Foundations of
Fuzzy Systems

Foundations of Fuzzy Systems

R. Kruse, J. Gebhardt and F. Klawonn

Computer Science Department
University of Braunschweig
Braunschweig
Germany

JOHN WILEY & SONS

Chichester · New York · Brisbane · Toronto · Singapore

ⓒ B. G. Teubner Stuttgart 1993: Kruse, R., Gebhardt, J./Klawonn, F: Fuzzy-Systeme. Translation arranged with approval of the publisher B. G. Teubner, Stuttgart, from the original German edition into English

Copyright ⓒ 1994 B. G. Teubner, Stuttgart

Published in 1994 by John Wiley & Sons Ltd.
Baffins Lane, Chichester
West Sussex PO19 1UD, England

National   Chichester (0243) 779777
International   (+44) 243 779777

Other Wiley Editorial Offices

John Wiley & Sons, Inc., 605 Third Avenue,
New York, NY 10158-0012, USA

Jacaranda Wiley Ltd, 33 Park Road, Milton,
Queensland 4064, Australia

John Wiley & Sons (Canada) Ltd, 22 Worcester Road,
Rexdale, Ontario M9W 1L1, Canada

John Wiley & Sons (SEA) Pte Ltd, 37 Jalan Pemimpin #05-04,
Block B, Union Industrial Building, Singapore 2057

Library of Congress Cataloging-in-Publication Data

Kruse, Rudolf.
    [Fuzzy Systeme. English]
    Foundations of fuzzy systems / R. Kruse, J. Gebhardt, F. Klawonn.
      p.   cm.
    Includes bibliographical references and index.
    ISBN 0 471 94243 X
    1. Fuzzy systems.   2. Fuzzy sets.   I. Gebhardt, J.   II. Klawonn,
F.   III. Title.
QA248.K74313   1994
003.7—dc20   94-9046
                                                            CIP

British Library Cataloguing in Publication Data

A catalogue record for this book is available from the British Library
ISBN 0 471 94243 X

Produced from camera-ready copy supplied by the authors using LaTeX
Printed and bound in Great Britain by Biddles Ltd, Guildford and Kings Lynn

Contents

Foreword   vii

Preface   xi

1 Introduction   1
   1.1   Fuzzy Systems   1
   1.2   Modelling Vague, Imprecise, and Uncertain Information   2
   1.3   About this Book   3

2 Elements of Fuzzy Set Theory   7
   2.1   Fuzzy Sets: An Elementary Introduction   7
   2.2   Simple Forms of Representing Fuzzy Sets   13
   2.3   Basic Operations on Fuzzy Sets   20
   2.4   The Extension Principle   28
   2.5   Efficient Operations on Fuzzy Sets   31
   2.6   Semantics of Fuzzy Sets   42
     2.6.1   Interpretation of Vague Concepts   43
     2.6.2   Interpretation of Vague Environments   47
   2.7   Fuzzy Logic   56
   2.8   Supplementary Remarks and References   65
     2.8.1   Historical Development: Fuzzy Systems   65
     2.8.2   Fuzzy Sets and their Semantics   68
     2.8.3   The Acquisition of Degrees of Membership   70
     2.8.4   Fuzzy Logic   71
     2.8.5   The SOLD System — An Implementation   73
     2.8.6   Exercises   77

3 Possibilistic Reasoning   81
   3.1   Possibility Distributions and Uncertainty Measures   82
   3.2   Concept of a Focusing Expert System for Possibilistic Data   91
   3.3   Interpretation of Possibilistic Inference Rules   106

3.4 Knowledge Representation and Propagation Using Hypergraphs 114
3.5 Logic-based Inference Mechanisms   129
     3.5.1 Possibilistic Logic   132
     3.5.2 Truth-functional Logic   139
3.6 Supplementary Remarks and References   142
     3.6.1 Historical Development: Approximate Reasoning   142
     3.6.2 Possibility Distributions   144
     3.6.3 Fuzzy Measures   149
     3.6.4 Extensions of Logic-based Mechanisms of Inference   151
     3.6.5 POSSINFER – An Implementation   152
     3.6.6 Exercises   155

4 Fuzzy Control   **159**
4.1 Knowledge-based versus Classical Models   160
4.2 Two Approaches to Fuzzy Control   164
     4.2.1 The Approach of Mamdani   165
     4.2.2 The Approach of Takagi and Sugeno   176
4.3 On the Design Parameters of Fuzzy Controllers   181
     4.3.1 Identification of Input and Control Variables   181
     4.3.2 The Domains of the Input and Control Variables   181
     4.3.3 Fuzzy Partitions of the Domains   182
     4.3.4 Linguistic Rules   182
     4.3.5 Evaluation of Linguistic Rules   184
     4.3.6 The Choice of the Defuzzification Strategy   185
     4.3.7 Optimizing a Fuzzy Controller   185
4.4 Fuzzy Control as Interpolation in the Presence of Imprecision 185
     4.4.1 Equality Relations as a Basis for Fuzzy Control   186
     4.4.2 Interpretation of a Mamdani Controller Based on
            Equality Relations   194
4.5 Fuzzy Control and Relational Equations   209
4.6 Supplementary Remarks and References   216
     4.6.1 Historical Development: Fuzzy Control   216
     4.6.2 Fuzzy Control   217
     4.6.3 Realization of Fuzzy Controllers   218
     4.6.4 Control of the Engine Idle Speed of a Spark Ignition
            Engine — An Implementation   222
     4.6.5 Exercises   233

Symbols   **237**

References   **241**

Index   **263**

Foreword

As a treatise on fuzzy set theory and its applications, "Foundations of Fuzzy Systems," by R. Kruse, J. Gebhardt and F. Klawonn, has few equals. Succinct, authoritative and up-to-date, it covers the basic theory very thoroughly and precisely, with emphasis on those aspects of the theory which play an important role in its applications. This is especially true of the chapters dealing with the calculus of fuzzy if-then rules — a subset of fuzzy set theory which plays a central role in the applications relating to the conception and design of both control and knowledge-based systems.

To view the contents of "Foundations of Fuzzy Systems" in a proper perspective, a digression is in order.

First, it is important to recognize that any crisp theory X can be fuzzified — and hence generalized to fuzzy X — by replacing the concept of a crisp set in X by that of a fuzzy set. In application to basic fields such as arithmetic, topology, graph theory, probability theory, etc., fuzzification leads to fuzzy arithmetic, fuzzy topology, fuzzy graph theory, fuzzy probability theory, etc. Similarly, in application to applied fields such as system theory, neural network theory, stability theory, pattern recognition, mathematical programming, etc., fuzzification leads to fuzzy system theory, fuzzy neural network theory, fuzzy stability theory, fuzzy pattern recognition, fuzzy mathematical programming, etc. What is gained through fuzzification is greater generality, higher expressive power, an enhanced ability to model real world problems and a methodology for exploiting the tolerance for imprecision and uncertainty to achieve tractability, robustness and low solution cost. For these reasons, it is inevitable that, eventually, fuzzy set theory will pervade most scientific theories and will have a particularly strong impact in the realms of systems analysis, artificial intelligence, intelligent control, signal processing, numerical analysis, optimization techniques, diagnostics, linguistics, information processing, decision analysis, cognitive science and related fields. In a very basic sense, this is the leitmotiv of the "Foundations of Fuzzy Systems."

Fuzzification applies not only to theories but, even more fundamentally, to concepts. One such concept — a concept which plays a central role in mathematics and its applications — is that of a theorem. In its traditional sense, a theorem may be viewed as a crisp if-then rule in which the premises form the antecedent. A fuzzy theorem, then, is a fuzzy if-then rule. A simple example of a fuzzy theorem is: if X and Y are large numbers then $X + Y$ is a large number. A fuzzy theorem is categorical if the truth value of the consequent is greater than or equal to that of the antecedent. In this sense, a categorical fuzzy theorem is reducible to a crisp theorem.

A dispositional fuzzy theorem is a usuality-qualified fuzzy if-then rule, that is, a rule which holds usually but not necessarily universally. For example, if X and Y are large numbers then usually $X + Y$ is a very large number. In this case, the underlying assumption is that the truth value of the consequent is usually — but not necessarily always — greater than or equal to that of the antecedent. In the case of dispositional fuzzy theorems, usuality qualification is frequently implicit rather than explicit and the expressions for the antecedent and consequent are informal rather than formal.

Viewed in this perspective, in the first three chapters of "Foundations of Fuzzy Systems" most of the results may be interpreted as categorical fuzzy theorems. By contrast, in the last chapter on fuzzy control, the rules are categorical but the conclusions regarding system performance are in effect dispositional fuzzy theorems — even though they are not presented as such. What this points to is that in the realm of practical applications the concept of a dispositional fuzzy theorem plays a dominant role albeit in a disguised rather than explicit form.

"Foundations of Fuzzy Systems" contains much that is worthy of praise. The exposition is tight and yet reader-friendly. The historical notes at the end of each chapter are illuminating; the inclusion of exercises enhances the pedagogical value of the text; and the description of software tools should appeal strongly to those who are interested in implementation issues.

There is a minor point of semantics which I do not see in the same light as the authors. Specifically, in my view — which was articulated in some of my early papers on the application of fuzzy sets to linguistics — the term vague is not coextensive with fuzzy, although it is frequently used in this sense in the literature of philosophy and linguistics. Thus, in my perception, vague and fuzzy are distinct concepts, with vagueness pertaining to insufficient specificity whereas fuzziness relates to unsharpness of boundaries. For example, "I will see you sometime," is vague and fuzzy while "I will meet you at approximately 5 pm" is fuzzy but not vague.

Putting aside minor matters, "Foundations of Fuzzy Systems" is a truly outstanding work. Much of the material is new and what is not new is treated with precision, clarity and concern for ease of understanding. The exposition

— as stated earlier — is authoritative, succinct and up-to-date. The authors deserve our thanks and congratulations for writing a book that contributes so importantly to the advancement of fuzzy set theory and its applications to fuzzy systems and related fields.

Lotfi A. Zadeh
Berkeley, California, March 1994

Preface

The primary aim of this monograph is to provide a robust framework for the study and development of fuzzy systems. Due to their impressive success in industrial applications, increasing interest has been directed to fuzzy systems. The research field of fuzzy systems has basically developed from set theory and statistics but it is now used in many applications for control tasks, expert systems, and other areas.

The fundamental objective of this textbook is to support one-semester graduate-level courses on the foundations of fuzzy systems, in which the different viewpoints of engineering and the natural sciences have been integrated.

The understanding of the text does not require specific prior knowledge of fuzzy systems theory, but we assume that the reader is familiar with the fundamental ideas of classical set theory, two-valued logic, and probability theory.

The material involved refers to several courses for students of computer science, mathematics, and electrical engineering that have been taught by the first author at the University of Braunschweig since 1986. But the book has also been written for professionals of various disciplines who will surely focus their attention on the supplementary remarks at the end of each chapter, which expand the tutorial by describing historical developments, giving main references, and emphasizing relationships to other subjects of research.

The chosen selection of topics reflects the two important application fields for fuzzy systems. These are approximate reasoning in knowledge-based systems and fuzzy control. We try not to restrict ourselves to the presentation of concepts and examples of their applicability, but the semantic background is also clarified in order to provide the reader with a well-founded development of fuzzy systems.

The book contains a number of original results which were obtained during cooperative contracts with Fraunhofer Gesellschaft, German Aerospace, Siemens AG, and Volkswagen AG. Furthermore, the CEC-ESPRIT III Basic Research Project 6156 DRUMS II (Defeasible Reasoning and Uncertainty

Management Systems) provided an important impulse to our scientific collaboration.

The book has benefited from our contacts with many colleagues in scientific research and partners in various projects. We particularly thank Hans Bandemer, Piero Bonissone, Didier Dubois, Patrik Eklund, Petr Hajek, Ulrich Höhle, Peter Klement, Ramon López de Mantaras, Abe Mamdani, Serafín Moral, Rainer Palm, Henri Prade, Michael M. Richter, Philippe Smets, Hideyuki Takagi, Enric Trillas, Lotfi Zadeh, and Hans-Jürgen Zimmermann for fruitful discussions about the material of the book, Janusz Kacprzyk for his valuable hints for improving the manuscript, Toshiro Terano for his creative suggestions during a stay of research at LIFE Institute (Laboratory for International Fuzzy Engineering Research, Yokohama, Japan), and Hans Bangen, Joachim Beckmann, Hans-Joachim Bohn, Horst Weber, and Hartmut Wolff for their professional support in the respective projects.

We received useful comments from students in our graduate courses on "Fuzzy Systems", and from the participants of several tutorials for professionals in industry who used the German version of this book (Fuzzy Systeme, Teubner, Stuttgart, 1993, ISBN 3–519–02130–7) as class notes.

Furthermore we thank Dieter Brökelmann, Günter Lehmann, Kai Michels, Detlef Nauck, and Herbert Toth for careful proofreading. We are grateful to Christian Borgelt and Maik Masuch for their excellent translation and typesetting of the manuscript. Our editors at John Wiley have been very helpful and cooperative in producing the final text.

R. Kruse
J. Gebhardt
F. Klawonn
Braunschweig, March 1994

1

Introduction

This chapter provides a brief introduction to the field of fuzzy systems and the construction of models for the management of imperfect information. Additionally the basic concepts, the resulting structure, and the arrangement of the material in this book are explained.

1.1 Fuzzy Systems

In engineering and science, complex physical systems are usually described by mathematical models. Typical examples are models that follow the laws of physics (e.g. systems of differential equations), stochastic models (e.g. Markov chains), but also models which have emerged from mathematical logic (e.g. the entity-relationship model in the theory of database systems). A general difficulty that arises when one tries to construct a model is that considerable idealizations are often necessary to get from a given problem to a suitable mathematical model. Although advances in computer technology have made it possible, in principle, to manage systems that are more and more complex, this has led at the same time to constantly growing conceptual demands to keep large software packages understandable. In the same way large databases lose their applicability if their users are no longer in a position to extract the relevant information in a sensible way and to have it presented in a proper form.

One method to simplify complex systems is to tolerate a reasonable amount of imprecision, vagueness, and uncertainty during the modelling phase. Certainly the resulting systems are not perfect, but in many cases they are capable of solving the modelling problem in an appropriate way. The principle of accepting a certain loss of information has already turned out to be satisfactory in knowledge-based systems, in which the imperfection of experts has to be taken into account while constructing the model.

In fuzzy systems, we attempt to take advantage of a reduced complexity (compared to other systems), which is achieved by the judicious use of imperfect information. Of course, if the models are too coarse, they bear the danger that relevant features of the real world are neglected. On the other hand the quality of a model cannot be measured by the degree of precision of the included information alone, but only with regard to criteria like correctness, completeness, adequacy, efficiency, and convenience of use. Thus it is not surprising that a model which gives a complexity-reduced description of the system under consideration, where the reduction in complexity is brought about by the systematic inclusion of imperfect information, can actually be better — with regard to several criteria of quality   than a model which is more difficult to handle because it demands precise information only.

The use of fuzzy systems has recently yielded impressive success, especially in the domain of control engineering. Therefore fuzzy methods are the subject matter of this textbook. They are treated in depth with regard to their applicability to the modelling of information which is afflicted with imprecision, vagueness, and uncertainty. But before we enter into details, we wish to give a brief explanation of what we understand by the above-mentioned types of imperfect information.

1.2   Modelling Vague, Imprecise, and Uncertain Information

In daily life *vagueness* is consciously accepted to facilitate communication and to restrict information to that portion which is essential for appropriate action in a given situation. The use of the adjective 'fast' to characterize, for example, the speed of an approaching car, is entirely sufficient to signal the necessity of an evasive movement. The precise velocity of the car at this moment is of subordinate consequence.

Even if we disregard that vague terms like 'fast' have different meanings if used by different persons and must be interpreted with respect to the observed environment (the velocity of a fast car is about equal to the velocity of a slow plane), a certain person under well-defined conditions cannot always reliably classify the movement of a car as 'fast' or 'not fast' with respect to given values of velocity. For some values the classification is certainly obvious, but for others it is difficult to find, because human reasoning recognizes different shades of meaning between the two concepts 'fast' and 'not fast.' In the following, such nuances are described with the help of so-called *fuzzy sets*. The next chapter is dedicated to a survey on this subject matter.

Imprecision of information can be traced back to the fact that in many cases it is not possible to observe or to measure with an arbitrary degree of accuracy. The most simple form of this kind of imperfect knowledge is a statement like 'The plane is between 50.6 km and 50.8 km away.' Such information is usually described by (non-stochastic) intervals of error [Moore66], which can be regarded as a special case of fuzzy sets.

The most familiar form of *uncertainty* is related to mechanisms of chance as they are well-known, for instance, from playing dice or roulette. To characterize such phenomena, probabilistic models are usually applied [Feller66]. In other cases, however, one has to deal with uncertainty caused by subjective estimation, such as that reflected in qualitatively rated rules stated by experts in medicine. In this sense the rule 163 of the expert system MYCIN [Shortliffe75, Buchanan84] for the diagnosis of bacteriogenic infectious diseases reads:

> **If** the morphology of the organism is rod, and
> the stain of the organism is gramneg, and
> the identity of the organism is not known with certainty, and
> the patient has had a genito-urinary manipulative procedure,
> **then** there is weakly suggestive evidence that
> the identity of the organism is pseudomonas.

The expert held the opinion that the conclusion can only be inferred with strength 'weakly suggestive evidence.'

In MYCIN the *calculus of certainty factors* [Shortliffe75] was employed. Yet this calculus can turn out to be inconsistent in other applications [Heckerman88a]. On this account stochastic methods are often used to model uncertainty [Kruse91a, Shafer90, Smithson89].

The approach that is presented in this book is based on the epistemic interpretation of fuzzy sets by means of possibility distributions, which are the topic of Chapter 3.

The types of imperfect knowledge mentioned above can also appear combined: for example, if an aircraft is selected by chance from a given set of aircraft (uncertainty), its speed is measured by an observer (imprecision), and it is classified to be a fast or a slow aircraft (vagueness). A uniform calculus that integrates all three phenomena would of course facilitate the modelling. We shall see that an important advantage of fuzzy systems (e.g. fuzzy controllers, which are presented in Chapter 4) is that they can accomplish just this integration.

1.3 About this Book

The aim of this book is to give a methodically ordered introductory course in fuzzy systems. The use of fuzzy methods in knowledge-based systems and for problem solving in control engineering are discussed in depth. With regard to the size of this book, however, we had to refrain from presenting other important domains in which fuzzy systems can be employed.

The material is partitioned into three chapters. Chapter 2 (theory of fuzzy sets) is elementary, while the two subsequent chapters (possibilistic reasoning and fuzzy control) can be studied independently of each other. Each chapter is divided into several subject-specific sections, the last of which contains supplementary remarks and many references. Historical remarks are also located here, recent developments are touched upon, cross-references to other domains of research are established, possible implementations are pointed out in case studies, and there are also some exercises. These supplementary sections should help the reader to deepen his/her knowledge and to extend it beyond the contents of this book in a well-directed way.

Now that we have described the overall concept of this textbook, we will give an overview of motivation and contents of particular chapters. Figure 1 shows the dependencies between the sections.

Chapter 2 deals with fundamental concepts of the theory of fuzzy sets and its relations to fuzzy logic. We start with a problem-oriented introduction of the notion of a fuzzy set, its formal definition, and its illustration. Thereafter we explain different representations of fuzzy sets, we show how mappings on ordinary sets are generalized to fuzzy sets (extension principle), and examine the conditions needed to operate on fuzzy sets efficiently. Finally we inspect the semantics of degrees of membership. This is often neglected, though it is very important from a theoretical point of view. Nevertheless we confine our discussion to a depth that is suitable for an introductory textbook and leave further considerations to the cited articles.

In Chapter 3 the fundamentals that have been learned are applied in a field that is of some interest in information science: approximate reasoning. We chose this domain, because in the first place it reflects an essential branch of research in fuzzy systems, and in the second place it brings about relations to knowledge-based systems, as they are known from artificial intelligence and the theory of database systems. At first we show, how fuzzy sets (in their epistemic interpretation as possibility distributions) can help to characterize uncertain data. In this context the elements of possibility theory and the measurement of uncertainty are discussed. Thereafter we examine how inference mechanisms and propagation algorithms can be realized in a multi-dimensional space of qualitative dependencies, that are specified in terms of hypergraphs, if general knowledge about a domain of application, provided by experts, and specific evidence, gained by observations, can be described by possibility distributions. Among other things we discuss the problem how rule bases, as they are known from expert systems, should be generalized to possibility distributions (instead of precise data) and how they should be interpreted. We choose a concise approach to possibility theory, though this approach is not mandatory. That is why we finally examine other approaches (e.g. possibilistic logic).

Chapter 4 deals with control engineering on the basis of fuzzy sets, explaining the elementary principles of fuzzy control based on the works of

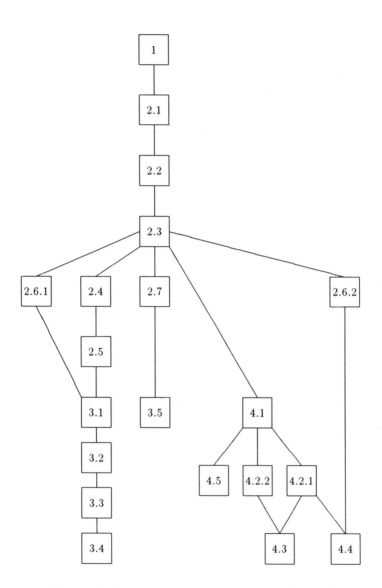

Figure 1 Dependencies between sections of the book

Mamdani and Takagi/Sugeno. In addition we attach great importance to a clean theoretical modelling based on the concepts presented in Chapter 2, including an evident motivation of the applied formalisms. Therefore no knowledge of control engineering is required. From this point of view fuzzy control is interpreted as a knowledge-based interpolation in a vague environment, which can be characterized by equality or similarity relations. In this sense fuzzy controllers are understood as real-time expert systems, which permit the control of systems with non-linear performance characteristics. Although we focus on control applications, this methodology of handling if-then rules can be easily adapted to other fields of research.

To increase the readability of the text, we have placed the reference citations in the supplementary remarks and have added a list of symbols in the appendix.

2

Elements of Fuzzy Set Theory

In this chapter we introduce the theory of fuzzy sets. We give a motivation for the basic concepts and discuss the representation and interpretation of fuzzy sets and especially how the semantics of operations on fuzzy sets can be substantiated.

The topic of Section 2.1 is to explain the notion of a membership function, which is used to characterize fuzzy sets. In Section 2.2 we consider an equivalent representation that employs systems of sets. In Section 2.3 the basic operations of set theory are extended to fuzzy sets with the help of the important concept of t-norms and t-conorms. By applying the extension principle, which is introduced in Section 2.4, it becomes possible to generalize mappings of real numbers to mappings of fuzzy sets. How these extended arithmetical operations can be carried out efficiently, using the set representation of fuzzy sets, is the subject of Section 2.5.

An understanding of the first five sections provides the necessary foundation for a clear account of the semantics of fuzzy sets in Section 2.6. In our book fuzzy sets are interpreted in two different ways: as contour functions of random sets and as extensions of ordinary sets induced by equality relations and t-norms. Section 2.7 describes important basic concepts of fuzzy logic. Finally Section 2.8 contains supplementary remarks and references.

2.1 Fuzzy Sets: An Elementary Introduction

In mathematics concepts and properties can often be described by specifying subsets of a given reference set. For example, if we take the set $\mathbf{N}$ of natural numbers to be our reference set, the property that a number has two digits can be expressed using the set

$$Z = \{10, 11, 12, \ldots, 97, 98, 99\} \subseteq \mathbf{N}.$$

Besides this representation, which is simple in structure but often not applicable, the same property can also be understood as a predicate, i.e.

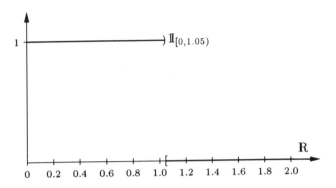

Figure 1   Characteristic function of the interval $[0, 1.05)$

two-digit-number(n) is true for a natural number n, if and only if the decimal representation of n consists of exactly two digits.

Therefore the set Z contains all natural numbers that satisfy the predicate *two-digit-number*:

$$Z = \{n \in \mathbf{N} \mid \textit{two-digit-number}(n)\}.$$

A third form to represent the datum *two-digit-number* is to use the concept of a *characteristic function*, which in our example is defined as:

$$\mathbb{I}_Z : \mathbf{N} \quad \rightarrow \quad \{0, 1\}$$
$$\mathbb{I}_Z(n) \quad \overset{\text{def}}{=} \quad \begin{cases} 1, & \text{if } \textit{two-digit-number}(n) \\ 0, & \text{otherwise.} \end{cases}$$

The members of Z are made to stand out from all other elements of the reference set by 'marking' them with the value 1. The two values 0 and 1 of the characteristic function are sometimes called *degrees of membership*.

Example 2.1 The characteristic function shown in Figure 1 describes the set of all non-negative real numbers smaller than 1.05. The point of discontinuity at 1.05 appears to be natural, if the characteristic function is interpreted as the set of body-heights of children who are shorter than 1.05 m.   □

The adequacy of characteristic functions with codomain $\{0, 1\}$ has to be questioned if, for instance, those body-heights are considered for which a child should be called *tall*. Even if we disregard that the word *tall* can have different meanings depending on the context in which it is used, a single physician in a context of four-year-old boys from Germany may have some difficulty to classify boys to be *tall* or *not tall*. Of course there are some body-heights which will always be classified as *tall* (e.g. 1.50 m) or *not tall* (e.g. 0.70 m), but for others (e.g. 1.10 m) a definite classification is hard to find.

The main issue of the above characterization problem is that a description with the help of ordinary sets requires us to fix a value of height in the way that a child, whose height exceeds this value, has to be called *tall*, and that no other value smaller than the fixed one has the same property. But the point of discontinuity, which then appears in the characteristic function, does not coincide with the conception that there is a smooth transition from the height of a child who has to be called *tall* to that of a child who has to be called *not tall*.

Other examples, in which a characterization with the help of ordinary sets appears to be unnatural, can be found in qualitative descriptions like 'Take a pinch of salt' or 'If the wall is not far away and the velocity of the car is high, then it is about time to apply the brakes.'

One method to model gradual transitions is to use other degrees of membership in addition to 0 (does certainly not belong to it) and 1 (does certainly belong to it). A mathematically simple way to introduce a gradual membership consists in taking the degrees of membership from the compact interval $[0, 1]$. Of course the semantics of these additional degrees remain to be defined.

Example 2.2 Figure 2 shows a 'generalized' characteristic function, which describes the vague predicate *tall* from the viewpoint of a physician who was asked for all values $x \in \mathbf{R}$ that body-height can assume in the context of four-year-old boys from Germany.

Every value x is related to a degree of membership, e.g. the height 1.20 m is related to 0.7. This means that the height 1.20 m satisfies the predicate *tall* with a degree of membership of 0.7 on a scale from 0 to 1. The more the degree of membership $\mu_{\text{tall}}(x)$ approaches 1, the more the value x satisfies the predicate *tall*.   □

The Examples 2.1 and 2.2 suggest how to formalize vaguely described data like 'sufficient supply of fuel', 'high throughput', and 'bright illumination' with the help of generalized characteristic functions, which are called fuzzy sets in this context.

Definition 2.3 *A* **fuzzy set** μ *of X is a function from the reference set X to the unit interval, i.e.*
$$\mu : X \to [0, 1].$$
$F(X)$ denotes the set of all fuzzy sets of X.

Each ordinary set A (in the context of fuzzy systems sometimes called *crisp* set) can be interpreted (via the characteristic function $\mathbb{1}_A$) as a special kind of a fuzzy set.

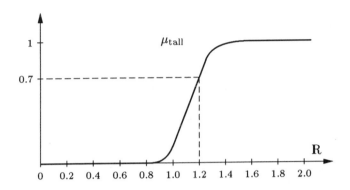

Figure 2 Generalized characteristic function μ_{tall}

In general, fuzzy sets merely have an intuitive basis as a formal description of vague data, since only few applications lay down a model which provides an unambiguous meaning for the degrees of membership. We return to this problem in Section 2.6 and confine our attention at present to a merely descriptive treatment of fuzzy set theory.

Example 2.4 A letting agency offers apartments of different sizes. Let one criterion for the quality of an apartment be the number of rooms. For a family of five the vague concept 'attractive number of rooms' can be described by the fuzzy set $\mu : \{1, \ldots, 8\} \to [0, 1]$, where the degrees of membership are fixed as $\mu(1) = 0$, $\mu(2) = 0.2$, $\mu(3) = 0.5$, $\mu(4) = 0.7$, $\mu(5) = 1$, $\mu(6) = 1$, $\mu(7) = 0.8$, and $\mu(8) = 0.2$. In subjectively fixing the degrees of membership it was taken into account that only apartments with up to eight rooms are offered. In this context five or six rooms would be preferred.   □

Example 2.5 Let x be the velocity of rotation of a hard disk in revolutions per minute. Since a statement like this is always afflicted with imprecision, it appears to be more realistic to use the statement

'The velocity of rotation is almost exactly equal to x'

and to interpret it in a suitable way.

If statistical data about the operation of the hard disk are available, one should prefer probabilistic approaches and should model the information 'almost exactly equal to x' using the known methods of calculating errors. But if such data are not available or if they are afflicted with imprecision, one can switch over to fuzzy sets, because they can often be specified by experts directly and in an intuitive way.

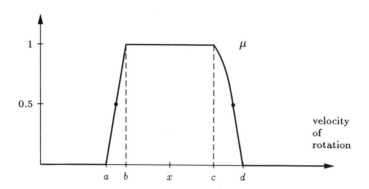

Figure 3   Fuzzy set μ characterizing the velocity of rotation of a harddisk

Let us assume that the expert chose the fuzzy set μ shown in Figure 3. It is regarded as being impossible that the velocity of rotation falls below a or exeeds d, whereas it is regarded as being highly certain that any value lying between b and c can occur. The interval $[a, d]$ is therefore called the *support* and $[b, c]$ the *core* of the fuzzy set.

In addition the expert provided the values of the velocity of rotation with a degree of membership of 0.5, and he chose a strictly increasing or decreasing course of the function in the remaining areas.   □

At this point we have to emphasize that the function in Figure 3 should *not* be interpreted as a density function in the sense of probability theory: from a solely formal point of view, the area beneath the graph of μ need not be equal to 1, and, from a semantical point of view, an important difference subsists in the fact that the definition of μ is based on assigning degrees of membership to all elements of $\mathbf{R}$ instead of stating probabilities for subsets of $\mathbf{R}$.

Example 2.6 An X-ray photograph can be digitized by scanning a raster of $m \times n$ pixels and assigning a shade of grey to each pixel. Let $P \stackrel{\text{def}}{=} \{x_1, \ldots, x_m\} \times \{y_1, \ldots, y_n\}$ be the set of all pixels and $G \stackrel{\text{def}}{=} \{g_0, \ldots, g_{\max}\}$ the set of all shades of grey ordered with respect to their darkness, so that g_0 corresponds to 'white' and $g_{\max}$ corresponds to 'black'.

Then the X-ray photograph can be represented as a grey scale picture that is formalized with the aid of a function

$$\rho : P \to G.$$

If every shade of grey is mapped to a number in the unit interval, e.g. by a function

$$\tau : G \to [0, 1],$$

satisfying $\tau(g_0) = 0$, $\tau(g_{max}) = 1$, and $\forall g, g' \in G : g \sqsubset g' \Longleftrightarrow \tau(g) < \tau(g')$, where $\sqsubset$ denotes the linear ordering of G, then the combination $\tau \circ \rho$ of the two functions ρ and τ can be regarded as the characteristic function of the grey scale picture. □

Remark 2.7 When using fuzzy sets, one has always to state exactly what is to be modelled by μ. There is a large variety of different interpretations of fuzzy sets in the literature (as it is the case also with probability distributions). In our book we distinguish mainly between two different interpretations.

A purely *physical interpretation* of a fuzzy set assumes that the fuzzy set under consideration describes a vague object, which is, for example, a grey scale picture, that is visible on a physically existing sheet of paper, and to be analysed with methods of mathematical morphology.

On the other hand, an *epistemic interpretation* of the fuzzy set characterizing the grey scale picture is used in the case of X-ray photographs, as mentioned in Example 2.6. The grey scale picture is regarded as a vague observation of an existing sharp object, and the X-ray photograph is to be read as a vague (in this case two-dimensional) perception of an existing three-dimensional sharp object.

In Example 2.5 the vague concept 'almost exactly equal to x' is described by a fuzzy set μ. μ has a 'physical' interpretation. But if it is intended to describe the velocity of rotation of a given hard disk at a certain point in time, then we are not dealing with a vague object, but rather with the uncertainty of an observer who cannot measure the velocity of rotation with exactitude. Fuzzy sets can also be applied to characterize this uncertainty, but then they are used in an epistemic interpretation, and in this context they are equated with *possibility distributions*. In Chapter 3 this interpretation of fuzzy sets is studied in detail. □

Remark 2.8 In Example 2.6 it was sensible to map the shades of grey into the interval $[0, 1]$, because the darkness of the shade of grey depends on the thickness of the X-rayed material alone. But there are also applications where an expert provides only qualitative assessments like 'I am nearly sure, that x belongs to it.' Then, by using the whole interval $[0, 1]$ as the range of degrees of membership, we appear to do an over-modelling, since, in the first place, a distinction between degrees of membership of 0.915 and 0.916 often cannot be motivated by any means. In the second place, an additivity of degrees of membership cannot be justified.

In such cases it is advisable to use an arbitrary lattice $(L, \sqcap, \sqcup)$ instead of the whole interval $[0, 1]$. Such a lattice expresses a merely qualitative order of the degrees of membership contained in L. If there is, for example, a linear ordering of L (i.e. for all $l, l' \in L$ it is $l \leq l'$ or $l > l'$), then $(L, \sqcap, \sqcup)$ is defined by

$$l \sqcap l' = l \iff l \leq l'$$
$$l \sqcup l' = l \iff l \geq l'.$$

These considerations lead us directly to the notion of an L-fuzzy set, which is a generalization of a fuzzy set. □

Definition 2.9 *Let $(L, \sqcap, \sqcup)$ be a lattice with l_{min} as smallest element and l_{max} as greatest. An **L-fuzzy set** η of X is a mapping from the reference set X to the set L, i.e.*

$$\eta : X \to L.$$

$L(X)$ denotes the set of all L-fuzzy sets of X.

Example 2.10 The so-called Sherman–Kent scale consists of 19 degrees of (qualitative) values for L as shown below:

l_{min} = impossible < highly doubtful < only a slight chance
< we believe not < unlikely < probably not
< chances are slightly less than even
< chances are about even
< chances are slightly better than even
< it's probable < chances are good < we estimate
< we believe < likely < highly likely
< highly probable < we are convinced
< virtually (almost) certain < certain = l_{max}. □

Examples like the assessment hierachy of the Sherman–Kent scale demonstrate that it is very important to take a close look at the semantics of L-fuzzy sets and fuzzy sets as special L-fuzzy sets. Before we deal with these aspects in Section 2.6, we now investigate a more intuitive approach to fuzzy sets.

2.2 Simple Forms of Representing Fuzzy Sets

In the following we discuss several examples of fuzzy sets as well as some possibilities to construct and represent the corresponding characteristic functions.

Any fuzzy set μ of X can obviously be described by assigning a degree of membership $\mu(x)$ to each element $x \in X$, but this is practically feasible only if the reference set X consists of a finite number of elements. If the number of elements in X is very large or even countably infinite or if a continuum is used for X (e.g. to measure temperature or velocity), then $\mu(x)$ is represented

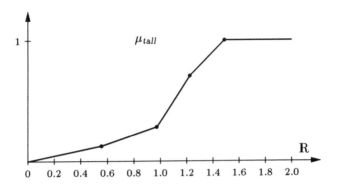

Figure 4   Interpretation of the vague concept 'tall'
by a piecewise linear function

best with the help of an appropriate, possibly parameterized term, where the
parameters can be adjusted according to the given problem.

If $X = \mathbf{R}$, then in most cases linguistic expressions like 'tall', 'about 10',
and 'approximately between a and b' turn up, which may be interpreted as
elements of suitable classes of parameterized fuzzy sets $\mu : X \rightarrow [0,1]$ that
satisfy the *normality condition* $\exists x \in X : \mu(x) = 1$ and are therefore called
normal fuzzy sets. In applications an interpretation of 'tall' as a monotonic
non-decreasing function like

$$\mu_{a,b}(x) \stackrel{\text{def}}{=} \begin{cases} 0, & \text{if } x \leq a \\ \frac{x-a}{b-a}, & \text{if } a \leq x \leq b \\ 1, & \text{if } x \geq b, \end{cases} \tag{2.1}$$

where $a < b$, is employed.

Other interpretations are piecewise linear functions, whose parameters are
the coordinates of the defining points. An example is shown in Figure 4.
Exponential functions like

$$\mu_{a,b}(x) \stackrel{\text{def}}{=} \begin{cases} 1 - e^{(-a(x-b))}, & \text{if } x \geq b \\ 0, & \text{if } x < b \end{cases} \tag{2.2}$$

where $a > 0$ and $b \in \mathbf{R}$, may also be used.

To interpret linguistic expressions like 'about 10', which are often
represented by the so-called *fuzzy numbers*, we most simply adopt symmetrical
triangular functions like

$$\mu_{m,d}(x) \stackrel{\text{def}}{=} \begin{cases} 1 - \left|\frac{m-x}{d}\right|, & \text{if } m - d \leq x \leq m + d \\ 0, & \text{if } x < m - d \text{ or } x > m + d, \end{cases} \tag{2.3}$$

where $d > 0$ and $m \in \mathbf{R}$, as well as bell-shaped Gaussian curves like

$$\mu_{a,m}(x) \stackrel{\text{def}}{=} e^{-a(x-m)^2} \tag{2.4}$$

where $a > 0$, $m \in \mathbf{R}$.

By analogy the expression 'approximately between b and c' is most simply characterized by the trapezoidal function

$$\mu_{a,b,c,d}(x) \stackrel{\text{def}}{=} \begin{cases} \frac{x-a}{b-a}, & \text{if } a \leq x < b \\ 1, & \text{if } b \leq x \leq c \\ \frac{x-d}{c-d}, & \text{if } c < x \leq d \\ 0, & \text{if } x < a \quad \text{or} \quad x > d, \end{cases} \tag{2.5}$$

where $a < b < c < d$ and $a, b, c, d \in \mathbf{R}$.

Note that real numbers t can be described by

$$\mu_t(x) \stackrel{\text{def}}{=} \mathbb{1}_{\{t\}}(x) = \begin{cases} 1, & \text{if } x = t \\ 0, & \text{if } x \neq t \end{cases} \tag{2.6}$$

and intervals $[a, b]$ by

$$\mu_{a,b}(x) \stackrel{\text{def}}{=} \mathbb{1}_{[a,b]}(x) = \begin{cases} 1, & \text{if } a \leq x \leq b \\ 0, & \text{if } x < a \quad \text{or} \quad x > b. \end{cases} \tag{2.7}$$

For this reason real numbers as well as intervals can be viewed as special cases of normal fuzzy sets.

In the examples considered up to now fuzzy sets were described exclusively by their characteristic function $\mu : X \to [0, 1]$. This representation corresponds to the usual view to functions, which map an element x, taken from the reference set X, to the value $\mu(x)$. With regard to the graphical representation of μ, this is called *vertical representation* of the corresponding fuzzy set.

Comparable to the motivation of L-fuzzy sets, which were introduced in Definition 2.9 and Example 2.10, the acquisition of fuzzy sets often happens in the following way: for all degrees of membership α that belong to a selected (finite) subset of the interval $[0, 1]$, an expert states which elements of the reference set X can be classed as belonging to the vague concept that is described by the fuzzy set with a degree at least equal to α. This is equivalent to the *horizontal representation* of fuzzy sets by their α-cuts.

Definition 2.11 *Let $\mu \in F(X)$ and $\alpha \in [0, 1]$. Then the set*

$$[\mu]_\alpha \stackrel{\text{def}}{=} \{x \in X \mid \mu(x) \geq \alpha\}$$

is called the α-cut or α-level set of μ.

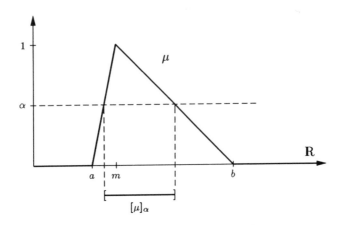

Figure 5 α-cut of a fuzzy set μ

Example 2.12 Let μ be the triangle function on **R** shown in Figure 5.

The α-cut of μ can be constructed by drawing a horizontal line parallel to the x-axis through the point $(0, \alpha)$ and projecting that section of the graph, which lies on this line and above, onto the x-axis. In this case we obtain

$$[\mu]_\alpha = \begin{cases} [a + \alpha(m - a), b - \alpha(b - m)], & \text{if} \quad 0 < \alpha \le 1 \\ \mathbf{R}, & \text{if} \quad \alpha = 0. \end{cases} \qquad \square$$

The following important properties of fuzzy sets can be gathered directly from Figure 5:

Theorem 2.13 Let $\mu \in F(X)$, $\alpha \in [0, 1]$ and $\beta \in (0, 1]$.

(a) $[\mu]_0 = X$,

(b) $\alpha < \beta \Longrightarrow [\mu]_\alpha \supseteq [\mu]_\beta$,

(c) $\bigcap_{\alpha : \alpha < \beta} [\mu]_\alpha = [\mu]_\beta$.

Proof:

(a), (b): obvious.

(c) Let $\beta \in (0, 1]$ be arbitrarily chosen, but fixed.
For all $\alpha \in [0, 1]$, where $\alpha < \beta$, we have $[\mu]_\alpha \supseteq [\mu]_\beta$ and therefore
$\bigcap_{\alpha : \alpha < \beta} [\mu]_\alpha \supseteq [\mu]_\beta$.
To show the inclusion in the opposite direction, we choose an

$x \in \bigcap\limits_{\alpha:\alpha<\beta} [\mu]_\alpha$. If $\alpha < \beta$, then $x \in [\mu]_\alpha$ and hence $\mu(x) \geq \alpha$.

It follows that $\mu(x) \geq \sup\{\alpha \mid \alpha < \beta\} = \beta$ and therefore $x \in [\mu]_\beta$. □

α-cuts are therefore important for the application of fuzzy sets, because any fuzzy set can be described by specifying its α-cuts. The following *representation theorem* holds:

Theorem 2.14 *Let $\mu \in F(X)$. Then*

$$\mu(x) = \sup_{\alpha \in [0,1]} \left\{\min(\alpha, \mathbb{1}_{[\mu]_\alpha}(x))\right\}.$$

Proof:
If $x \in X$ and $\alpha \in [0,1]$, then

$$\min(\alpha, \mathbb{1}_{[\mu]_\alpha}(x)) = \left\{ \begin{array}{ll} \alpha, & \text{if } \mu(x) \geq \alpha \\ 0, & \text{if } \mu(x) < \alpha. \end{array} \right.$$

It follows that $\mu(x) = \sup\{\alpha \mid \alpha \leq \mu(x)\} = \sup\limits_{\alpha \in [0,1]} \left\{\min(\alpha, \mathbb{1}_{[\mu]_\alpha}(x))\right\}.$ □

Geometrically speaking, a fuzzy set can be obtained as the upper envelope of its α-cuts, if we draw the α-cuts parallel to the horizontal axis in the height α. As we already indicated with respect to L-fuzzy sets, in applications it is recommended to select a finite subset $L \subseteq [0,1]$ of relevant degrees of membership, that are distinguishable semantically, and to fix the level sets of the fuzzy set to characterize only for these levels. In this manner a system of sets

$$\mathcal{A} = (A_\alpha)_{\alpha \in L}, \qquad L \subseteq [0,1], \text{ card}(L) \in \mathbf{N}, \tag{2.8}$$

results, which, for $\alpha, \beta \in L$, has to satisfy consistency conditions

$$\begin{array}{lll} \text{(a)} & 0 \in L \implies A_0 = X & \text{(fixing of the reference set),} \\ \text{(b)} & \alpha < \beta \implies A_\alpha \supseteq A_\beta & \text{(monotonicity)} \end{array} \tag{2.9}$$

and which induces the fuzzy set

$$\begin{array}{rcl} \mu_{\mathcal{A}} : X & \rightarrow & [0,1], \\ \mu_{\mathcal{A}}(x) & = & \sup_{\alpha \in L} \left\{\min(\alpha, \mathbb{1}_{A_\alpha}(x))\right\}. \end{array} \tag{2.10}$$

If L is not supposed to be finite, but can comprise the whole range of values ($L = [0,1]$), then μ has to satisfy

$$\text{(c)} \quad \bigcap\limits_{\alpha:\alpha<\beta} A_\alpha = A_\beta \qquad \text{(condition for continuity).} \tag{2.11}$$

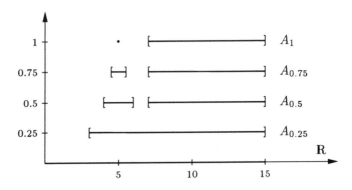

Figure 6   The system $\mathcal{A}$ that characterizes a fuzzy set
(horizontal view)

Definition 2.15 $\mathcal{FL}(X)$ *denotes the set of all families* $(A_\alpha)_{\alpha \in [0,1]}$ *of sets that
satisfy*

(F1) $A_0 = X$,

(F2) $\alpha < \beta \Longrightarrow A_\alpha \supseteq A_\beta$,

(F3) $\displaystyle\bigcap_{\alpha:\alpha<\beta} A_\alpha = A_\beta$.

Remark 2.16 Any family $\mathcal{A} = (A_\alpha)_{\alpha \in [0,1]}$ of sets of X that satisfies the
conditions (F1), (F2), and (F3), represents a fuzzy set

$$
\begin{aligned}
\mu_{\mathcal{A}} &\in F(X), \\
\mu_{\mathcal{A}}(x) &= \sup\{\alpha \mid \alpha \in [0,1] \wedge x \in A_\alpha\}.
\end{aligned}
\tag{2.12}
$$

In reverse, if there is a $\mu \in F(X)$, then the family $([\mu]_\alpha)_{\alpha \in [0,1]}$ of α-cuts
of μ satisfies the three conditions stated in Definition 2.15 according to
Theorem 2.13.   □

Example 2.17 Let $X = [0, 15]$. To characterize the vague datum 'approximately 5 or greater than or equal to 7', an expert chooses five degrees of membership $L = \{0, 0.25, 0.5, 0.75, 1\}$ as well as the α-cuts $A_0 = [0, 15]$, $A_{0.25} =$
$[3, 15]$, $A_{0.5} = [4, 6] \cup [7, 15]$, $A_{0.75} = [4.5, 5.5] \cup [7, 15]$, and $A_1 = \{5\} \cup [7, 15]$.
The family $(A_\alpha)_{\alpha \in L}$ of sets induces the fuzzy set $\mu_{\mathcal{A}}$ shown in Figure 6, where
$\mu_{\mathcal{A}}$ can be obtained graphically as the upper envelope of the family $\mathcal{A}$ of sets.

The difference between the horizontal (Figure 6) and the vertical view of
fuzzy sets (Figure 7) is obvious. To efficiently process fuzzy sets with the help
of the computer, the horizontal representation is often preferred. In addition,

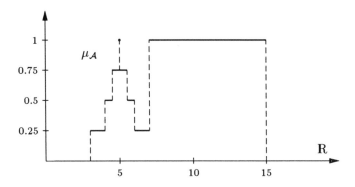

Figure 7   The fuzzy set μ_A induced by A
(vertical view)

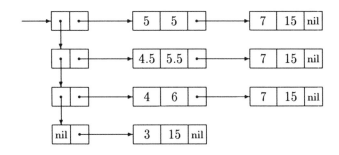

Figure 8   Representation of μ_A in a software tool

restricting the domain of the x-axis to a discrete set should be taken into consideration.

Whereas a discretization of the codomain always appears to be plausible, doing the same to the domain need not be sensible in any case. That is why, for example, in the interactive software system SOLD (Statistics On Linguistic Data) which allows us to statistically analyse vague data that are describable by fuzzy sets, only a partition into a reasonable finite number of equidistant levels can be chosen. The domain, on the other hand, always remains continuous. (The system SOLD is presented in greater detail in Section 2.8.5.) For storing fuzzy sets μ_A a data structure in the style of the set representation is used (see Figure 8.).

It is a chain of linear lists for each α-level, $\alpha \neq 0$, in which a finite union of closed intervals is stored by noting their bounds.

Having assumed that the α-cuts can be represented in this way, the

chosen data structure turns out to be appropriate, especially with regard
to arithmetic operations (cf. Section 2.5).   □

The horizontal view on fuzzy sets can be used to conveniently transfer and
to generalize several notions from ordinary sets to fuzzy sets. As an example,
let us take a look at set inclusion.

An ordinary set A is called a subset of A', if every element of A is also
an element of A'. Taking two families of sets, $(A_\alpha)_{\alpha \in [0,1]}$ and $(A'_\alpha)_{\alpha \in [0,1]}$, a
generalized subset notation should therefore satisfy $A_\alpha \subseteq A'_\alpha$ for all $\alpha \in [0,1]$.
It is easy to show that $(\forall \alpha \in [0,1] : A_\alpha \subseteq A'_\alpha)$ holds for the fuzzy sets μ and
μ' induced by these two families of sets, if $(\forall x \in X : \mu(x) \leq \mu'(x))$ is valid.
Hence we define:

Definition 2.18 *Let μ, μ' be two fuzzy sets of X. μ is called a* **subset** *of μ',
denoted by $\mu \subseteq \mu'$, if for all $x \in X$ the inequality $\mu(x) \leq \mu'(x)$ holds.*

Remark 2.19 Obviously $\mathbb{I}_\emptyset \subseteq \mu \subseteq \mathbb{I}_X$ and $(\mu \subseteq \mu' \wedge \mu' \subseteq \mu \implies \mu = \mu')$
are valid for all $\mu', \mu \in F(X)$. It follows that $\mathbb{I}_\emptyset$ (i.e. the empty set) is the
smallest and $\mathbb{I}_X$ (i.e. the whole reference set) the greatest element of $F(X)$
with respect to $\subseteq$. $\subseteq$ forms a partial ordering of the set $F(X)$. If A and B are
crisp sets, then $\mathbb{I}_A \subseteq \mathbb{I}_B$ holds, if and only if A is a subset of B.

Therefore $\subseteq$ can be regarded as a generalization of ordinary set inclusion
to fuzzy sets.   □

2.3 Basic Operations on Fuzzy Sets

In the following the basic operations of set theory — intersection, union, and
complement — are introduced in the framework of fuzzy sets. An inspiration
for such generalized set operations can be received from the example below.

Example 2.20 Let a cube C be made up of $4 \times 4 \times 4$ smaller cubes $C_{i,j,k}$,
$1 \leq i, j, k \leq 4$, of equal size. Let each small cube consist of glass either crystal
clear or homogeneously coloured grey. If C is exposed to sunlight, then a grey
scale picture appears on the surface beneath the cube. This picture shows five
different shades of grey, depending on the number of grey cubes the light has
to pass through (see Figure 9).

We can describe this grey scale picture as an L-fuzzy set or as a fuzzy
set $\mu_K : P \to G$, where $P = \{1, 2, 3, 4\}^2$ denotes the set of pixels (squares
on the surface beneath C), and $G = \{0, \frac{1}{4}, \frac{1}{2}, \frac{3}{4}, 1\}$ denotes the ordered
set of all shades of grey. Let K be the configuration of the small cubes
that determines the composition of the cube C under consideration, namely
$K = \{(i, j, k) \mid W_{i,j,k} \text{ is grey}\} \subseteq \{1, 2, 3, 4\}^3$.

If $\mathcal{K} = \{K \mid K \subseteq \{1, 2, 3, 4\}^3\}$ is the set of all possible configurations, and
$\pi : \mathcal{K} \to F(P)$ is the mapping which relates a configuration to the grey scale

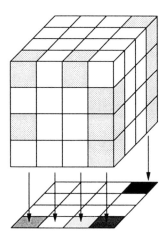

Figure 9   Emergence of a grey scale picture

picture $\mu_K = \pi(K)$ it produces, then $\pi(K)$ contains some information about the arrangement of the small cubes, but since π is not an injection, an exact identification of K is not possible in general. That is, $\pi(K)$ can be regarded as an information-compressed description of K.

In the light of the (epistemic) semantics of fuzzy sets adopted in this example, we now discuss the problem of intersecting two fuzzy sets.

Let $K_1, K_2 \in \mathcal{K}$ be the configurations of two cubes C_1 and C_2. The intersection of these, the configuration $K_1 \cap K_2$, has grey cubes in those positions where there are grey cubes in K_1 as well as in K_2, and it also produces a grey scale picture, namely $\mu_{K_1 \cap K_2} = \pi(K_1 \cap K_2)$.

In general the grey scale picture $\mu_{K_1 \cap K_2}$ cannot be calculated from the knowledge of the grey scale pictures μ_{K_1} and μ_{K_2} alone, since for all $(i,j) \in P$ we have:

$$\max(0, \mu_{K_1}(i,j) + \mu_{K_2}(i,j) - 1) \leq \mu_{K_1 \cap K_2}(i,j) \leq \min(\mu_{K_1}(i,j), \mu_{K_2}(i,j)).$$

But if additional information is available about K_1 and K_2, then there are some special cases, in which $\mu_{K_1 \cap K_2}$ can be calculated from μ_{K_1} and μ_{K_2}. We consider three examples:

1) It may be known that in every pile (i,j) of both configurations K_1 and K_2 the clear cubes lie on top of the grey ones. Then it follows:

$$\mu_{K_1 \cap K_2}(i,j) = \min(\mu_{K_1}(i,j), \mu_{K_2}(i,j)).$$

2) Given the information that in every pile (i,j) of configuration K_1 the clear cubes lie on top, yet in every pile of configuration K_2 they lie at the

bottom, we obtain:

$$\mu_{K_1 \cap K_2}(i,j) = \max(0, \mu_{K_1}(i,j) + \mu_{K_2}(i,j) - 1).$$

3) If we assume that in every pile (i,j) any arrangement of small cubes producing the shade of grey $\mu_{K_1}(i,j)$ or $\mu_{K_2}(i,j)$, respectively, is equally likely and independent of the pile (i,j) and the cube under consideration, then the 'expected grey scale picture' can be defined as follows:

In the pile (i,j) there are $\binom{4}{4\mu_{K_1}(i,j)}$ different arrangements in C_1 and $\binom{4}{4\mu_{K_2}(i,j)}$ in C_2. In all there are $N = \binom{4}{4\mu_{K_1}(i,j)}\binom{4}{4\mu_{K_2}(i,j)}$ different combinations, each of which is equally likely and therefore has the same probability $\frac{1}{N}$. If the combinations are ordered in an appropriate way, and if $a_n(i,j)$ is the number of grey cubes in pile (i,j) of the intersection of μ_{K_1} and μ_{K_2} induced by the n-th combination, then we get the expected value

$$\sum_{n=1}^{N} \frac{1}{N} \cdot a_n(i,j) = \mu_{K_1}(i,j) \cdot \mu_{K_2}(i,j).$$

In this case it is therefore sensible to define

$$\mu_{K_1 \cap K_2}(i,j) = \mu_{K_1}(i,j) \cdot \mu_{K_2}(i,j).$$

Unlike the general situation, where no additional facts about μ_{K_1} and μ_{K_2} are known, the three special cases discussed above deal with *truth functionality*, i.e. the value of $\mu_{K_1 \cap K_2}(i,j)$ can be inferred unambiguously from the values $\mu_{K_1}(i,j)$ and $\mu_{K_2}(i,j)$.   □

Taking Example 2.20 as a starting point, we now consider the question how the classic operations of set theory can be transferred to fuzzy sets. Starting with intersection, we assume, as a general principle, that the intersection $\mu \cap \mu'$ of two fuzzy sets $\mu, \mu' \in F(X)$, which we want to define, must be calculated *point by point*. Then there is a function $\sqcap : [0,1]^2 \to [0,1]$ which satisfies

$$(\mu \cap \mu')(x) = \sqcap(\mu(x), \mu'(x)).$$

To be acceptable as an operator for intersection, $\sqcap$ has to satisfy some axioms, that are stated in the definition below and that lead to the notion of a *t*-norm (*t* norms are known from multi-valued logic). However, for the time being, we will apply this notion to the elementary view on degrees of membership only without mentioning any relations to mathematical logic.

Definition 2.21 *A function* $\mathsf{T} : [0,1]^2 \to [0,1]$ *is called a* **t-norm** *if the following holds:*

(i)	$\mathsf{T}(a,1) = a$	*(unit element),*
(ii)	$a \leq b \Longrightarrow \mathsf{T}(a,c) \leq \mathsf{T}(b,c)$	*(monotonicity),*
(iii)	$\mathsf{T}(a,b) = \mathsf{T}(b,a)$	*(commutativity),*
(iv)	$\mathsf{T}(a,\mathsf{T}(b,c)) = \mathsf{T}(\mathsf{T}(a,b),c)$	*(associativity).*

It is evident that T is monotonic non-decreasing in both arguments and that $\mathsf{T}(a,0) = 0$.

The meaning of (i) is, that an intersection of a fuzzy set and an ordinary set may only lead to an exclusion of elements (the degrees of membership become 0) or to a conservation of the existing degree of membership.

Condition (ii) ensures that monotonic increasing degrees of membership as arguments cause monotonic increasing degrees of membership in the intersection of two fuzzy sets.

In addition it is obvious that we require commutativity (iii) and associativity (iv).

The *t*-norms that are most often used in applications have already been encountered in Example 2.20, namely

$$\mathsf{T}_{\min}(a,b) \stackrel{\text{def}}{=} \min\{a,b\},$$

$$\mathsf{T}_{\text{Luka}}(a,b) \stackrel{\text{def}}{=} \max\{0, a+b-1\}, \quad \text{and} \tag{2.13}$$

$$\mathsf{T}_{\text{prod}}(a,b) \stackrel{\text{def}}{=} a \cdot b.$$

Dual to the notion of *t*-norms, the notion of *t*-conorms is used, which is likely to define different generalized union-operators.

Definition 2.22 *A function* $\perp : [0,1]^2 \to [0,1]$ *is called a* **t-conorm**, *if and only if* $\perp$ *is commutative, associative, monotonic non-decreasing in both arguments, and has 0 as unit element.*

Well-known examples of *t*-conorms are

$$\perp_{\min}(a,b) \stackrel{\text{def}}{=} \max\{a,b\},$$

$$\perp_{\text{Luka}}(a,b) \stackrel{\text{def}}{=} \min\{a+b,1\}, \quad \text{and} \tag{2.14}$$

$$\perp_{\text{prod}}(a,b) \stackrel{\text{def}}{=} a+b-ab.$$

t-norms and *t*-conorms induce connectives for fuzzy sets by means of

$$(\mu \cap_\mathsf{T} \mu')(x) \stackrel{\text{def}}{=} \mathsf{T}(\mu(x), \mu'(x)) \quad \text{and}$$

$$(\mu \cup_\perp \mu')(x) \stackrel{\text{def}}{=} \perp(\mu(x), \mu'(x)). \tag{2.15}$$

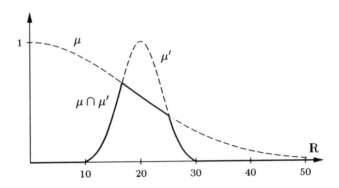

Figure 10   Intersection of two fuzzy sets

Example 2.23 In his paper 'Fuzzy Sets', written in 1965, L.A. Zadeh used
the operators

$$(\mu \cap \mu')(x) \quad \overset{\text{def}}{=} \quad \min\{\mu(x), \mu'(x)\} \qquad \text{(intersection)},$$

$$(\mu \cup \mu')(x) \quad \overset{\text{def}}{=} \quad \max\{\mu(x), \mu'(x)\} \qquad \text{(union)},$$

$$\overline{\mu}(x) \quad \overset{\text{def}}{=} \quad 1 - \mu(x) \qquad \text{(complement)}.$$

In particular the pair of t-norm *min* and t-conorm *max* has favourable
algebraic qualities like the distributive laws

$$\mu \cap (\mu' \cup \mu'') = (\mu \cap \mu') \cup (\mu \cap \mu'') \quad \text{and}$$
$$\mu \cup (\mu' \cap \mu'') = (\mu \cup \mu') \cap (\mu \cup \mu''). \tag{2.16}$$

The laws of DeMorgan

$$\overline{\mu \cup \mu'} = \overline{\mu} \cap \overline{\mu'} \quad \text{and} \quad \overline{\mu \cap \mu'} = \overline{\mu} \cup \overline{\mu'} \tag{2.17}$$

hold not only for $\cap$ and $\cup$, but also for all pairs of t-norms und t-conorms.
$(F(X), \cap, \cup)$ forms a complete distributive lattice with zero element $\mathbb{1}_\emptyset$ and
unit element $\mathbb{1}_X$.

However, we do not get a Boolean algebra, since there is no complement $^{-}$,
for which $\mu \cap \overline{\mu} = \mathbb{1}_\emptyset$ and $\mu \cup \overline{\mu} = \mathbb{1}_X$ are true in any case.

From a practical point of view, the pair (min, max) is comfortable to work
with arithmetically as well as graphically. For example, if we take $\mu, \mu' \in F(\mathbb{R})$
to describe the linguistic values 'young' and 'about 20 years old', then $\mu \cap \mu'$
is obtained as shown in Figure 10. $\square$

(min, max) is the only pair of distributive and thus absorbing and idempotent t-norms and t-conorms. Therefore, if one wants to preserve the equalities

$$\mu \cap \mu = \mu \quad \text{and} \quad \mu \cup \mu = \mu$$

for fuzzy sets μ, one has to turn to this pair of operators. But since in many applications fuzzy sets affect each other when united or intersected — even if they have the same characterization — then pairs of parameterized operators are often used instead of (min, max).

Example 2.24 Let $(T_\lambda, \perp_\lambda)$, $\lambda \in (-1, \infty)$, be the parameterized *Weber-family* of t-norms and t-conorms, i.e.

$$\begin{aligned}
T_\lambda(a,b) &= \max\left\{\frac{a+b-1+\lambda ab}{1+\lambda}, 0\right\} \quad \text{and} \\
\perp_\lambda(a,b) &= \min\left\{a+b-\frac{\lambda ab}{1+\lambda}, 1\right\}.
\end{aligned} \tag{2.18}$$

Particularly for $\lambda = 0$ we have the pair $(T_{\text{Luka}}, \perp_{\text{Luka}})$, for $\lambda \to \infty$ the pair $(T_{\text{prod}}, \perp_{\text{prod}})$, and for $\lambda \to -1$ the pair $(T_{-1}, \perp_{-1})$ with

$$T_{-1}(a,b) = \begin{cases} a, & \text{if } b = 1 \\ b, & \text{if } a = 1 \\ 0, & \text{otherwise} \end{cases} \quad \text{(drastic product)}$$

and

$$\perp_{-1}(a,b) = \begin{cases} a, & \text{if } b = 0 \\ b, & \text{if } a = 0 \\ 1, & \text{otherwise} \end{cases} \quad \text{(drastic sum).} \qquad \square$$

Since any t-norm $T : [0,1]^2 \to [0,1]$ defines a t-conorm dual to itself by

$$\begin{aligned}
\perp : [0,1]^2 &\to [0,1], \\
\perp(a,b) &= 1 - T(1-a, 1-b),
\end{aligned} \tag{2.19}$$

sensible pairs of t-norms and t-conorms are easily determined if the t-norms are given. Further families of such pairs are for instance:

Hamacher-family

$$\begin{aligned}
T'_\gamma(a,b) &= \frac{ab}{\gamma + (1-\gamma)(a+b-ab)}, \\
\perp'_\gamma(a,b) &= \frac{a+b-ab-(1-\gamma)ab}{1-(1-\gamma)ab}, \quad \gamma > 0,
\end{aligned} \tag{2.20}$$

Yager-family

$$T_p''(a, b) = 1 - \min\left\{[(1 - a)^p + (1 - b)^p]^{\frac{1}{p}}, 1\right\},$$

$$\perp_p''(a, b) = \min\left\{[a^p + b^p]^{\frac{1}{p}}, 1\right\}, \quad p > 0. \tag{2.21}$$

In the Yager-family, for $p \to 0$ we obtain the pair $(T_{-1}, \perp_{-1})$, and for $p \to \infty$ the pair (min, max).

By this the Yager-family covers a broad range of t-norms and t-conorms, since it can be shown for all t-norms T and t-conorms $\perp$ that the following holds:

$$\begin{aligned} T_{-1}(a, b) &\leq T(a, b) \leq \min\{a, b\} \quad \text{and} \\ \max\{a, b\} &\leq \perp(a, b) \leq \perp_{-1}(a, b). \end{aligned} \tag{2.22}$$

Archimedian t-norms and t-conorms are especially attractive from a mathematical point of view.

Definition 2.25 *Let* $T : [0, 1]^2 \to [0, 1]$ *and* $\perp : [0, 1]^2 \to [0, 1]$ *be two functions.*

a) T *is called an* **Archimedian t-norm***, if and only if* T *is a continuous t-norm and the inequality* $T(a, a) < a$ *is valid for all* $a \in (0, 1)$.

b) $\perp$ *is called* **Archimedian t-conorm***, if and only if* $\perp$ *is a continuous t-conorm and the inequality* $\perp(a, a) > a$ *is valid for all* $a \in (0, 1)$.

Archimedian t-norms can be represented as follows:

Theorem 2.26 *A function* $T : [0, 1]^2 \to [0, 1]$ *is an Archimedian t-Norm, if and only if there is a strictly decreasing continuous function* $f : [0, 1] \to [0, \infty]$ *with* $f(1) = 0$ *and*

$$T(a, b) = f^{(-1)}(f(a) + f(b)),$$

where $f^{(-1)}$ *is the* **pseudo-inverse** *of* f, *i.e.*

$$f^{(-1)}(y) \overset{\text{def}}{=} \begin{cases} \{x \in [0, 1] \mid f(x) = y\}, & \text{if} \quad y \in [0, f(0)] \\ 0, & \text{if} \quad y \in [f(0), \infty]. \end{cases}$$

If $f(0) = \infty$, *then* T *is strictly increasing in both arguments.*

Proof:
A proof of this theorem can be found in [Schweizer61]. $\square$

By analogy any Archimedian t-conorm $\perp$ can be described by a strictly increasing continuous function $g : [0, 1] \to [0, \infty]$ with $g(0) = 0$,

$$\perp(a, b) = g^{(-1)}(g(a) + g(b))$$

and the pseudo-inverse

$$g^{(-1)}(y) = \begin{cases} \{x \in [0,1] \mid g(x) = y\}, & \text{if } y \in [0, g(1)] \\ 1, & \text{if } y \in [g(1), \infty]. \end{cases}$$

If $g(1) = \infty$, then $\perp$ is strictly increasing in both arguments. The mentioned functions f and g are uniquely determined with the exception of a positive factor.

Example 2.27

(a) $f_p : [0,1] \to [0,1]$, $f_p(x) \overset{\text{def}}{=} (1-x)^p$,

$g_p : [0,1] \to [0,1]$, $g_p(x) \overset{\text{def}}{=} x^p$, $p > 0$.

f_p is strictly decreasing with $f_p(1) = 0$ and $f_p(0) = 1$,

g_p on the other hand is strictly increasing with $g_p(0) = 0$ and $g_p(1) = 1$.

The pairs of t-norm and t-conorm induced by f_p and g_p form the Yager-family.

(b) $f : [0, \infty] \to [0,1]$, $f(x) \overset{\text{def}}{=} \begin{cases} -\ln x, & \text{if } x \in (0,1] \\ \infty, & \text{if } x = 0, \end{cases}$

$g : [0,1] \to [0, \infty]$, $g(x) \overset{\text{def}}{=} \begin{cases} -\ln(1-x), & \text{if } x \in [0,1) \\ \infty, & \text{if } x = 1. \end{cases}$

f and g satisfy the preconditions of Theorem 2.26. It follows that:

$$\begin{aligned} \mathsf{T}(a,b) &= f^{(-1)}(f(a) + f(b)) \\ &= f^{(-1)}(-\ln a - \ln b) & \text{(algebraic product)}, \\ &= e^{\ln(a)+\ln(b)} = a \cdot b \end{aligned}$$

$$\begin{aligned} \perp(a,b) &= g^{(-1)}(g(a) + g(b)) \\ &= g^{(-1)}(-\ln(1-a) - \ln(1-b)) & \text{(algebraic sum)}. \\ &= 1 - e^{\ln(1-a)+\ln(1-b)} = a + b - ab \end{aligned}$$ □

To form a complement of fuzzy sets, analogous considerations are made with regard to axiomatization. A function $n : [0,1] \to [0,1]$ is called a *negation*, if $n(0) = 1$, $n(1) = 0$, and $n(a) \geq n(b)$ is valid for all $a \leq b$. The most frequently used negation is the one suggested by Zadeh, which is

$$n(x) = 1 - x.$$

It is *strict*, that means strictly monotonic and continuous.

Using a strict negation, any t-norm T is related to a dual t-conorm $\perp_n$, for which

$$\perp_n(a,b) = n(\mathsf{T}(n(a), n(b))) \tag{2.23}$$

holds. For instance, if $\mathsf{T} = \min$, then for $n(x) = 1 - x$ we obtain the dual t-conorm $\perp_n = \max$.

2.4   The Extension Principle

In the previous section we only discussed how generalizations to fuzzy sets of the basic operations on ordinary sets can be realized. Now we turn to the motivation of a general method, by which mappings of the form $\phi : X^n \to Y$ can be extended to mappings of the kind $\hat{\phi} : F(X)^n \to F(Y)$. This motivation will take place on an intuitive level without going into details concerning the relations between fuzzy set theory and multi-valued logic. These are postponed till Section 2.7, where they are considered in more depth.

Let $\mu \in F(\mathbf{R})$ be the fuzzy set interpretation of a vague concept named with the linguistic value 'about 2'. The *degree of membership* $\mu(2.2)$, with which the number 2.2 belongs to the fuzzy set μ, can also be viewed as the *acceptance degree* regarding the correctness of the statement '2.2 is about equal to 2'.

In addition, let $\mu' \in F(\mathbf{R})$ be the fuzzy set interpretation of the vague concept 'old'. Then, in reverse, the acceptance degree of the statement '2.2 is about equal to 2, and 2.2 is old' can be interpreted as the degree of membership of 2.2 with respect to the vague concept 'about 2 and old', i.e. to the fuzzy set $\mu \cap \mu'$.

Now the question arises, how we can operate on these 'degrees of acceptance', which are generalized truth values. In principle, to represent conjunction, any t-norms can be used, or to represent disjunction, any t-conorms. However, in this section we restrict ourselves to the pair (min, max).

Let $\mathcal{P}$ be a set of vague statements, which can be combined by the connectives *and* and *or*. The mapping acc $: \mathcal{P} \to [0, 1]$ may assign an acceptance degree $\mathrm{acc}(a)$ to every statement $a \in \mathcal{P}$. $\mathrm{acc}(a) = 0$ means that a is definitely false, $\mathrm{acc}(a) = 1$, on the other hand, that it is definitely true. However, if $\mathrm{acc}(a) \in (0, 1)$, then we can only speak of a *gradual truth* of the statement a.

If we combine two statements $a, b \in P$, their combination is rated according to the following scheme:

$$\begin{aligned} \mathrm{acc}(a \text{ and } b) &= \mathrm{acc}(a \wedge b) = \min\{\mathrm{acc}(a), \mathrm{acc}(b)\}, \\ \mathrm{acc}(a \text{ or } b) &= \mathrm{acc}(a \vee b) = \max\{\mathrm{acc}(a), \mathrm{acc}(b)\}. \end{aligned} \quad (2.24)$$

For an infinite number of statements a_i, $i \in I$, we define analogously:

$$\begin{aligned} \mathrm{acc}(\forall i \in I : a_i) &= \inf\{\mathrm{acc}(a_i) \mid i \in I\}, \\ \mathrm{acc}(\exists i \in I : a_i) &= \sup\{\mathrm{acc}(a_i) \mid i \in I\}. \end{aligned} \quad (2.25)$$

With the help of the concept of acceptance degrees of vague statements we now show how a function $\phi : X^n \to Y$, which maps the tuples $(x_1, \ldots, x_n)$ of X^n to the crisp value $\phi(x_1, \ldots, x_n)$ of Y, can be extended in a suitable way to a function $\hat{\phi} : F(X)^n \to F(Y)$, which maps vague descriptions $(\mu_1, \ldots, \mu_n) \in F(X)^n$ of $(x_1, \ldots, x_n)$ to the fuzzy value $\hat{\phi}(\mu_1, \ldots, \mu_n)$.

Example 2.28 Let $+ : \mathbf{R}^2 \to \mathbf{R}$ be the addition of real numbers. If the exact values x_1 and x_2 are not accessible, then in an *ad hoc* modelling one would try to work with intervals A_1 and A_2 that satisfy $x_1 \in A_1$ and $x_2 \in A_2$ with certainty.

Therefore the addition of A_1 and A_2 is defined as

$$A_1 + A_2 \stackrel{\text{def}}{=} \{a_1 + a_2 \mid a_1 \in A_1 \wedge a_2 \in A_2\}.$$

This procedure is well known from the arithmetic of intervals. For instance,

$$[4, 5] + [8, 10] = [12, 15],$$

i.e. by merely adding the bounds a new interval is obtained.   □

The extension of a function $\phi : X^n \to Y$ to power sets, as carried out in the example above, in the general case leads to

$$\bar{\phi} : (\mathfrak{P}(X))^n \quad \to \quad \mathfrak{P}(Y),$$
$$\bar{\phi}(A_1, \ldots, A_n) \quad = \quad \{y \in Y \mid \exists (x_1, \ldots, x_n) \in A_1 \times \ldots \times A_n : \\ \phi(x_1, \ldots, x_n) = y\}.$$

If we determine the acceptance degree of the statement 'y belongs to the image of $(A_1, \ldots, A_n)$', we get

$$\mathrm{acc}(y \text{ belongs to the image of } (A_1, \ldots, A_n))$$
$$= \quad \mathrm{acc}(\exists (x_1, \ldots, x_n) \in A_1 \times \ldots \times A_n : \phi(x_1, \ldots, x_n) = y)$$
$$= \quad \begin{cases} 1, & \text{if } \exists (x_1, \ldots, x_n) \in A_1 \times \ldots \times A_n : \phi(x_1, \ldots, x_n) = y \\ 0, & \text{otherwise.} \end{cases}$$

A further generalization of the function $\phi : X^n \to Y$ to $\hat{\phi} : F(X)^n \to F(Y)$ can be obtained proceeding in an analogous way: The acceptance degree of the statement 'y belongs to the image of $(\mu_1, \ldots, \mu_n)$' with $\mu_i \in F(X)$, $i = 1, \ldots, n$, then is

$$\mathrm{acc}(y \text{ belongs to the image of } (\mu_1, \ldots, \mu_n))$$

$$= \mathrm{acc}(\exists (x_1, \ldots, x_n) \in X^n : \quad x_1 \text{ belongs to } \mu_1 \text{ and} \\ x_2 \text{ belongs to } \mu_2 \text{ and} \\ \cdots \qquad \qquad \text{and} \\ x_n \text{ belongs to } \mu_n \text{ and} \\ y = \phi(x_1, \ldots, x_n))$$

$$= \sup_{(x_1,\dots,x_n)\in X^n} \{\mathrm{acc}(x_1 \text{ belongs to } \mu_1 \text{ and}$$
$$\cdots \qquad \text{and}$$
$$x_n \text{ belongs to } \mu_n \text{ and}$$
$$y = \phi(x_1,\dots,x_n))\}$$

$$= \sup_{(x_1,\dots,x_n)\in X^n} \{\min\{\mathrm{acc}(x_1 \text{ belongs to } \mu_1),$$
$$\cdots$$
$$\mathrm{acc}(x_n \text{ belongs to } \mu_n),$$
$$\mathrm{acc}(y = \phi(x_1,\dots,x_n))\}\}$$

$$= \sup_{\substack{(x_1,\dots,x_n)\in X^n: \\ y=\phi(x_1,\dots,x_n)}} \{\min\{\mu_1(x_1),\dots,\mu_n(x_n)\}\},$$

since the acceptance degree of x_i belonging to the vague concept characterized by μ_i *per definitionem* is equal to the acceptance degree $\mu_i(x_i)$, and since it is evident that

$$\mathrm{acc}(y = \phi(x_1,\dots,x_n)) = \begin{cases} 1, & \text{if } \phi(x_1,\dots,x_n) = y \\ 0, & \text{otherwise} \end{cases}$$

holds.

In all we get the extended mapping

$$\hat{\phi} : F(X)^n \rightarrow F(Y)$$

with

$$\hat{\phi}(\mu_1,\dots,\mu_n) : Y \rightarrow [0,1],$$
$$\hat{\phi}(\mu_1,\dots,\mu_n)(y) = \sup\{\min\{\mu_1(x_1),\dots,\mu_n(x_n)\} \mid \qquad (2.26)$$
$$(x_1,\dots,x_n) \in X^n$$
$$\text{and } y = \phi(x_1,\dots,x_n)\}.$$

Example 2.29 Let the fuzzy set interpretation of the vague concept 'approximately 2' be defined as

$$\mu \in F(\mathbf{R}), \quad \mu(x) = \begin{cases} x - 1, & \text{if } 1 \leq x \leq 2 \\ 3 - x, & \text{if } 2 \leq x \leq 3 \\ 0, & \text{otherwise.} \end{cases}$$

Then the extension of the mapping $\phi : \mathbf{R} \rightarrow \mathbf{R}$, $x \mapsto x^2$ to fuzzy sets on $\mathbf{R}$ provides us with

$$\hat{\phi}(\mu) : \mathbf{R} \rightarrow [0,1],$$
$$\hat{\phi}(\mu)(y) = \sup\{\mu(x) \mid x \in \mathbf{R} \wedge x^2 = y\}$$
$$= \begin{cases} \sqrt{y} - 1, & \text{if } 1 \leq y \leq 4 \\ 3 - \sqrt{y}, & \text{if } 4 \leq y \leq 9 \\ 0, & \text{otherwise.} \end{cases} \qquad \square$$

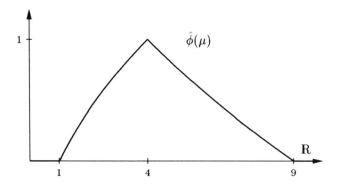

Figure 11   Graph of $\hat{\phi}(\mu)$ of Example 2.29

The method described above, by which a function $\phi : X^n \to Y$ can be extended to $\hat{\phi} : F(X)^n \to F(Y)$, is called the *extension principle*.

Definition 2.30 *Let $\phi : X^n \to Y$ be a mapping. The **extension** of ϕ is given by*

$$\hat{\phi} : (F(X))^n \quad \to \quad F(Y) \quad with$$
$$\hat{\phi}(\mu_1, \ldots, \mu_n)(y) \quad \overset{\text{def}}{=} \quad \sup\{\min\{\mu_1(x_1), \ldots, \mu_n(x_n)\} \mid$$
$$(x_1, \ldots, x_n) \in X^n$$
$$\text{and } y = \phi(x_1, \ldots, x_n)\}$$

assuming that $\sup \emptyset \overset{\text{def}}{=} 0.$

As a first application of the extension principle, that can be taken as a basis for 'fuzzifying' whole theories, we look at arithmetical operations on fuzzy intervals in the next section.

2.5   Efficient Operations on Fuzzy Sets

In this section we restrict ourselves to the examination of special classes of fuzzy sets of real numbers.

Definition 2.31
(a) $F_N(\mathbf{R}) \overset{\text{def}}{=} \{\mu \in F(\mathbf{R}) \mid \exists x \in \mathbf{R} : \mu(x) = 1\},$

(b) $F_C(\mathbf{R}) \overset{\text{def}}{=} \{\mu \in F_N(\mathbf{R}) \mid \forall \alpha \in (0,1] : [\mu]_\alpha \text{ compact}\},$

(c) $F_I(\mathbf{R}) \overset{\text{def}}{=} \{\mu \in F_N(\mathbf{R}) \mid \forall a, b, c \in \mathbf{R} :$

$$a \leq c \leq b \Rightarrow \mu(c) \geq \min\{\mu(a), \mu(b)\}\}.$$

The elements of $F_N(\mathbf{R})$ are called *normal fuzzy sets*. The normality condition is sensible in all cases where a fuzzy set $\mu \in F_N(\mathbf{R})$ is used to characterize an existing yet not directly accessible real-valued quantity. In such cases it would not be plausible to assign 1 as the maximum acceptance degree to not a single number.

The fuzzy sets contained in $F_C(\mathbf{R})$ are upper semicontinuous. This leads to a simplification of the arithmetic functions applied to them, which remain to be examined in greater detail below. A special role play the fuzzy sets of $F_I(\mathbf{R})$. They are normal and *fuzzy convex* (i.e. all α-cuts are convex). They are called *fuzzy intervals*, since their core is an ordinary interval. We emphasize that fuzzy convexity should not be confused with the convexity of the characteristic function. If they are interpreted as vague characterization of real numbers, the elements of $F_I(\mathbf{R})$ are often called *fuzzy numbers*.

Before we can do calculations with fuzzy numbers, we have to define the basic arithmetic operations. With the help of the extension principle, for the sum $\mu \oplus \mu'$, the product $\mu \odot \mu'$, and the reciprocal value $\text{rec}(\mu)$ of arbitrary fuzzy sets $\mu, \mu' \in F(\mathbf{R})$, we lay down the following:

$$(\mu \oplus \mu')(t) = \sup\{\min\{\mu(x_1), \mu'(x_2)\} \mid x_1, x_2 \in \mathbf{R} \wedge x_1 + x_2 = t\},$$
$$(\mu \odot \mu')(t) = \sup\{\min\{\mu(x_1), \mu'(x_2)\} \mid x_1, x_2 \in \mathbf{R} \wedge x_1 x_2 = t\},$$
$$\text{rec}(\mu)(t) = \sup\{\mu(x) \mid x \in \mathbf{R}\backslash\{0\} \wedge \tfrac{1}{x} = t\}, \qquad \text{for all } t \in \mathbf{R}.$$

Other operations can be defined in an analogous way.

Example 2.32 Figure 12 shows three normal fuzzy sets $\mu_1, \mu_2, \mu_3 \in F_C(\mathbf{R})$; μ_1 and μ_3 are even fuzzy intervals, and, in particular, μ_3 is the characterization of the ordinary set $\{2\}$.

Considering the graphs of $\mu_1 \oplus \mu_2$, $\mu_1 \odot \mu_3$, and $\text{rec}(\mu_1)$, the first two coincide with what is expected intuitively. Considering $\text{rec}(\mu_1)$ one has to take into account that $\mu_1(0) = 0.5$, and that a division by zero is not allowed. Therefore an asymptotic course of the function near a degree of membership of 0.5 results:

$$\text{rec}(\mu_1)(t) = \sup\{\mu_1(x) \mid x \in \mathbf{R} \wedge \tfrac{1}{x} = t\}$$

$$= \begin{cases} \frac{1}{2t} + \frac{1}{2}, & \text{if} \quad t \leq -1 \quad \vee \quad t \geq 1 \\ 2 - \frac{1}{t}, & \text{if} \quad \frac{1}{2} \leq t \leq 1 \\ 0, & \text{if} \quad -1 \leq t \leq \frac{1}{2}. \end{cases} \qquad \square$$

In general operations on fuzzy sets are much more complicated, especially if fuzzy convexity is not given, and if the vertical instead of the horizontal

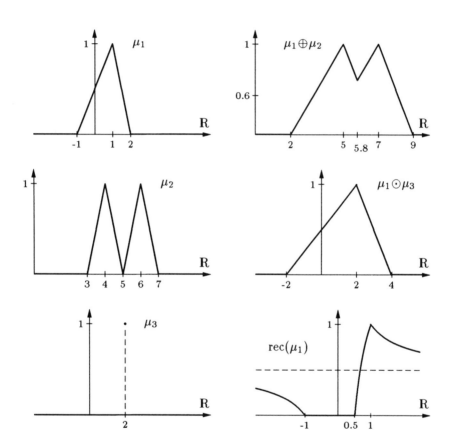

Figure 12 Arithmetic operations on fuzzy sets

representation in the sense of Definition 2.30 is applied. That is why it is desirable to reduce fuzzy arithmetics to ordinary set arithmetics and to take advantage of using elementary operations of *interval arithmetics*. In this connection we return to the concept introduced in Definition 2.15. The condition for continuity (F3) described in that definition can be used in a proof to establish a correspondence between fuzzy sets on $\mathbf{R}$ and the families of sets in $\mathcal{FL}(\mathbf{R})$. Thus the mapping

$$\phi : \mathcal{FL}(\mathbf{R}) \quad \rightarrow \quad F(\mathbf{R}),$$
$$\phi\left((A_\alpha)_{\alpha \in [0,1]}\right) \quad \overset{\text{def}}{=} \quad \mu \tag{2.27}$$

with $\mu(x) = \sup\{\alpha \mid \alpha \in [0, 1] \wedge x \in A_\alpha\}$ is a one-to-one-mapping.

But if one wants to represent normal fuzzy sets with families $\mathcal{A} = (A_\alpha)_{\alpha \in (0,1)}$ of sets, that are as general as possible, the condition for continuity (F3) is omitted as well as the condition (F1), which is rendered superfluous by $\alpha \neq 0$ anyhow. Merely using (F2) and the additional condition $\bigcap_{\alpha \in (0,1)} A_\alpha \neq \emptyset$, which ensures normality, the fuzzy set $\mu_\mathcal{A}$ induced by $\mathcal{A}$ can be specified by

$$\mu_\mathcal{A}(t) = \sup\{\alpha \mid \alpha \in (0, 1) \wedge t \in A_\alpha\}.$$

This motivates the following definition:

Definition 2.33 *A family* $(A_\alpha)_{\alpha \in (0,1)}$ *of sets is called a* **set representation** *of* $\mu \in F_N(\mathbf{R})$, *if*

(a) $0 < \alpha < \beta < 1 \Longrightarrow A_\beta \subseteq A_\alpha \subseteq \mathbf{R}$, *and*

(b) $\mu(t) = \sup\{\alpha \mid \alpha \in (0, 1) \wedge t \in A_\alpha\}$

holds, where $\sup \emptyset \overset{\text{def}}{=} 0$.

The next theorem shows that a fuzzy set can have several set representations.

Theorem 2.34 *Let* $\mu \in F_N(\mathbf{R})$. *The family* $(A_\alpha)_{\alpha \in (0,1)}$ *of sets is a set representation of* μ, *if and only if*

$$\{t \in \mathbf{R} \mid \mu(t) > \alpha\} \subseteq A_\alpha \subseteq \{t \in \mathbf{R} \mid \mu(t) \geq \alpha\}$$

is valid for all $\alpha \in (0, 1)$.

Proof:
Let $[\mu]_{\underline{\alpha}} \overset{\text{def}}{=} \{t \in \mathbf{R} \mid \mu(t) > \alpha\}$, $\alpha \in (0, 1)$.

'$\Longrightarrow$': Starting from a set representation $(A_\gamma)_{\gamma \in (0,1)}$ of μ and an arbitrary, but fixed $\alpha \in (0,1)$, the following holds for all $t \in \mathbf{R}$:

$$t \in [\mu]_{\underline{\alpha}} \implies \mu(t) = \sup\{\gamma \mathbb{I}_{A_\gamma}(t) \mid \gamma \in (0,1)\} > \alpha$$
$$\implies \exists \beta \in (\alpha, 1) : t \in A_\beta$$
$$\implies t \in A_\alpha.$$

$t \in A_\alpha$ can be weakened to $t \in [\mu]_\alpha$ by using $\mu(t) \geq \alpha \cdot \mathbb{I}_{A_\alpha}(t) = \alpha$, i.e. $[\mu]_{\underline{\alpha}} \subseteq A_\alpha \subseteq [\mu]_\alpha$.

'$\Longleftarrow$': Let the inclusion $[\mu]_{\underline{\gamma}} \subseteq A_\gamma \subseteq [\mu]_\gamma$ be satisfied for all $\gamma \in (0,1)$. We have to show that $(\overline{A_\gamma})_{\gamma \in (0,1)}$ is a set representation of μ.

(a) If $0 < \alpha < \beta < 1$, then $t \in [\mu]_\beta \implies \mu(t) \geq \beta > \alpha \implies t \in [\mu]_{\underline{\alpha}}$ and therefore $[\mu]_\beta \subseteq [\mu]_{\underline{\alpha}}$. Because of the prerequisite $[\mu]_{\underline{\gamma}} \subseteq A_\gamma \subseteq [\mu]_\gamma$, $\gamma \in (0,1)$, the chain of inclusions $A_\beta \subseteq [\mu]_\beta \subseteq [\mu]_{\underline{\alpha}} \subseteq A_\alpha$ is valid.

(b) To show $\mu(t) = \sup\{\alpha \mid \alpha \in (0,1) \wedge t \in A_\alpha\}$, we begin with $\mu(t) > 0$. Applying Theorem 2.14 and taking care of the precondition we infer as follows:

$$\begin{aligned} \mu(t) &= \sup_{\alpha \in [0,1]} \left\{\min\{\alpha, \mathbb{I}_{[\mu]_\alpha}(t)\}\right\} \\ &= \sup\{\alpha \mid \alpha \in [0,1] \wedge t \in [\mu]_\alpha\} \\ &= \sup\{\alpha \mid \alpha \in (0,1) \wedge t \in [\mu]_\alpha\} \\ &\geq \sup\{\alpha \mid \alpha \in (0,1) \wedge t \in A_\alpha\} \\ &\geq \sup\{\alpha \mid \alpha \in (0,1) \wedge t \in [\mu]_{\underline{\alpha}}\} \\ &= \sup\{\alpha \mid \alpha \in (0,1) \wedge \mu(t) > \alpha\} \\ &= \mu(t). \end{aligned}$$

If $\mu(t) = 0$, then we have $t \notin A_\alpha$ for all $\alpha \in (0,1)$ and therefore *per definitionem*
$$\sup\{\alpha \mid \alpha \in (0,1) \wedge t \in A_\alpha\} = \sup \emptyset = 0 = \mu(t). \qquad \square$$

Remark 2.35 Let $\mu \in F_N(\mathbf{R})$. In Definition 2.11 we introduced $[\mu]_\alpha = \{x \in \mathbf{R} \mid \mu(x) \geq \alpha\}$, $\alpha \in [0,1]$, as the α-cut of μ.
The sets $[\mu]_{\underline{\alpha}} \overset{\text{def}}{=} \{x \in \mathbf{R} \mid \mu(x) > \alpha\}$, $\alpha \in [0,1]$, are often called the *strict α-cuts* of μ.
The families $([\mu]_{\underline{\alpha}})_{\alpha \in (0,1)}$ of sets and $([\mu]_\alpha)_{\alpha \in (0,1)}$, induced by these strict α-cuts, are set representations of μ and — according to Theorem 2.34 — they obviously form the upper and lower bound of all set representations of μ. $\square$

The next theorem demonstrates how for arbitrary mappings $\phi : \mathbf{R}^n \to \mathbf{R}$ a set representation of their extension $\hat{\phi}(\mu_1, \ldots, \mu_n)$ can be obtained with the help of set representations $((A_i)_\alpha)_{\alpha \in (0,1)}$ of normal fuzzy sets $\mu_i \in F_N(\mathbf{R})$, $i = 1, 2, \ldots, n$.

The result achieved can be employed, for example, to carry out arithmetic operations on fuzzy sets efficiently.

Theorem 2.36 *Let $\mu_1, \mu_2, \ldots, \mu_n$ be normal fuzzy sets of $\mathbf{R}$ and $\phi : \mathbf{R}^n \to \mathbf{R}$ be a mapping. Then the following holds:*

(a) $\forall \alpha \in [0, 1) : [\hat{\phi}(\mu_1, \ldots, \mu_n)]_{\underline{\alpha}} = \phi([\mu_1]_{\underline{\alpha}}, \ldots, [\mu_n]_{\underline{\alpha}})$,

(b) $\forall \alpha \in (0, 1] : [\hat{\phi}(\mu_1, \ldots, \mu_n)]_\alpha \supseteq \phi([\mu_1]_\alpha, \ldots, [\mu_n]_\alpha)$,

(c) *if $((A_i)_\alpha)_{\alpha \in (0,1)}$ is a set representation of μ_i for $i = 1, 2, \ldots, n$, then*
$(\phi((A_1)_\alpha, \ldots, (A_n)_\alpha))_{\alpha \in (0,1)}$ *is a set representation of $\hat{\phi}(\mu_1, \ldots, \mu_n)$.*

Proof:

(a) Let $\alpha \in [0, 1)$ and $t \in \mathbf{R}$ be arbitrary, but fixed.

$\quad t \in [\hat{\phi}(\mu_1, \ldots, \mu_n)]_{\underline{\alpha}}$

$\quad \Longleftrightarrow \quad t \in \{y \in \mathbf{R} \mid \hat{\phi}(\mu_1, \ldots, \mu_n)(y) > \alpha\}$

$\quad \Longleftrightarrow \quad t \in \{y \in \mathbf{R} \mid \sup\{\min\{\mu_1(x_1), \ldots, \mu_n(x_n)\} \mid$
$\qquad\qquad\qquad\qquad (x_1, \ldots, x_n) \in \mathbf{R}^n \wedge \phi(x_1, \ldots, x_n) = y\} > \alpha\}$

$\quad \overset{(*)}{\Longleftrightarrow} \quad \exists (x_1, \ldots, x_n) \in \mathbf{R}^n :$
$\qquad\qquad (\phi(x_1, \ldots, x_n) = t \wedge \min\{\mu_1(x_1), \ldots, \mu_n(x_n)\} > \alpha)$

$\quad \Longleftrightarrow \quad \exists (x_1, \ldots, x_n) \in \mathbf{R}^n :$
$\qquad\qquad (\phi(x_1, \ldots, x_n) = t \wedge (x_1, \ldots, x_n) \in \times_{i=1}^n [\mu_i]_{\underline{\alpha}})$

$\quad \Longleftrightarrow \quad t \in \phi([\mu_1]_{\underline{\alpha}}, \ldots, [\mu_n]_{\underline{\alpha}})$.

(b) Just as in (a), yet at step $(*)$ the supremum term on the left entails that only '$\Longleftarrow$' is valid.

(c) This statement will be a trivial inference from Theorem 2.34, if we have proved

$$[\hat{\phi}(\mu_1, \ldots, \mu_n)]_{\underline{\alpha}} \subseteq \phi((A_1)_\alpha, \ldots, (A_n)_\alpha) \subseteq [\hat{\phi}(\mu_1, \ldots, \mu_n)]_\alpha.$$

Since $((A_i)_\alpha)_{\alpha \in (0,1)}$, $i = 1, 2, \ldots, n$, are set representations of μ_i, by applying Theorem 2.34 we get $[\mu_i]_{\underline{\alpha}} \subseteq (A_i)_\alpha \subseteq [\mu_i]_\alpha$. We make use of (a) and (b) and at last arrive at

$$\begin{aligned}
&[\hat{\phi}(\mu_1, \ldots, \mu_n)]_{\underline{\alpha}} \\
&= \phi([\mu_1]_{\underline{\alpha}}, \ldots, [\mu_n]_{\underline{\alpha}}) \\
&\subseteq \phi((A_1)_\alpha, \ldots, (A_n)_\alpha) \\
&\subseteq \phi([\mu_1]_\alpha, \ldots, [\mu_n]_\alpha) \\
&\subseteq [\hat{\phi}(\mu_1, \ldots, \mu_n)]_\alpha.
\end{aligned}$$

$\qquad\qquad\qquad\qquad\qquad\qquad\qquad\qquad\qquad\qquad\qquad\qquad\qquad\quad \square$

Example 2.32 (continuation) For $\mu_1, \mu_2 \in F_C(\mathbf{R})$ with

$$\mu_1(x) = \begin{cases} \frac{x+1}{2}, & \text{if } -1 \leq x \leq 1 \\ 2 - x, & \text{if } \quad 1 \leq x \leq 2 \\ 0, & \text{otherwise} \end{cases}$$

and

$$\mu_2(x) = \begin{cases} x - 3, & \text{if } 3 \leq x \leq 4 \\ 5 - x, & \text{if } 4 \leq x \leq 5 \\ x - 5, & \text{if } 5 \leq x \leq 6 \\ 7 - x, & \text{if } 6 \leq x \leq 7 \\ 0, & \text{otherwise} \end{cases}$$

we obtain the set representations $([\mu_1]_\alpha)_{\alpha \in (0,1)}$ and $([\mu_2]_\alpha)_{\alpha \in (0,1)}$, where

$$\begin{aligned} [\mu_1]_\alpha &= [2\alpha - 1, 2 - \alpha], \\ [\mu_2]_\alpha &= [\alpha + 3, 5 - \alpha] \cup [\alpha + 5, 7 - \alpha]. \end{aligned}$$

Let add $: \mathbf{R}^2 \to \mathbf{R}$, $\text{add}(x, y) \overset{\text{def}}{=} x + y$. Then, according to Theorem 2.36, $(A_\alpha)_{\alpha \in (0,1)}$, defined as $A_\alpha \overset{\text{def}}{=} \text{add}([\mu_1]_\alpha, [\mu_2]_\alpha)$, is a set representation of $\mu_1 \oplus \mu_2$. We calculate

$$\begin{aligned} A_\alpha &= \text{add}([\mu_1]_\alpha, [\mu_2]_\alpha) = [3\alpha + 2, 7 - 2\alpha] \cup [3\alpha + 4, 9 - 2\alpha] \\ &= \begin{cases} [3\alpha + 2, 7 - 2\alpha] \cup [3\alpha + 4, 9 - 2\alpha], & \text{if } \alpha \geq 0.6 \\ [3\alpha + 2, 9 - 2\alpha], & \text{if } \alpha \leq 0.6, \end{cases} \end{aligned}$$

and from this it follows that

$$(\mu_1 \oplus \mu_2)(x) = \begin{cases} \frac{x-2}{3}, & \text{if } 2 \ \ \leq x \leq 5 \\ \frac{7-x}{2}, & \text{if } 5 \ \ \leq x \leq 5.8 \\ \frac{x-4}{3}, & \text{if } 5.8 \leq x \leq 7 \\ \frac{9-x}{2}, & \text{if } 7 \ \ \leq x \leq 9 \\ 0, & \text{otherwise.} \end{cases}$$

$\square$

Remark 2.37 The continuation of Example 2.32 makes clear that a set representation $(A_\alpha)_{\alpha \in (0,1)}$ of the sum $\mu_1 \oplus \mu_2$ of two fuzzy sets of $F_C(\mathbf{R})$ can be found by calculating the level sets A_α by adding the α-cuts of μ_1 and μ_2. In an analogous way determining set representations of other combinations of fuzzy sets of $F_C(\mathbf{R})$ can often be reduced to executing simple operations of interval arithmetics.

If we consider the fundamental operations of arithmetics, we obtain the following interval operations (for $a, b, c, d \in \mathbf{R}$):

$$[a, b] + [c, d] = [a + c, b + d],$$
$$[a, b] - [c, d] = [a - d, b - c],$$

$$[a, b] \cdot [c, d] = \begin{cases} [ac, bd], & \text{for} \quad a \geq 0 \wedge c \geq 0 \\ [bd, ac], & \text{for} \quad b < 0 \wedge d < 0 \\ [\min\{ad, bc\}, \max\{ad, bc\}], & \text{for} \quad ab \geq 0 \wedge cd \geq 0 \\ & \qquad\qquad \wedge\ ac < 0 \\ [\min\{ad, bc\}, \max\{ac, bd\}], & \text{for} \quad ab < 0 \vee cd < 0 \end{cases}$$

$$\frac{1}{[a, b]} = \begin{cases} \left[\frac{1}{b}, \frac{1}{a}\right], & \text{if } 0 \notin [a, b] \\ \left[\frac{1}{b}, \infty\right) \cup \left(-\infty, \frac{1}{a}\right], & \text{if } a < 0 \wedge b > 0 \\ \left[\frac{1}{b}, \infty\right), & \text{if } a = 0 \wedge b > 0 \\ \left(-\infty, \frac{1}{a}\right], & \text{if } a < 0 \wedge b = 0. \end{cases}$$

Note that the addition, subtraction, or multiplication of two fuzzy intervals always yields another fuzzy interval, but that division does not. Also, in contrast to division, the other operations do not leave $F_C(\mathbf{R})$.

Many laws of calculation known from the fundamental operations of arithmetics with real numbers can be transferred to the corresponding extensions for fuzzy sets and consequently to interval operations. Thus the commutative and the associative laws hold for the addition of fuzzy sets of $\mathbf{R}$, i.e. for $\mu_1, \mu_2, \mu_3 \in F(\mathbf{R})$

$$\mu_1 \oplus \mu_2 = \mu_2 \oplus \mu_1,$$
$$\mu_1 \oplus (\mu_2 \oplus \mu_3) = (\mu_1 \oplus \mu_2) \oplus \mu_3.$$

The distributive law does not always hold, but one can show that

$$\mu_1 \odot (\mu_2 \oplus \mu_3) \subseteq (\mu_1 \odot \mu_2) \oplus (\mu_1 \odot \mu_3)$$

is valid.

It should be recognized that $F(\mathbf{R})$ does not form a group with respect to addition. With $\mathbb{1}_{\{0\}}$ there is a zero element, but generally there is no additive inverse. This can be inferred easily from the fact that the equation

$$[2, 3] + A = \{0\}$$

has no solution in the domain of ordinary sets.   □

In Theorem 2.36(b) it was proved that the set representation of α-cuts of extensions $\hat{\phi}(\mu_1, \ldots, \mu_n)$, which are very important for application, in general cannot be determined directly from the α-cuts of the fuzzy sets involved. A very simple example demonstrating this is:

$$\phi : \mathbf{R} \to \mathbf{R}, \qquad \phi(x) \stackrel{\text{def}}{=} \mathbb{1}_{\{1\}},$$

$$\mu \in F_C(\mathbf{R}), \qquad \mu(x) = \begin{cases} x, & \text{if } x \in [0,1] \\ 0, & \text{otherwise,} \end{cases}$$

where $[\hat{\phi}(\mu)]_1 = \{0,1\} \supset \phi([\mu]_1) = \{1\}$.

But if we consider continuous functions ϕ and fuzzy sets $\mu_1, \ldots, \mu_n$ that are upper semicontinuous, then equality can be achieved in Theorem 2.36(b).

Theorem 2.38 *Let* $(\mu_1, \mu_2, \ldots, \mu_n) \in [F_C(\mathbf{R})]^n$ *and* $\phi : \mathbf{R}^n \to \mathbf{R}$ *be a continuous mapping. Then:*

$$\forall \alpha \in (0,1] : \hat{\phi}(\mu_1, \ldots, \mu_n)]_\alpha = \phi([\mu_1]_\alpha, \ldots, [\mu_n]_\alpha).$$

Proof:
The inclusion '$\supseteq$' is a trivial conclusion from Theorem 2.36 (b), since $F_C(\mathbf{R}) \subseteq F_N(\mathbf{R})$ holds. To show '$\subseteq$', let $\alpha \in (0,1]$ and $t \in \mathbf{R}$ be arbitrary, but fixed. We define

$$X_\alpha(t) \stackrel{\text{def}}{=} \{(x_1, \ldots, x_n) \in \mathbf{R}^n \mid \phi(x_1, \ldots, x_n) = t \wedge \forall i \in \{1, \ldots, n\} : x_i \in [\mu_i]_\alpha\}$$

as well as

$$M(t) \stackrel{\text{def}}{=} \{\min\{\mu_1(x_1), \ldots, \mu_n(x_n)\} \mid (x_1, \ldots, x_n) \in \mathbf{R}^n \wedge \phi(x_1, \ldots, x_n) = t\}.$$

Case 1. $X_\alpha(t) \neq \emptyset$. There is a $(x_1, \ldots, x_n) \in X_\alpha(t)$ with $x_i \in [\mu_i]_\alpha$, $i = 1, 2, \ldots, n$, and $\phi(x_1, \ldots, x_n) = t$. Then

$$t \in [\hat{\phi}(\mu_1, \ldots, \mu_n)]_\alpha \Longleftrightarrow t \in \phi([\mu_1]_\alpha, \ldots, [\mu_n]_\alpha).$$

Case 2. $X_\alpha(t) = \emptyset$. On account of

$$[\hat{\phi}(\mu_1, \ldots, \mu_n)]_\alpha$$
$$= \{y \in \mathbf{R} \mid \hat{\phi}(\mu_1, \ldots, \mu_n)(y) \geq \alpha\}$$
$$= \{y \in \mathbf{R} \mid \sup\{\min\{\mu_1(x_1), \ldots, \mu_n(x_n)\} \mid (x_1, \ldots, x_n) \in \mathbf{R}^n$$
$$\wedge \phi(x_1, \ldots, x_n) = y\} \geq \alpha\}$$
$$= \{y \in \mathbf{R} \mid \sup M(y) \geq \alpha\}$$

we obtain $\sup M(t) \geq \alpha$, if $t \in [\hat{\phi}(\mu_1, \ldots, \mu_n)]_\alpha$ holds.

Since we assumed $X_\alpha(t) = \emptyset$, then $m < \alpha$ for all $m \in M(t)$. Therefore α is a point of accumulation of $M(t)$, i.e. there is a strictly increasing sequence

$(\alpha_k)_{k \in \mathbf{N}}$ of elements $\alpha_k \in M(t)$ with $\lim\limits_{k \to \infty} \alpha_k = \alpha$. The property $\alpha_k \in M(t)$ is connected to the existence of tuples $(x_1^k, \ldots, x_n^k) \in \mathbf{R}^n$, which satisfy the conditions $\phi(x_1^k, \ldots, x_n^k) = t$ and $\min\{\mu_1(x_1^k), \ldots, \mu_n(x_n^k)\} = \alpha_k$, hence $(x_1^k, \ldots, x_n^k) \in \times_{i=1}^{n}[\mu_i]_{\alpha_k}$.

Since we have assumed $(\mu_1, \ldots, \mu_n) \in [F_C(\mathbf{R})]^n$, we have compact, hence closed and bounded sets of real numbers for $i = 1, \ldots, n$ and $k \in \mathbf{N}$ with respect to $[\mu_i]_{\alpha_k}$. Consequently $\times_{i=1}^{n}[\mu_i]_{\alpha_k}$ for $k \in \mathbf{N}$ is compact, too. We conclude that the sequence $\left((x_1^k, \ldots, x_n^k)\right)_{k \in \mathbf{N}}$ has a point of accumulation $(x_1, \ldots, x_n)$, and especially that it contains a partial sequence $\left((x_1^{i_k}, \ldots, x_n^{i_k})\right)_{k \in \mathbf{N}}$ convergent to $(x_1, \ldots, x_n)$. From the compactness of $\times_{i=1}^{n}[\mu_i]_{\beta}$ for $\beta \in (0,1]$ it follows that

$$(x_1, \ldots, x_n) \in \bigcap_{k=1}^{\infty} \times_{i=1}^{n}[\mu_i]_{\alpha_k} = \times_{i=1}^{n}[\mu_i]_{\alpha}.$$

The function ϕ is continuous on the set $\times_{i=1}^{n}[\mu_i]_{\alpha_1}$.

Per definitionem and according to the inclusion $[\mu_i]_{\alpha_k} \supseteq [\mu_i]_{\alpha_{k+1}}$, $k \in \mathbf{N}$, $i = 1, \ldots, n$, we obtain:

$$\lim_{k \to \infty} (x^{i_k} n_1, \ldots, x^{i_k} n) = (x_1, \ldots, x_n)$$
$$\implies \lim_{k \to \infty} \phi(x^{i_k} n_1, \ldots, x^{i_k} n) = \phi(x_1, \ldots, x_n).$$

For all $k \in \mathbf{N}$, $\phi(x_1^k, \ldots, x_n^k) = t$, hence $\phi(x_1, \ldots, x_n) = t$ and $(x_1, \ldots, x_n) \in X_\alpha(t)$. But this is a contradiction to the assumption $X_\alpha(t) = \emptyset$ stated at the beginning.

Therefore only the first case occurs, i.e. the assertion is proved.   □

Theorem 2.38 shows an advantage of the horizontal representation of fuzzy sets compared to the vertical representation: determining the values of extensions is, in general, simplified considerably compared to a direct application of Definition 2.30. But unfortunately there is also a drawback, namely that the term of the membership function of a fuzzy set can be represented in a computer, but normally none of its set representations. In applications one is therefore forced to confine to a finite number of α-cuts.

To store fuzzy sets in a computer and to develop efficient algorithms to process them, the following sub-classes of $F_C(\mathbf{R})$ are especially suited. If they are employed, one can dispense with the condition for continuity, which was required to hold for the mapping ϕ in Theorem 2.38.

Definition 2.39 *The fuzzy set* $\mu \in F_N(\mathbf{R})$ *belongs to the set* $F_{D_k}(\mathbf{R})$ *with* $k \in \mathbf{N}$, *if and only if:*

(a) $[\mu]_\beta = [\mu]_{i/k}$ *holds for all* $i \in \{1, 2, \ldots, k\}$ *and* $\beta \in (\frac{i-1}{k}, \frac{i}{k}]$, *and*

(b) for all $i \in \{1, 2, \ldots, k\}$ *the set* $[\mu]_{i/k}$ *is a finite union of closed intervals.*

Theorem 2.40 *From* $k, n \in \mathbf{N}$, $(\mu_1, \ldots, \mu_n) \in [F_{D_k}(\mathbf{R})]^n$ *and* $\phi : \mathbf{R}^n \to \mathbf{R}$ *it follows that*

$$\forall \alpha \in (0,1] : [\hat{\phi}(\mu_1, \ldots, \mu_n)]_\alpha = \phi([\mu_1]_\alpha, \ldots, [\mu_n]_\alpha).$$

Proof:
Property (b) of Definition 2.39 is not needed. With (a), for an arbitrary $\alpha \in (0,1]$, we can find two numbers $i_\alpha \in \{1, 2, \ldots, k\}$ and $\varepsilon_\alpha > 0$ such that $\alpha, \alpha - \varepsilon_\alpha \in (\frac{i_\alpha - 1}{k}, \frac{i_\alpha}{k}]$.

Let $y \in [\hat{\phi}(\mu_1, \ldots, \mu_n)]_\alpha$. Then there is a tuple $(x_1, \ldots, x_n) \in \mathbf{R}^n$ with $\phi(\mu_1, \ldots, \mu_n) = y$ and $\min\{\mu_1(x_1), \ldots, \mu_n(x_n)\} \geq \alpha - \varepsilon_\alpha$. Consequently $(x_1, \ldots, x_n) \in \times_{i=1}^n [\mu_i]_{\alpha - \varepsilon_\alpha}$, $y \in \phi([\mu_1]_{\alpha - \varepsilon_\alpha}, \ldots, [\mu_n]_{\alpha - \varepsilon_\alpha})$ and therefore $[\hat{\phi}(\mu_1, \ldots, \mu_n)]_\alpha \subseteq \phi([\mu_1]_{\alpha - \varepsilon_\alpha}, \ldots, [\mu_n]_{\alpha - \varepsilon_\alpha})$.

ε_α was fixed, so that $[\mu_i]_\alpha = [\mu_i]_{\alpha - \varepsilon_\alpha}$ holds for $i = 1, 2, \ldots, n$, as well as $\phi([\mu_1]_{\alpha - \varepsilon_\alpha}, \ldots, [\mu_n]_{\alpha - \varepsilon_\alpha}) = \phi([\mu_1]_\alpha, \ldots, [\mu_n]_\alpha)$.

Finally we make use of statement (b) of Theorem 2.36, and by

$$
\begin{aligned}
\phi([\mu_1]_{\alpha - \varepsilon_\alpha}, \ldots, [\mu_n]_{\alpha - \varepsilon_\alpha}) &= \phi([\mu_1]_\alpha, \ldots, [\mu_n]_\alpha) \\
&\subseteq [\hat{\phi}(\mu_1, \ldots, \mu_n)]_\alpha \\
&\subseteq \phi([\mu_1]_{\alpha - \varepsilon_\alpha}, \ldots, [\mu_n]_{\alpha - \varepsilon_\alpha})
\end{aligned}
$$

we obtain the assertion to be proved. $\square$

Besides the fuzzy sets of $F_{D_k}(\mathbf{R})$, for which all fundamental operations of arithmetics are supported and carried out level by level in the software system SOLD (Statistics On Linguistic Data, cf. Section 2.7), the class of fuzzy sets of the L-R-type is often chosen to make calculations efficient:

Definition 2.41 *Let* $L, R : [0, \infty) \to \mathbf{R}$ *be two functions that are strictly decreasing with regard to their support and that satisfy* $L(0) = R(0) = 1$.

A fuzzy set $\mu \in F_I(\mathbf{R})$ *is called a* **fuzzy interval of the L-R-type**, *if* $M_1, M_2, l, r \in \mathbf{R}$ *with* $M_1 \leq M_2$, $l > 0$ *and* $r > 0$ *exist, so that*

$$
\mu(x) = \begin{cases}
L\left(\frac{M_1 - x}{l}\right), & \text{if} \quad x \leq M_1 \\
R\left(\frac{x - M_2}{r}\right), & \text{if} \quad x \geq M_2 \\
1, & \text{if} \quad M_1 \leq x \leq M_2
\end{cases}
$$

holds for all $x \in \mathbf{R}$.

Example 2.42 Let $L : [0, \infty) \to [0, 1]$ and $R : [0, \infty) \to [0, 1]$ be defined by

$$L(x) \; = \; \begin{cases} 1 - x, & \text{if } x \in [0, 1] \\ 0, & \text{otherwise,} \end{cases}$$

$$R(x) \; = \; \begin{cases} 1 - x, & \text{if } x \in [0, 1] \\ 0, & \text{otherwise.} \end{cases}$$

Two fuzzy intervals μ and μ' of the L-R-type can be characterized stating their four parameters, namely

$$\mu \; \overset{\wedge}{=} \; (M_1, M_2, l, r),$$
$$\mu' \; \overset{\wedge}{=} \; (M_1', M_2', l', r').$$

The addition of μ and μ' can be carried out directly with the help of these parameters. We obtain

$$\mu \oplus \mu' \overset{\wedge}{=} (M_1 + M_1', M_2 + M_2', l + l', r + r'),$$

i.e.

$$\mu \oplus \mu'(x) = \begin{cases} 1 - \frac{M_1 + M_1' - x}{l + l'}, & \text{if } M_1 + M_1' - l - l' \le x \le M_1 + M_1' \\ 1 - \frac{x - M_2 - M_2'}{r + r'}, & \text{if } M_2 + M_2' \le x \le M_2 + M_2' + r + r' \\ 1, & \text{if } M_1 + M_1' \le x \le M_2 + M_2' \\ 0, & \text{otherwise.} \end{cases}$$

The product and the quotient of fuzzy intervals of the L-R-type, however, leave the L-R-class considered here.   □

2.6   Semantics of Fuzzy Sets

If mathematicians try to develop and to analyse new concepts, they often prefer the *axiomatic* method: a suitable mathematical base structure is chosen, which can be extended and analysed subsequently. Thus in books on probability theory the approach which prevails is based on the axioms of Kolmogorov and their formal embedding in measure theory. We followed a similar course when we introduced the classical theory of fuzzy sets, and we will adhere to it, when we substantiate fuzzy sets using multivalued logic in Section 2.7.

But since there are many different interpretations of probability and since acquiring subjective probabilities raises some difficulties, it is very important to explain the *semantics* of the concepts under consideration. In the works of de Finetti and Savage [de Finetti74, Savage72] probabilities are derived from concepts like usefulness and preference. Such concepts assign a precise meaning to subjective probabilities and, in this way, often facilitate their acquisition. Comparable semantical foundations for fuzzy sets are seldom dealt with in the literature, but in the following we examine two promising attempts.

In the first approach fuzzy sets serve as a compressed description of imprecise, generally contradictory pieces of information. We employ a refined model, based on the theory of random sets [Matheron75, Dempster68].

In contrast to this interpretation of fuzzy sets, the second approach considered here rests on a phenomenon which is known from quantum physics, namely that the assumption of an exact quantity that cannot be measured precisely is not consistent with all phenomena occuring in nature. One method, which has been proposed to deal with such phenomena, refers to 'Unschärfemengen' [Ludwig90], that are collections of such quantities, which cannot be distinguished. We, however, employ the theory of equality relations to achieve a more precise modelling of vague concepts. The approach used here was also employed in [Ruspini90, Trillas84, Höhle91]

Note that in both approaches fuzzy sets are viewed as *induced* concepts, derived from other basic notations.

2.6.1   Interpretation of Vague Concepts

The basic assumption of the model we deal with in the following is that vagueness of data is caused by the presence of set-valued information in a system of competing contexts.

Example 2.43 Four persons are asked to give their opinion at which body-heights women from London should be called 'tall'. With the help of the information obtained from their answers it is intended to represent the linguistic value 'tall' as compactly as possible. The result of the inquiry may be that the first person regards all women to be tall if their height exceeds 160 cm. The corresponding heights stated by the second to fourth persons may be 165 cm, 170 cm, and 180 cm, respectively. Each person describes the concept 'tall' by a subset of R.   □

In this example vagueness emerges from the fact that competing, possibly contradicting statements have been given. If the observer is able to state preferences between the interrogated persons or even to assign ratings to them, he can carry out a more refined analysis. If the ratings are additive like probabilities, then we can define the following notion of a *random set*.

Definition 2.44 *Let $(C, \mathfrak{P}(C), P)$ be a finite probability space, X an arbitrary non-empty reference set, and $\Gamma : C \to \mathfrak{P}(X)$ a set-valued mapping. The pair (P, Γ) is called a* **random set**. *The sets $\Gamma(c)$, $c \in C$, are called the* **focal sets** *of (P, Γ).*

We call an element $c \in C$ a *context*. c can be an interrogated person as in the example above, or, more generally, a point of view, that has to be examined separately with respect to the considered application. The probability measure $P : \mathfrak{P}(C) \to [0, 1]$ is exclusively used to model the weights of the contexts. On the other hand no frequentistic interpretation of the measure space $(C, \mathfrak{P}(C), P)$ is needed and especially no random experiment with results in C. But fixing the weights can nevertheless be guided by frequentistic methods. $\Gamma(c)$ consists of those elements of the reference set X that are regarded as being possible from the point of view of the context $c \in C$. Note, that these semantics are slightly different from those of ordinary random set theory [Matheron75, Stoyan87].

Example 2.43 (continuation) Each of the four interrogated persons represents a 'context', i.e. we define $C \stackrel{\text{def}}{=} \{c_1, c_2, c_3, c_4\}$. Since no preferences have been stated, we choose $P(\{c_i\}) = 0.25$, $i = 1, 2, 3, 4$. The mapping $\Gamma : C \to \mathfrak{P}(\mathbf{R})$ is defined by $c_1 \mapsto (160, \infty)$, $c_2 \mapsto (165, \infty)$, $c_3 \mapsto (170, \infty)$, and $c_4 \mapsto (180, \infty)$.

The question of whether a woman from London with a body-height of 168 cm should be called 'tall', obviously has to be answered with 'yes' in the contexts c_1 and c_2, but with 'no' in the contexts c_3 and c_4, respectively. The combined answer 'In two out of four contexts she is tall' should be adequate. This compression of information can be carried out for each particular value that the body-height can assume by assigning to any value $x \in \mathbf{R}$ the weight

$$\frac{|(\{c \in C \mid x \in \Gamma(c)\})|}{|(C)|} = \sum_{c \in C : x \in \Gamma(c)} P(\{c\})$$

of the confirming contexts and representing it in the form of a generalized histogram (see Figure 13).

Formally, a fuzzy set $\mu_\Gamma : \mathbf{R} \to [0, 1]$ emerges, but in addition μ_Γ has a fixed meaning.   □

In general, for every random set (P, Γ), such a compressed representation can be found in an analogous way. This representation is sometimes called the *contour function of Γ*.

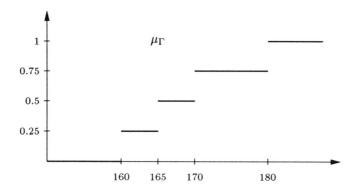

Figure 13 Fuzzy set induced by the random set (P, Γ)

Definition 2.45 *If (P, Γ) with $\Gamma : C \to \mathfrak{P}(X)$ is a random set, then*

$$\mu_\Gamma : \mathbf{R} \quad \to \quad [0, 1],$$
$$x \quad \mapsto \quad P(\{c \in C \mid x \in \Gamma(c)\})$$

is called the **contour function of Γ**.

Note that in this approach it is *not* assumed that the sets $\Gamma(c)$ are nested, i.e. that $\Gamma(c) \subseteq \Gamma(c')$ or $\Gamma(c') \subseteq \Gamma(c)$ holds for all $c, c' \in C$.

If a fuzzy set is interpreted as a contour function, there are of course many methods of acquisition, which are based on statistical procedures. A decisive advantage of this interpretation is seen in the fact that it can motivate some of the t-norms and t-conorms used for intersection and union in practice.

To illustrate this, let (P, Γ_1) and (P, Γ_2) be two fuzzy sets with respect to their common finite context measure space $(C, \mathfrak{P}(C), P)$. The canonical extension of the operations intersection, union, and complement to random sets can be obtained by combining the sets $\Gamma_1(c), \Gamma_2(c)$ for all contexts $c \in C$ applying the corresponding operations of set theory.

Definition 2.46 *Let (P, Γ_1) and (P, Γ_2) with $\Gamma_1, \Gamma_2 : C \to \mathfrak{P}(X)$ be two random sets. Then:*

(a) $\quad (P, \Gamma_1) \cap (P, \Gamma_2) \quad \overset{\text{def}}{=} \quad (P, \Gamma_1 \cap \Gamma_2),$
$\quad\quad \Gamma_1 \cap \Gamma_2 : C \to \mathfrak{P}(X), \quad\quad c \mapsto \Gamma_1(c) \cap \Gamma_2(c) \quad\quad$ *(intersection)*,

(b) $\quad (P, \Gamma_1) \cup (P, \Gamma_2) \quad \overset{\text{def}}{=} \quad (P, \Gamma_1 \cup \Gamma_2),$
$\quad\quad \Gamma_1 \cup \Gamma_2 : C \to \mathfrak{P}(X), \quad\quad c \mapsto \Gamma_1(c) \cup \Gamma_2(c) \quad\quad$ *(union)*,

(c) $\quad \overline{(P, \Gamma_1)} \quad\quad\quad \overset{\text{def}}{=} \quad (P, \overline{\Gamma_1}),$
$\quad\quad \overline{\Gamma_1} \quad\quad : C \to \mathfrak{P}(X), \quad\quad c \mapsto \overline{\Gamma_1(c)} \quad\quad\quad$ *(complement)*.

Theorem 2.47 *With respect to the contour functions which are induced by the random sets stated in Definition 2.46, for all $x \in X$, the following (in-)equalities hold:*

(a) $\max\{0, \mu_{\Gamma_1}(x) + \mu_{\Gamma_2}(x) - 1\} \leq \mu_{\Gamma_1 \cap \Gamma_2}(x) \leq \min\{\mu_{\Gamma_1}(x), \mu_{\Gamma_2}(x)\}$,

(b) $\max\{\mu_{\Gamma_1}(x), \mu_{\Gamma_2}(x)\} \leq \mu_{\Gamma_1 \cup \Gamma_2}(x) \leq \min\{1, \mu_{\Gamma_1}(x) + \mu_{\Gamma_2}(x)\}$,

(c) $\mu_{\overline{\Gamma_1}}(x) = 1 - \mu_{\Gamma_1}(x)$.

Proof:

(a) $\max\{0, \mu_{\Gamma_1}(x) + \mu_{\Gamma_2}(x) - 1\}$

$$
\begin{aligned}
&= \max\{0, 1 - ((1 - \mu_{\Gamma_1}(x)) + (1 - \mu_{\Gamma_2}(x)))\} \\
&\leq 1 - P(\{c \in C \mid x \notin \Gamma_1(c) \vee x \notin \Gamma_2(c)\}) \\
&= P(\{c \in C \mid x \in \Gamma_1(c) \wedge x \in \Gamma_2(c)\}) \\
&= P(\{c \in C \mid x \in (\Gamma_1 \cap \Gamma_2)(c)\}) \\
&= \mu_{\Gamma_1 \cap \Gamma_2}(x) \\
&\leq \min\{P(\{c \in C \mid x \in \Gamma_1(c)\}), P(\{c \in C \mid x \in \Gamma_2(c)\})\} \\
&= \min\{\mu_{\Gamma_1}(x), \mu_{\Gamma_2}(x)\},
\end{aligned}
$$

(b) $\max\{\mu_{\Gamma_1}(x), \mu_{\Gamma_2}(x)\}$

$$
\begin{aligned}
&= \max\{P(\{c \in C \mid x \in \Gamma_1(c)\}), P(\{c \in C \mid x \in \Gamma_2(c)\})\} \\
&\leq \mu_{\Gamma_1 \cup \Gamma_2}(x) \\
&= P(\{c \in C \mid x \in (\Gamma_1 \cup \Gamma_2)(c)\}) \\
&= P(\{c \in C \mid x \in \Gamma_1(c) \vee x \in \Gamma_2(c)\}) \\
&\leq \min\{P(\{c \in C \mid x \in \Gamma_1(c)\}) + P(\{c \in C \mid x \in \Gamma_2(c)\}), 1\} \\
&= \min\{\mu_{\Gamma_1}(x) + \mu_{\Gamma_2}(x), 1\},
\end{aligned}
$$

(c) $\mu_{\overline{\Gamma_1}}(x)$

$$
\begin{aligned}
&= P(\{c \in C \mid x \in \overline{\Gamma_1}(c)\}) \\
&= 1 - P(\{c \in C \mid x \in \Gamma_1(c)\}) \\
&= 1 - \mu_{\Gamma_1}(x). \qquad \qquad \square
\end{aligned}
$$

If fuzzy sets are interpreted as contour functions, then the theorem above shows that an unambiguous definition of intersection and union cannot be achieved on the basis of the membership function alone, but that one can choose from a set of possibilities, which underlie the restrictions of Theorem 2.47. To obtain a membership function for set-theoretic combinations of fuzzy sets, at least 'qualitative' information about the specification of the contexts is required. Thus Example 2.20 in Section 2.3 can be easily explained in the scope of the above-mentioned context model. The intersection operations discussed there are applicable for the development of fuzzy expert systems.

2.6.2   Interpretation of Vague Environments

In contrast to the context model treated above we assume in the second approach discussed here that vagueness is caused by the fact that it is sometimes impossible to distinguish certain objects with regard to a given context. To achieve an adequate formalism, we start with an axiomatic introduction of the notion of equality or similarity, which is essential in this connection.

Definition 2.48 *Let* $\top$ *be a t-norm. A mapping* $E : X \times X \rightarrow [0, 1]$ *is called an* **equality relation** $\top$, *if the following axioms are satisfied:*

(a) $E(x, x) = 1$ *(reflexivity)*,

(b) $E(x, x') = E(x', x)$ *(symmetry)*,

(c) $\top(E(x, x'), E(x', x'')) \leq E(x, x'')$ *(transitivity)*.

The value $E(x, x')$ is interpreted as the degree of acceptance that x and x' are equal or indistinguishable.

The most simple example for an equality relation is the ordinary crisp equality on X, namely

$$E(x, x') = \begin{cases} 1, & \text{if} \quad x = x' \\ 0, & \text{otherwise.} \end{cases}$$

The following examples motivate the use of equality relations with respects to the t-norms $\top_{\min}$ and $\top_{\text{Luka}}$.

Example 2.49 Assume X is a set of objects which can only be observed through cloudy glasses of a thickness ranging from 0 to 1 (for this interpretation of vague data see also [Kruse91a]). Let $x, y \in X$ be two objects. We observe the isolated object x through a cloudy glass of thickness α. After that we observe the isolated object y through a cloudy glass of thickness α. If we cannot decide whether we have observed different objects x and y or the same object x twice, then we say that x and y are α-indistinguishable ($\alpha \in [0, 1]$). Define

$$E(x, y) = \sup\{1 - \alpha \mid x \text{ and } y \text{ are } \alpha\text{-indistinguishable}\},$$

where $\sup \emptyset = 0$. Obviously, $E(x, x) = 1$ and $E(x, y) = E(y, x)$ hold. Let x and y be α-indistinguishable and let y and z be β-indistinguishable. Let $\gamma = \max\{\alpha, \beta\}$. Since $\gamma \geq \alpha$ holds, x and y are γ-indistinguishable. The same argument applies to y and z. This means that if we observe x, y, and z through a cloudy glass of thickness γ we can neither distinguish x from y nor y from z, which implies that we cannot distinguish x from z (except in the case when we have to deal with Poincaré's paradox [Poincaré02, Poincaré04], i.e. $A = B$,

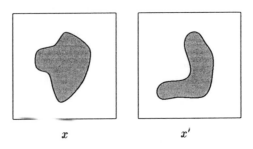

Figure 14   Two black-and-white pictures

$B = C$, but $A \neq C$, which we will not consider here). In other words, x and z are γ-indistinguishable. Therefore, we obtain

$$
\begin{aligned}
E(x, z) \;=\;& \sup\{1 - \gamma \mid x \text{ and } z \text{ are } \gamma\text{-indistinguishable}\} \\
\geq\;& \sup\{1 - \gamma \mid x \text{ and } y \text{ are } \gamma\text{-indistinguishable and} \\
& \qquad\qquad y \text{ and } z \text{ are } \gamma\text{-indistinguishable}\} \\
\geq\;& \sup\{1 - \max\{\alpha, \beta\} \mid x \text{ and } y \text{ are } \alpha\text{-indistinguishable and} \\
& \qquad\qquad\qquad y \text{ and } z \text{ are } \beta\text{-indistinguishable}\} \\
=\;& \sup\{\min\{1 - \alpha, 1 - \beta\} \mid x \text{ and } y \text{ are } \alpha\text{-indistinguishable and} \\
& \qquad\qquad\qquad\qquad y \text{ and } z \text{ are } \beta\text{-indistinguishable}\} \\
=\;& \min\{E(x, y), E(y, z)\}
\end{aligned}
$$

which means that E is an equality relation with respect to the t-norm $\mathsf{T}_{\min}$.
□

Example 2.50 Let X be a set of black-and-white pictures on squares of edge length 1 (see Figure 14 for an example).

We assume that for each picture x the set of black points $S(x) \subseteq [0, 1]^2$ is a (Borel) measureable subset of the unit square. We define

$$
\begin{aligned}
E_2(x, x') \;=\;& m\left[\{(s, t) \in [0, 1]^2 \mid (s, t) \in (S(x) \cap S(x'))\right. \\
& \qquad\qquad\qquad \left. \cup([0, 1]^2 - (S(x) \cup S(x')))\}\right].
\end{aligned}
$$

where m is the Borel measure on the unit square.

The degree of acceptance $E_2(x, x')$, that two pictures x and x' are equal, is the measure of the set of points, where the two pictures coincide with each other.

Obviously we have again $E_2(x, x) = 1$ and $E_2(x, x') = E_2(x', x)$. In addition E_2 satisfies the inequality

$$E_2(x, x') + E_2(x', x'') - 1 \leq E_2(x, x''),$$

so that E_2 is an equality relation with regard to the t-norm T_{Luka}.   □

Example 2.51 In the domains of physics and engineering the problem often arises, of how errors of measurement can be incorporated in the model. The most simple approach is to consider the interval $[x_0 - \varepsilon, x_0 + \varepsilon]$ instead of the measured value x_0, where $\varepsilon > 0$ is an upper bound for the error of measurement. With a fixed error bound ε the actual value $x_0^{(\text{actual})}$ can be any value in the interval $[x_0 - \varepsilon, x_0 + \varepsilon]$, if x_0 has been measured. Therefore we consider the relation

$$R_\varepsilon = \{(x, y) \in \mathbf{R}^2 \mid |x - y| \leq \varepsilon\}. \tag{2.28}$$

If x_0 has been measured, then $(x_0, x_0^{(\text{actual})}) \in R_\varepsilon$ holds for the actual value $x_0^{(\text{actual})}$. R_ε is a reflexive and symmetrical relation. R_ε is *not* transitive. If R_ε was transitive, it would be convenient to look at the quotient space $\mathbf{R}/R_\varepsilon$ instead of $\mathbf{R}$, in which those values are identified that are indistinguishable with respect to a given error of measurement. In other words the non-transitivity of R_ε means the following: Although a and b as well as b and c, respectively, are indistinguishable according to the imprecise measurements (i.e. we have $|a - b| \leq \varepsilon$ and $|b - c| \leq \varepsilon$) a can be distinguished from c (i.e. $|a - c| > \varepsilon$). This phenomenon, equivalent to the situation '$a = b$', '$b = c$', but '$a \neq c$', is called the *Poincaré paradox*.

We now consider the relation R_ε for all error bounds $\varepsilon \in [0, 1]$ and define the mapping

$$\begin{aligned} E : \mathbf{R} \times \mathbf{R} &\longrightarrow [0, 1], \\ (x, y) &\longmapsto 1 - \inf\{\varepsilon \in [0, 1] \mid (x, y) \in R_\varepsilon\} \end{aligned} \tag{2.29}$$

where $\inf \emptyset \overset{\text{def}}{=} 1$.

Obviously E is an equality relation with respect to the t-norm T_{Luka} and for all $\varepsilon \in [0, 1]$

$$(x, y) \in R_\varepsilon \quad \Leftrightarrow \quad E(x, y) \geq 1 - \varepsilon.$$

holds. In addition it follows $[x_0 - \varepsilon, x_0 + \varepsilon] = \{x \in \mathbf{R} \mid E(x_0, x) \geq 1 - \varepsilon\}$. In this sense an equality relation describes which identifications we have to make, if we measure with an error bound $\varepsilon \in [0, 1]$. If an equality relation E on X is given and if we examine the point $x_0 \in X$, then the set $\{x \in X \mid E(x_0, x) \geq 1 - \varepsilon\}$ is the set of all points which are indistinguishable from x_0 with respect to the error bound ε. Therefore the set

$$R_\varepsilon^{(E)} = \{(x, y) \in X \times X \mid E(x, y) \geq 1 - \varepsilon\} \tag{2.30}$$

contains all pairs (x, y) that are indistinguishable with respect to an error bound ε. $\square$

Example 2.52 In Example 2.51 we considered the equality relation induced by the standard metric on $\mathbf{R}$. This metric is not always appropriate, since it does not take any scaling factors into account. For instance, if we want to compute the degree to which the ages of two persons are equal, the result depends on the unit (years, months, days, etc.) which we use for measuring the age. To amend this, we might introduce a scaling factor $c \geq 0$ and define the equality relation $E^{(c)}$

$$E^{(c)}(x, x') = 1 - \min\{|c \cdot x - c \cdot x'|, 1\}. \tag{2.31}$$

Depending on the unit we use for measuring, we have to choose an adequate scaling factor c. $\square$

Example 2.53 In some applications it is reasonable to modify the idea of a scaling factor as described in Example 2.52. Let us assume that we consider the real interval $[a, b]$ as possible values (our set for the vague environment). Instead of one scaling factor for all values, we can specify different scaling factors for different ranges.

A vague environment induced by a measuring instrument that works with high precision in the range of $[-1, 1] \subseteq [a, b]$, and with lower precision outside this region, might be characterized by choosing the scaling factor 3 for the range $[-1, 1]$ and the scaling factor 0.5 for the region outside $[-1, 1]$. This means that we magnify the interval $[-1, 1]$ by the factor 3 and we shorten the intervals $[-1, a]$ and $[1, b]$ by the factor 2 (i.e. we 'magnify' them by the factor 0.5). To illustrate this idea, let us assume $a = -3$ and $b = 5$. In order to 'measure' the distance between two elements $x, x' \in [-3, 5]$, we map the interval $[-3, 5]$ (piecewise linearly) to the interval $[0, 9]$, where $[-3, -1]$, $[-1, 1]$, $[1, 5]$ is mapped linearly to $[0, 1]$, $[1, 7]$, $[7, 9]$, respectively. This piecewise linear transformation is given by the mapping

$$f : [-3, 5] \quad \to \quad [0, 9],$$
$$x \quad \mapsto \quad \begin{cases} 0.5 \cdot (x + 3), & \text{if } -3 \leq x \leq -1 \\ 3 \cdot (x + 1) + 1, & \text{if } -1 \leq x \leq 1 \\ 0.5 \cdot (x - 1) + 7, & \text{if } 1 \leq x \leq 5. \end{cases}$$

The equality relation on $[-3, 5]$ induced by this transformation is defined by

$$E : [-3, 5] \times [-3, 5] \quad \to \quad [0, 1],$$
$$(x, x') \quad \mapsto \quad 1 - \min\{|f(x) - f(x')|, 1\}.$$

This equality relation is intended to model the vague environment induced by a measuring instrument that does not measure with the same exactness throughout the range $[-3,5]$. Therefore, the equality relation reflects indistinguishability.

It is also reasonable to define such an equality relation to represent similarity. For example, if we want to regulate the room temperature, we may use a thermometer with a certain exactness for temperatures between $0°$ C and $35°$ C. But for our purposes we are not interested in a precise value for a temperature below $15°$ C or above $27°$ C, since we consider these temperatures as much too cold or warm, respectively, so that we have to heat or cool the room as much as possible. Temperatures between $15°$ C and $19°$ C or between $23°$ C and $27°$ C are also considered as too cold or too warm. But heating or cooling should be carried out moderately in these cases. For temperatures between $19°$ C and $23°$ C we are interested in more exact measurements, since these temperatures are near the optimal value for the room temperature and the adjustment has to be chosen carefully. In order to reflect this vague environment we might use the transformation

$$f : [0,35] \longrightarrow [0,8],$$

$$x \longmapsto \begin{cases} 0, & \text{if} \quad 0 \le x \le 15 \\ 0.25 \cdot (x-15), & \text{if} \quad 15 \le x \le 19 \\ 1.5 \cdot (x-19)+1, & \text{if} \quad 19 \le x \le 23 \\ 0.25 \cdot (x-23)+7, & \text{if} \quad 23 \le x \le 27 \\ 8, & \text{if} \quad 27 \le x \le 35, \end{cases}$$

leading to the equality relation $E(x,x') = 1 - \min\{|f(x) - f(x')|, 1\}$. $\quad\square$

Example 2.54 In Example 2.53 we considered vague environments where we specified different factors $c \ge 0$ for different intervals. A scaling factor $c > 1$ for an interval means that we 'look at this interval through a magnifying glass' and the indistinguishability or similarity between values of this interval is low. On the contrary, a scaling factor $c < 1$ implies that the values of this interval show great indistinguishability or similarity. We now consider a more general approach where we associate to each element of our vague environment a scaling factor, describing the magnifying factor with which we look at the neighbourhood of the element. If the interval $[a,b]$ is the underlying set of our vague environment, then we can represent the idea of scaling factors by a mapping $c : [a,b] \rightarrow [0,\infty)$. Assuming that the mapping c is integrable, the corresponding transformation is given by

$$f : [a,b] \longrightarrow [0,\infty),$$

$$x \longmapsto \int_a^x c(t)dt.$$

Again, we obtain the equality relation, characterizing our vague environment by

$$E : [a, b] \times [a, b] \quad \rightarrow \quad [0, 1],$$
$$(x, x') \quad \mapsto \quad 1 - \min\{|f(x) - f(x')|, 1\}. \qquad \square$$

Definition 2.55 *Let E be an equality relation on X with respect to the t-norm T. A fuzzy set μ of X, that satisfies the condition*

$$\mathsf{T}(\mu(x), E(x, y)) \leq \mu(y) \qquad \text{(extensionality)}$$

is called **extensional**.

Only those fuzzy sets are extensional, that behave well with respect to the equality relation. This means that the membership degrees of two elements that are similar according to the equality relation must not differ significantly.

Fuzzy sets that are particularly simple are the so-called *singletons* μ_{x_0} where $x_0 \in X$ is fixed:

$$\mu_{x_0}(x) \overset{\text{def}}{=} E(x, x_0), \qquad x \in X.$$

The fuzzy set μ_{x_0} is the smallest extensional fuzzy set for which $\mu_{x_0}(x_0) = 1$ holds.

Example 2.56 Let $X = \mathbf{R}$, $\delta : \mathbf{R} \times \mathbf{R} \to \mathbf{R}$, $(x, y) \mapsto |x - y|$ be the standard metric on $\mathbf{R}$ and E_δ the equality relation induced by the t-norm T_{Luka} with respect to δ:

$$E_\delta(x, y) \overset{\text{def}}{=} 1 - \min\{\delta(x, y), 1\}.$$

In this case the singleton μ_{x_0} is described by the membership function

$$\mu_{x_0}(x) = 1 - \min\{|x_0 - x|, 1\},$$

i.e. the triangular function depicted in Figure 15. $\qquad \square$

If a fuzzy set is seen as a singleton and assuming that the corresponding equality relation is induced by a scaling function as in Example 2.54, then we obtain a concrete interpretation of membership degrees: The first derivative of a fuzzy set at the point x determines the scaling factor or distinguishability in the neighbourhood of x.

It is not only single elements of X that induce fuzzy sets in the form of singletons with respect to an equality relation. Any subset of X can be associated with the smallest extensional fuzzy set containing it.

Definition 2.57 *Let E be a fuzzy set on X with respect to the continuous t-norm T. The* **extensional hull** *of the set $M \subseteq X$ is the fuzzy set*

$$\mu_M : X \to [0, 1], \qquad x \mapsto \sup\{E(x, x') \mid x' \in M\}.$$

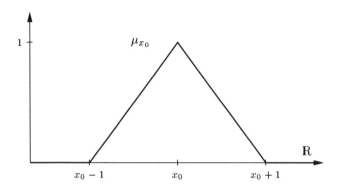

Figure 15   Singleton induced by x_0

μ_M is the smallest extensional fuzzy set for which $\forall x \in M : \mu_M(x) = 1$ holds. Definition 2.57 describes how fuzzy sets can emerge from ordinary sets. If an equality relation is defined on X, which describes the indistinguishability of elements of X in the sense of the Examples 2.49 or 2.50, then an ordinary set can only be observed as a fuzzy set according to this indistinguishability.

Example 2.58 We consider the same equality relation on $\mathbf{R}$ as in Example 2.56. Let $M = [a, b]$ be an interval. The extensional hull of M is a trapezodial function (see Figure 16).

$$\mu_M(x) = \begin{cases} 1, & \text{if} \quad a \quad \leq x \leq b \\ x - a + 1, & \text{if} \quad a - 1 \leq x \leq a \\ b - x + 1, & \text{if} \quad b \quad \leq x \leq b + 1 \\ 0, & \text{otherwise.} \end{cases} \qquad \square$$

By using extensional fuzzy sets the indistinguishability of objects induced by equality relations is taken into account. Mappings to sets, on which equality relations are defined, should respect these equality relations too.

Definition 2.59 *Let E and F be two equality relations on the sets X and Y, respectively. A mapping $\phi : X \rightarrow Y$ is called **extensional** with respect to E and F, if the condition*

$$E(x, x') \leq F(\phi(x), \phi(x'))$$

is satisfied.

Extensionality of a mapping ϕ means that the degree of equality of the images of two elements under ϕ is at least as great as the degree of equality of the elements themselves. This means that the images of two elements cannot be better distinguished than the original elements.

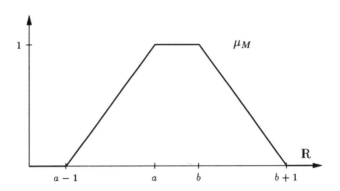

Figure 16   The extensional hull of the interval $[a, b]$

Example 2.60 We interpret the equality relation in the sense of Example 2.51 and examine what it means for a mapping $\phi : X \to Y$ to be extensional. Let the equality relations E and F be given on X and Y, respectively.

ϕ is extensional $\iff$ $\forall x, x' \in X : E(x, x') \leq F(\phi(x), \phi(x'))$

$\qquad\qquad\qquad\iff$ $\forall x, x' \in X, \forall \varepsilon \in [0, 1] :$
$\qquad\qquad\qquad\qquad (E(x, x') \geq 1 - \varepsilon \implies F(\phi(x), \phi(x')) \geq 1 - \varepsilon)$

$\qquad\qquad\qquad\iff$ $\forall x, x' \in X, \forall \varepsilon \in [0, 1] :$
$\qquad\qquad\qquad\qquad ((x, x') \in R_\varepsilon^{(E)} \implies (\phi(x), \phi(x')) \in R_\varepsilon^{(F)})$

$\qquad\qquad\qquad\iff$ $\forall \varepsilon \in [0, 1] : (\phi \times \phi)(R_\varepsilon^{(E)}) \subseteq R_\varepsilon^{(F)}.$

Therefore, the extensionality of ϕ means that ϕ preserves error bounds, i.e. if x and x' are indistinguishable in X with respect to a given error bound ε then $\phi(x)$ and $\phi(x')$ are indistinguishable in Y with respect to the same error bound.   □

If we take a look at the Cartesian product $X \times Y$ of two sets X and Y, on which the equality relations E and F, respectively, are defined, then the question arises as to what an equality relation on $X \times Y$ may look like. The next theorem provides the answer:

Theorem 2.61 *Let $E_1, E_2, \ldots, E_n$ be equality relations on $X_1, X_2, \ldots, X_n$, respectively, with respect to the t-norm $\top$. We define*

$$E : \quad (X_1 \times \ldots \times X_n)^2 \quad \to \quad [0, 1],$$
$$((x_1, \ldots, x_n), (y_1, \ldots, y_n)) \quad \mapsto \quad \min\{E_1(x_1, y_1), \ldots, E_n(x_n, y_n)\}.$$

Then the following holds:

(a) E is an equality relation on the set $X_1 \times \ldots \times X_n$ with respect to the t-norm T.

(b) For all $i \in \{1, \ldots, n\}$: the projection $\pi_i : X_1 \times \ldots \times X_n \rightarrow X_i$, $(x_1, \ldots, x_n) \mapsto x_i$, is extensional with respect to E and E_i.

(c) If E' is an equality relation on $X_1 \times \ldots \times X_n$ with respect to T so that all projections π_i $(i = 1, \ldots, n)$ are extensional, then $E' \leq E$ follows.

Proof:

(a) Obviously E satisfies conditions (a) and (b) of Definition 2.48 for equality relations. To show condition (c) we need the monotonicity of T:

$$\mathsf{T}(E((x_1, \ldots, x_n), (y_1, \ldots, y_n)), E((y_1, \ldots, y_n), (z_1, \ldots, z_n)))$$

$$= \mathsf{T}(\min_{i=1,\ldots,n} \{E_i(x_i, y_i)\}, \min_{i=1,\ldots,n} \{E_i(y_i, z_i)\})$$

$$\leq \min_{i=1,\ldots,n} \{\mathsf{T}(E_i(x_i, y_i), E_i(y_i, z_i))\}$$

$$\leq \min_{i=1,\ldots,n} \{E_i(x_i, z_i)\}$$

$$= E((x_1, \ldots, x_n), (z_1, \ldots, z_n)).$$

(b)
$$E((x_1, \ldots, x_n), (y_1, \ldots, y_n)) = \min_{j=1,\ldots,n} \{E_j(x_j, y_j)\}$$
$$\leq E_i(x_i, y_i)$$
$$= E_i(\pi_i(x_1, \ldots, x_n), \pi_i(y_1, \ldots, y_n)).$$

(c) Because of the projection π_i is extensional with respect to E' and E_i it follows for all $i = 1, \ldots, n$ that

$$E'((x_1, \ldots, x_n), (y_1, \ldots, y_n)) \leq E_i(x_i, y_i)$$

and therefore

$$E'((x_1, \ldots, x_n), (y_1, \ldots, y_n)) \leq \min_{i=1,\ldots,n} \{E_i(x_i, y_i)\}$$
$$= E((x_1, \ldots, x_n), (y_1, \ldots, y_n)). \qquad \square$$

Theorem 2.61 shows that E is the coarsest equality relation on $X_1 \times \ldots \times X_n$ for which all projections are extensional.

The foundations of equality relations and the results presented in this section will be applied in the domain of fuzzy control in Chapter 4. It should be emphasized that especially in applications, where measurements in the form of real numbers are considered, equality relations with respect to the t-norm T_{Luka} play the central role, since they are suitable for modelling imprecision according to Examples 2.51 and 2.60.

ϕ	ψ	$\phi \wedge \psi$
w	w	w
w	f	f
f	w	f
f	f	f

Table 1   The truth table for conjunction

2.7   Fuzzy Logic

The concepts for fuzzy sets described in the previous sections are based on the assumption alone that the relation 'is element of' ($\in$) is not simply true or false, but can assume arbitrary values from the unit interval. If $\in$ is conceived as a predicate, then in the theory of fuzzy sets it is not required — at least for the predicate $\in$ — that only one of the two truth values 'true' and 'false' can be assigned to a proposition. All other propositions and predicates — for example equality — remain two-valued still. Set theory is based on the classical first order predicate calculus. If classical two-valued logic is replaced by a multi-valued one, then more than two truth values are possible — not only for the predicate $\in$ alone, but also for all other predicates. If instead of the truth values t for true and f for false the unit interval $[0,1]$ is admitted as the range of truth values, then logical connectives like $\wedge$ (and), $\vee$ (or), $\rightarrow$ (implication), $\leftrightarrow$ (biimplication), or $\neg$ (negation) can no longer be represented by a truth table such as the one shown in Table 1.

Rather, each logical connective is associated with a mapping $[0,1]^2 \rightarrow [0,1]$ (or $[0,1] \rightarrow [0,1]$, respectively, for negation) by which the truth value of a combined proposition can be determined from the truth values of the atomic propositions.

By means of t-norms and t-conorms we have already made acquaintance with truth functions for conjunction and disjunction. Table 2 shows some examples for implications, where $[\![\phi]\!] \in [0,1]$ denotes the truth value of the proposition ϕ.

For negation mostly $[\![\neg\phi]\!] = 1 - [\![\phi]\!]$ is used. The biimplication generally depends on the implication:

$$[\![\phi \leftrightarrow \psi]\!] \;=\; [\![(\phi \rightarrow \psi) \wedge (\psi \rightarrow \phi)]\!].$$

In this equation also a conjunction appears. But if we assume that for implication

$$[\![\phi]\!] \leq [\![\psi]\!] \;\;\Longrightarrow\;\; [\![\phi \rightarrow \psi]\!] = 1$$

name	$[\![\phi \to \psi]\!]$
Lukasiewicz	$\min\{1 - [\![\phi]\!] + [\![\psi]\!], 1\}$
Gödel	$\begin{cases} 1, & \text{if } [\![\phi]\!] \leq [\![\psi]\!] \\ [\![\psi]\!], & \text{otherwise} \end{cases}$
Goguen	$\begin{cases} 1, & \text{if } [\![\phi]\!] = 0 \\ \min\{1, \frac{[\![\psi]\!]}{[\![\phi]\!]}\}, & \text{otherwise} \end{cases}$
Kleene–Dienes	$\max\{1 - [\![\phi]\!], [\![\psi]\!]\}$
Zadeh	$\max\{1 - [\![\phi]\!], \min\{[\![\phi]\!], [\![\psi]\!]\}\}$
Reichenbach	$1 - [\![\phi]\!] + [\![\phi]\!] \cdot [\![\psi]\!]$

Table 2   Some evaluation functions for implication

holds, we have either $[\![\phi \to \psi]\!] = 1$ or $[\![\psi \to \phi]\!] = 1$. If the conjunction is assigned a truth function by means of a t-norm,¡ we obtain

$$[\![\phi \leftrightarrow \psi]\!] = \min\{[\![\phi \to \psi]\!], [\![\psi \to \phi]\!]\}$$

independent of $\top$.

Similar to classical logic the connectives of multi-valued logic can be expressed by other connectives. In Section 2.3 we have already demonstrated that by

$$[\![\phi \vee \psi]\!] = [\![\neg(\neg\phi \wedge \neg\psi)]\!]$$

the t-conorm dual to $\top$ can be obtained for disjunction if the conjunction $\wedge$ is associated with the t-norm $\top$ and negation $\neg$ is defined as $[\![\neg\phi]\!] = 1 - [\![\phi]\!]$. But expressions that are equivalent to those of classical logic can lead to different definitions if the truth values can range over the unit interval. For example, with the Lukasiewicz implication, two different disjunctions can be defined:

$$[\![\phi \vee_1 \psi]\!] = [\![\neg\phi \to \psi]\!] = \min\{[\![\phi]\!] + [\![\psi]\!], 1\}, \tag{2.32}$$

$$[\![\phi \vee_2 \psi]\!] = [\![\neg\phi \to \neg(\psi \to \phi)]\!] = \max\{[\![\phi]\!], [\![\psi]\!]\}. \tag{2.33}$$

In classical (two-valued) logic the formulae $\neg\phi \to \psi$ and $\neg\phi \to \neg(\psi \to \phi)$ are equivalent to a disjunction of ϕ and ψ. Whereas in classical logic the implication can be stated equivalently with the help of negation and disjunction, the Lukasiewicz implication can only be expressed with negation and disjunction (2.32), but not with (2.33).

A standard technique which generates an implication from a t-norm is that of residuation.

Definition 2.62 *Let* T *be a continuous t-norm. A mapping* $R : [0,1] \times [0,1] \to$ $[0,1]$ *is called a* **residuum** *of* T, *if the condition*

$$\mathsf{T}(\alpha, \beta) \le \gamma \Leftrightarrow \alpha \le R(\beta, \gamma) \tag{2.34}$$

holds.

Theorem 2.63 *Let* T *be a continuous t-norm. There is exactly one residuum for* T, *namely*

$$
\begin{aligned}
\vec{\mathsf{T}} : [0,1] \times [0,1] &\to [0,1], \\
(\beta, \gamma) &\mapsto \sup\{\alpha \in [0,1] \mid \mathsf{T}(\alpha, \beta) \le \gamma\}.
\end{aligned}
$$

Proof:
First, we show that $\vec{\mathsf{T}}$ is a residuum. From $\mathsf{T}(\alpha, \beta) \le \gamma$ we derive, according to the definition of $\vec{\mathsf{T}}$, the inequality $\vec{\mathsf{T}}(\beta, \gamma) \ge \alpha$. If, on the contrary, $\alpha \le \vec{\mathsf{T}}(\beta, \gamma)$ holds, we obtain because of the continuity and the monotonicity of T, the inequality $\mathsf{T}(\alpha, \beta) \le \gamma$. Therefore, $\vec{\mathsf{T}}$ is a residuum of T.

If R is another residuum of T, we consequently have for all $\alpha \in [0,1]$: $\alpha \le \vec{\mathsf{T}}(\beta, \gamma) \Leftrightarrow \mathsf{T}(\alpha, \beta) \le \gamma \Leftrightarrow \alpha \le R(\beta, \gamma)$, and therefore $\vec{\mathsf{T}} = R$. $\qquad\square$

According to the uniqueness of the residuum proved in Theorem 2.63, we will refer in the following to the residuum of the continuous t-norm always as $\vec{\mathsf{T}}$.

A t-norm represents a truth function of a conjunction. A residuum can be interpreted as a truth function of an implication. To explain this let us take a look at classical two-valued logic (consisting of the truth values 0 and 1). Obviously, we have for all propositions φ, ψ and χ

$$[\![\varphi \wedge \psi]\!] \le [\![\chi]\!] \Longleftrightarrow [\![\varphi]\!] \le [\![\psi \to \chi]\!]. \tag{2.35}$$

We derive formula (2.34) from formula (2.35) by using the t-norm T as truth function for the conjunction and the residuum $R = \vec{\mathsf{T}}$ as truth function for the implication. Consequently, every t-norm interpreted as a conjunction, induces a corresponding implication.

Based on a t-norm as conjunction, we are able to define — with assistance of the residuum as implication — a logical equivalence, the biimplication.

Definition 2.64 *Let* T *be a continuous t-norm.*

$$
\begin{aligned}
\overleftrightarrow{\mathsf{T}} : [0,1] \times [0,1] &\to [0,1], \\
(\alpha, \beta) &\mapsto \mathsf{T}(\vec{\mathsf{T}}(\alpha, \beta), \vec{\mathsf{T}}(\beta, \alpha))
\end{aligned}
$$

is called the **biimplication** *induced by* T.

This definition is motivated by the equivalence

$$\alpha \leftrightarrow \beta \Longleftrightarrow (\alpha \to \beta) \wedge (\beta \to \alpha),$$

where we evaluate the conjunction by the t-norm T and the implication by the residuum $\overrightarrow{\mathsf{T}}$.

Theorem 2.65 *Let* T *be a continuous* t-*norm,* $\alpha, \beta \in [0,1]$. *Then*

$$\overleftrightarrow{\mathsf{T}}(\alpha, \beta) = \overrightarrow{\mathsf{T}}(\max\{\alpha, \beta\}, \min\{\alpha, \beta\})$$

holds.

Proof:
Let us first consider the case $\alpha \le \beta$. Consequently, we have $\mathsf{T}(\alpha, 1) = \alpha \le \beta$ and therefore

$$
\begin{aligned}
\overleftrightarrow{\mathsf{T}}(\alpha, \beta) &= \mathsf{T}(\sup\{\gamma \in [0,1] \mid \mathsf{T}(\alpha, \gamma) \le \beta\}, \overrightarrow{\mathsf{T}}(\beta, \alpha)) \\
&= \mathsf{T}(1, \overrightarrow{\mathsf{T}}(\beta, \alpha)) \\
&= \overrightarrow{\mathsf{T}}(\beta, \alpha).
\end{aligned}
$$

For the case $\beta \le \alpha$, we obtain

$$\overleftrightarrow{\mathsf{T}}(\alpha, \beta) = \overrightarrow{\mathsf{T}}(\alpha, \beta). \qquad \square$$

Example 2.66
(a) Assume $\mathsf{T} = \mathsf{T}_{\text{Luka}}$. This leads to the Lukasiewicz implication as the residuum to T.

$$
\begin{aligned}
\overrightarrow{\mathsf{T}}(\alpha, \beta) &= \sup\{\gamma \in [0,1] \mid \max\{\alpha + \gamma - 1, 0\} \le \beta\} \\
&= \sup\{\gamma \in [0,1] \mid \alpha + \gamma - 1 \le \beta\} \\
&= \sup\{\gamma \in [0,1] \mid \gamma \le 1 - \alpha + \beta\} \\
&= \min\{1 - \alpha + \beta, 1\}.
\end{aligned}
$$

For the biimplication we obtain according to Theorem 2.65:

$$
\begin{aligned}
\overleftrightarrow{\mathsf{T}}(\alpha, \beta) &= \min\{1 - \max\{\alpha, \beta\} + \min\{\alpha, \beta\}, 1\} \\
&= 1 - |\alpha - \beta|.
\end{aligned}
$$

(b) For $\mathsf{T} = \mathsf{T}_{\min}$ we get the Gödel implication as the residuum:

$$
\begin{aligned}
\overrightarrow{\mathsf{T}}(\alpha, \beta) &= \sup\{\gamma \in [0,1] \mid \min\{\alpha, \gamma\} \le \beta\} \\
&= \begin{cases} 1, & \text{if } \alpha \le \beta \\ \beta, & \text{otherwise.} \end{cases}
\end{aligned}
$$

The corresponding biimplication is

$$\overleftrightarrow{T}(\alpha, \beta) = \overrightarrow{T}(\max\{\alpha, \beta\}, \min\{\alpha, \beta\})$$

$$= \begin{cases} 1, & \text{if } \alpha = \beta \\ \min\{\alpha, \beta\}, & \text{otherwise.} \end{cases}$$

(c) In case of $T = T_{\text{prod}}$ we have the Goguen implication as the residuum:

$$\overrightarrow{T}(\alpha, \beta) = \sup\{\gamma \in [0, 1] \mid \alpha \cdot \gamma \leq \beta\}$$

$$= \begin{cases} \frac{\beta}{\alpha}, & \text{if } \beta < \alpha \\ 1, & \text{otherwise.} \end{cases}$$

and as biimplication

$$\overleftrightarrow{T}(\alpha, \beta) = \overrightarrow{T}(\max\{\alpha, \beta\}, \min\{\alpha, \beta\})$$

$$= \begin{cases} 1, & \text{if } \alpha = \beta \\ \dfrac{\min\{\alpha, \beta\}}{\max\{\alpha, \beta\}}, & \text{otherwise.} \end{cases} \qquad \square$$

The truth values of statements that contain quantifiers are normally determined as follows:

$$[\![\forall x : P(x)]\!] = \inf\{[\![P(x)]\!] \mid x\}, \tag{2.36}$$

$$[\![\exists x : P(x)]\!] = \sup\{[\![P(x)]\!] \mid x\}. \tag{2.37}$$

Equation (2.36) for the evaluation of the universal quantifier yields the smallest truth value possible for the statement $P(x)$ about an arbitrary individuum x.

The extension principle, which was presented in Section 2.4, allows us to extend ordinary mappings to fuzzy sets. It is based on the fact that an element $y \in Y$ belongs to the image of the set $A_1 \times \ldots \times A_n \subseteq X^n$ under the mapping $\phi : X^n \rightarrow Y$, if and only if

$$\exists (x_1, \ldots, x_n) \in X^n : (y = \phi(x_1, \ldots, x_n) \wedge x_1 \in A_1 \wedge \ldots \wedge x_n \in A_n)$$

holds. Applying the extension principle produces a truth value in the unit interval for the statement '$y \in \hat{\phi}(\mu_1, \ldots, \mu_n)$'. The extension principle yields truth values from the unit interval for the predicate $\in$ only.

A fundamental question which arises, when using fuzzy sets, is whether or not the extension of the range of truth values for the predicate $\in$ from the set $\{0, 1\}$ to the interval $[0, 1]$ entails the extension of the range of truth values for all propositions and predicates.

Example 2.67 We consider equality as a two-place predicate. Two sets M and N are equal, if and only if

$$\forall x : (x \in M \leftrightarrow x \in N)$$

holds. In the following we examine the statement '$x \in \mu$' for fuzzy sets μ. It is assigned the truth value $[\![x \in \mu]\!] = \mu(x)$. For two fuzzy sets μ and ν of X we define the truth value of the statement '$\mu = \nu$' by

$$[\![\mu = \nu]\!] = [\![\forall x : (x \in \mu \leftrightarrow x \in \nu)]\!] \tag{2.38}$$

$$= [\![\forall x : ((x \in \mu \to x \in \nu) \wedge (x \in \nu \to x \in \mu))]\!]. \tag{2.39}$$

The value of $[\![\mu = \nu]\!]$ depends on the truth functions chosen for the implication $\to$ and the conjunction $\wedge$. If the conjunction is evaluated using an arbitrary t-norm, employing the Lukasiewicz implication yields

$$[\![\mu = \nu]\!] = \inf\{1 - |\mu(x) - \nu(x)| \mid x \in X\}, \tag{2.40}$$

whereas the Gödel implication leads to

$$[\![\mu = \nu]\!] = \inf\{\text{göd}(\mu(x), \nu(x)) \mid x \in X\} \tag{2.41}$$

where

$$\text{göd}: \quad [0,1]^2 \quad \to \quad [0,1],$$
$$(\alpha, \beta) \quad \mapsto \quad \begin{cases} 1, & \text{if} \quad \alpha = \beta \\ \min\{\alpha, \beta\}, & \text{otherwise.} \end{cases} \qquad \square$$

If instead of exact equality for fuzzy sets the equality induced by (2.40) is used, then two similar fuzzy sets are regarded as being approximately equal. This is demonstrated by the next example.

Example 2.68 Let the fuzzy sets μ and ν on $\mathbf{R}$ be defined as

$$\mu(x) = \begin{cases} 1, & \text{if} \quad 1 \le x \\ x, & \text{if} \quad 0 \le x \le 1 \\ 0, & \text{if} \quad\quad x \le 0 \end{cases}$$

and

$$\nu(x) = \begin{cases} 1, & \text{if} \quad 1.1 \le x \\ \frac{10}{11}x, & \text{if} \quad\ 0 \le x \le 1.1 \\ 0, & \text{if} \quad\quad x \le 0 \end{cases}$$

respectively (see Figure 17). Both, μ and ν could be interpreted as the fuzzy sets of values that are considerably greater than zero.

From (2.40) we obtain $[\![\mu = \nu]\!] = \frac{10}{11}$. $\qquad\qquad\qquad\qquad\qquad \square$

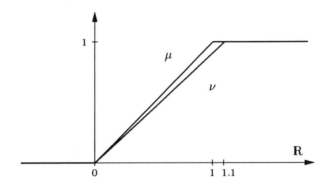

Figure 17　Two fuzzy sets with a degree of equality of $\frac{10}{11}$

The next theorem shows that fuzzy equality as defined in (2.40) is an equality relation on $F(X)$ with respect to the t-norm T_{Luka}.

Theorem 2.69 *The mapping*

$$E_{\text{Luka}}^{(X)} : F(X) \times F(X) \quad \rightarrow \quad [0,1],$$
$$(\mu, \nu) \quad \mapsto \quad \inf\{1 - |\mu(x) - \nu(x)| \mid x \in X\}$$

is an equality relation on $F(X)$ with respect to the t-norm T_{Luka}.

Remark: if it is obvious, which reference set X is referred to, we write E_{Luka} instead of $E_{\text{Luka}}^{(X)}$.

Proof:
Let $\mu, \nu, \eta \in F(X)$. Obviously $E_{\text{Luka}}(\mu, \mu) = 1$ and $E_{\text{Luka}}(\mu, \nu) = E_{\text{Luka}}(\nu, \mu)$ holds. Furthermore

$$\mathsf{T}_{\text{Luka}} \left(E_{\text{Luka}}(\mu, \nu), E_{\text{Luka}}(\nu, \eta) \right)$$
$$= \quad \max \Big\{ \inf_{x \in X} \{1 - |\mu(x) - \nu(x)|\}$$
$$+ \inf_{x \in X} \{1 - |\nu(x) - \eta(x)|\} - 1, 0 \Big\}. \tag{2.42}$$

Since we have $E_{\text{Luka}}(\mu, \eta) \geq 0$, we have to examine only the case in which the maximum in (2.42) is greater than zero.

$$\inf_{x \in X} \{1 - |\mu(x) - \nu(x)|\} + \inf_{x \in X} \{1 - |\nu(x) - \eta(x)|\} - 1$$
$$\leq \quad \inf_{x \in X} \{1 - (|\mu(x) - \nu(x)| + |\nu(x) - \eta(x)|)\}$$

$$\leq \quad \inf_{x \in X} \{1 - |\mu(x) - \eta(x)|\}$$

$$= \quad E_{\text{Luka}}(\mu, \eta). \qquad \qquad \square$$

The extension principle introduced in Section 2.4 extends mappings $\phi : X^n \to Y$ to $\hat{\phi} : F(X)^n \to F(Y)$. The mapping $\hat{\phi}$ is extensional with respect to the equality relation stated in (2.40). To prove this, we define the equality relation

$$E_{\text{Luka}}^{(n)} ((\mu_1, \ldots, \mu_n), (\nu_1, \ldots, \nu_n))$$
$$= \quad \inf_{(x_1, \ldots, x_n) \in X^n} \{1 - | \min\{\mu_1(x_1), \ldots, \mu_n(x_n)\}$$
$$- \min\{\nu_1(x_1), \ldots, \nu_n(x_n)\}|\}$$

on the set $F(X)^n$. Using the mapping

$$(\mu_1, \ldots, \mu_n) \mapsto \mu \in F(X^n)$$

where

$$\mu(x_1, \ldots, x_n) = \min \{\mu_1(x_1), \ldots, \mu_n(x_n)\}$$

we embed the set $F(X)^n$ into the set $F(X^n)$ by identifying the tuple $(\mu_1, \ldots, \mu_n)$ of fuzzy sets with their Cartesian product. We examine the equality relation $E_{\text{Luka}}^{(X^n)}$ on $F(X^n)$. Then $E_{\text{Luka}}^{(n)}$ is the restriction of the equality relation $E_{\text{Luka}}^{(X^n)}$ to the set $F(X)^n$.

Theorem 2.70 *Let $\phi : X^n \to Y$ be a mapping. Then the mapping $\hat{\phi} : F(X)^n \to F(Y)$ is extensional with respect to the equality relations $E_{\text{Luka}}^{(n)}$ on $F(X)^n$ and $E_{\text{Luka}}^{(Y)}$ on $F(Y)$.*

Proof:
Let $\mu_i, \nu_i \in F(X)$, $i = 1, \ldots, n$. We show that for any ε, $0 < \varepsilon \leq 1$, the condition

$$E_{\text{Luka}}^{(Y)} \left(\hat{\phi}(\mu_1, \ldots, \mu_n), \hat{\phi}(\nu_1, \ldots, \nu_n) \right) < \varepsilon \qquad (2.43)$$

implies the inequality

$$E_{\text{Luka}}^{(n)} ((\mu_1, \ldots, \mu_n), (\nu_1, \ldots, \nu_n)) < \varepsilon, \qquad (2.44)$$

from which it follows that $\hat{\phi}$ is extensional. Let an arbitrary ε, $0 < \varepsilon \leq 1$, be given, so that (2.43) is satisfied. By definition of $E_{\text{Luka}}^{(Y)}$ and $\hat{\phi}$ we have

$$\inf_{y \in Y} \left\{ 1 - \left| \sup_{\substack{(x_1, \ldots, x_n) \in X^n : \\ \phi(x_1, \ldots, x_n) = y}} \{\min\{\mu_1(x_1), \ldots, \mu_n(x_n)\}\} \right. \right.$$

$$\left. \left. - \sup_{\substack{(x_1, \ldots, x_n) \in X^n : \\ \phi(x_1, \ldots, x_n) = y}} \{\min\{\nu_1(x_1), \ldots, \nu_n(x_n)\}\} \right| \right\} < \varepsilon.$$

That means, that there is an $y_\varepsilon \in Y$ with

$$1 \ - \ \Big| \sup_{\substack{(x_1,\ldots,x_n)\in X^n: \\ \phi(x_1,\ldots,x_n)=y_\varepsilon}} \{\min\{\mu_1(x_1),\ldots,\mu_n(x_n)\}\}$$
$$- \sup_{\substack{(x_1,\ldots,x_n)\in X^n: \\ \phi(x_1,\ldots,x_n)=y_\varepsilon}} \{\min\{\nu_1(x_1),\ldots,\nu_n(x_n)\}\} \Big| < \varepsilon. \tag{2.45}$$

Because $\varepsilon \leq 1$ it follows that the two suprema in (2.45) cannot assume the same value. Therefore, without loss of generality, let

$$\sup_{\substack{(x_1,\ldots,x_n)\in X^n: \\ \phi(x_1,\ldots,x_n)=y_\varepsilon}} \{\min\{\nu_1(x_1),\ldots,\nu_n(x_n)\}\}$$
$$< \sup_{\substack{(x_1,\ldots,x_n)\in X^n: \\ \phi(x_1,\ldots,x_n)=y_\varepsilon}} \{\min\{\mu_1(x_1),\ldots,\mu_n(x_n)\}\}.$$

Hence there exists an $(x_1^{(\varepsilon)},\ldots,x_n^{(\varepsilon)}) \in X^n$ with $\phi(x_1^{(\varepsilon)},\ldots,x_n^{(\varepsilon)}) = y_\varepsilon$, so that

$$1 \ - \ \min\left\{\mu_1(x_1^{(\varepsilon)}),\ldots,\mu_n(x_n^{(\varepsilon)})\right\}$$
$$+ \sup_{\substack{(x_1,\ldots,x_n)\in X^n: \\ \phi(x_1,\ldots,x_n)=y_\varepsilon}} \{\min\{\nu_1(x_1),\ldots,\nu_n(x_n)\}\} < \varepsilon$$

holds. This implies especially

$$1 - \min\left\{\mu_1(x_1^{(\varepsilon)}),\ldots,\mu_n(x_n^{(\varepsilon)})\right\} + \min\left\{\nu_1(x_1^{(\varepsilon)}),\ldots,\nu_n(x_n^{(\varepsilon)})\right\} < \varepsilon.$$

Therefore we obtain

$$E_{\text{Luka}}^{(n)}\left((\mu_1,\ldots,\mu_n)(\nu_1,\ldots,\nu_n)\right)$$
$$= \inf_{(x_1,\ldots,x_n)\in X^n} \Big\{1 - \big| \min\{\mu_1(x_1),\ldots,\mu_n(x_n)\}$$
$$- \min\{\nu_1(x_1),\ldots,\nu_n(x_n)\}\big|\Big\}$$
$$\leq \ 1 - \big|\min\{\mu_1(x_1^{(\varepsilon)}),\ldots,\mu_n(x_n^{(\varepsilon)})\} - \min\{\nu_1(x_1^{(\varepsilon)}),\ldots,\nu_n(x_n^{(\varepsilon)})\}\big|$$
$$= \ 1 - \min\{\mu_1(x_1^{(\varepsilon)}),\ldots,\mu_n(x_n^{(\varepsilon)})\} + \min\{\nu_1(x_1^{(\varepsilon)}),\ldots,\nu_n(x_n^{(\varepsilon)})\}$$
$$< \ \varepsilon. \qquad \square$$

We demonstrated that by interpreting the implication in (2.38) in the sense of Lukasiewicz an equality relation with respect to the t-norm T_{Luka} is induced on the fuzzy sets of X. The extension principle leads to an extensional mapping with respect to this equality relation. If instead of the Lukasiewicz implication the Gödel implication is used in (2.38), we obtain an equality relation on $F(X)$ with respect to the t-norm $\mathsf{T}_{\min}$. For this equality relation a theorem analogous to 2.70 can be proved, i.e. the extension principle is extensional with respect to this equality relation, too.

In Section 2.6.2 we showed how an equality relation induces a fuzzy set as an extensional hull of an ordinary set (cf. Examples 2.56, 2.58). Conversely we can now record the result that on the set of fuzzy sets of X an equality relation is defined in a canonical way by Theorem 2.70. This equality relation first reflects the idea that 'similar' fuzzy sets are 'more or less equal' also. In the second place concepts like the extension principle can be assessed with the help of this equality relation. 'Similar' fuzzy sets should lead to 'similar' results, which means nothing else but that concepts for fuzzy sets should possess properties of extensionality with respect to this equality relation.

2.8   Supplementary Remarks and References

In this section we give additional comments on the historical development of fuzzy systems, on fuzzy sets and their semantics, on the problem how to choose suitable membership functions, and on fuzzy logic, and finally we describe a software tool for statistical analysis of vague data. In addition the publications relevant for Chapter 2 and further references to recommendable supplementary literature can be found here.

2.8.1   Historical Development: Fuzzy Systems

The concept of a fuzzy set was suggested first in the mid sixties by L.A. Zadeh to achieve a simplified modelling of complex systems. Zadeh made significant contributions to control engineering (e.g. [Desoer63]). Puzzled by the increasing complexity of systems he stated in [Zadeh73]

'...that the conventional quantitative techniques of system analysis are intrinsically unsuited for dealing with humanistic systems or, for that matter, any system whose complexity is comparable to that of humanistic systems. The basis of this contention rests on what might be called the *principle of incompatibility*. Stated informally, the essence of this principle is that as the complexity of a system increases, our ability to make precise and yet significant statements about its behaviour diminishes until a threshold is reached beyond which precision and significance (or relevance) become almost mutually exclusive characteristics.'

In this paper he advocates an alternative approach [Zadeh73]:

'...based on the premise that the key elements in human thinking
are not numbers, but labels of fuzzy sets, that is, classes of objects
in which the transition from membership to non-membership is
gradual rather than abrupt. Indeed the pervasiveness of fuzziness
in human thought processes suggests that much of the logic behind
human reasoning is not the traditional two-valued or even multi-
valued logic, but a logic with fuzzy truths, fuzzy connectives, and
fuzzy rules of inference.'

The publications that appeared shortly after were dedicated mostly to a
theoretical analysis of promising applications of fuzzy sets, which on the whole
resulted from the fuzzification of known theories. A thoroughly commented
bibliography that covers the first decade of this domain of research was
written by B.R. Gaines and L.J. Kohout [Gaines77]. As representative of the
publications that appeared in the domain of fuzzy sets and fuzzy systems up to
1978 the book [Dubois80a] by D. Dubois and H. Prade should be mentioned.
The most important treatises of Zadeh, who has influenced the development of
the whole theory up to now in an authorative way, are collected in [Yager87].
In [Dubois93b] important papers are compiled and discussed.

Since about 1975, besides a further theoretical deepening, there were
attempts to develop applications for laboratories as well as for large
scale technical utilization. As a pioneer in the domain of fuzzy control
E.H. Mamdani [Mamdani76] is well respected. His control system for a
cement kiln [Holmblad82] was the first large scale application that was
realized successfully. In Japan the possibilities opened up by fuzzy control
were recognized and promoted very early. This not only led to spectacular
applications like the underground of Sendai, but also to more simple
realizations in consumer goods, that were extremely successful [Terano91].
An additional field for employing fuzzy methods was seen in the domains of
knowledge processing, artificial intelligence [Smets88] and analysis of fuzzy
data [Miyamoto90, Bezdek81, Bandemer92].

In Europe and in the U.S.A. Zadeh's basic concepts and even some first
results in control engineering and mathematics met with only modest interest.
R.E. Kalman remarked in 1972 at a presentation given by Zadeh:

'His proposals could be severely, ferociously, even brutally
criticized from a technical point of view [...] Professor Zadeh's
fears of unjust criticism can be mitigated by recalling that the
alchemists were not prosecuted for their beliefs but because they
failed to produce gold.'

S. MacLane wrote as late as 1986 in an assessment of fuzzy sets:

'It was hoped that this ingenious notion would lead to all sorts of
fruitful applications, to fuzzy automata, fuzzy decision theory and

elsewhere. However, as yet most of the intended applications turn out to be just extensive exercises, not actually applicable.'

Whereas the idea of fuzzy control was rejected point-blank by engineers, good mathematicians were deterred by the poor quality [Arbib77] of some of the treatises on fuzzy systems (P.T. Johnstone remarked in 1988: 'Fuzzy mathematics is not an excuse for fuzzy thinking') and by the fact that many authors confined their attention to the application of the extension principle within the theories they were interested in. This ensured that the word 'fuzzy' was avoided in numerous applications of fuzzy sets. One of the most controversial topics was the question 'Is fuzziness just probability in clever disguise' and the claim by several promoters of probability theory — 'anything you can do [with fuzzy sets] I can do [with probability] better' [Lindley87]. See the special issue on fuzziness vs. probability of the IEEE Transactions of Fuzzy Systems, Feb. 94, Vol. 2, No. 1, for an uptodate discussion.

Reports about the success of fuzzy systems in Japan ('Matsushita's turnover in fuzzy products came to 1 billion US $ in 1990,' 'In Japan 'fuzzy' was elected word of the year 1990'), the public demand for consumer goods (e.g. the use of fuzzy methods in cameras), and the registration of patents (e.g. fuzzy ABS for motor vehicles) led to an increased interest in fuzzy systems again.

Most of the successful industrial applications are developed in the area of fuzzy control (cf. Chapter 4) and to some extent in the field of fuzzy expert systems (cf. Chapter 3). Other industrial application fields for fuzzy systems not considered in this book are especially fuzzy clustering, pattern classification and recognition [Bellman66, Ruspini69, Ruspini70, Bocklisch86, Bezdek81, Bezdek92a, Bezdek92b, Krishnapuram93a], fuzzy image processing and computer vision [Keller85, Krishnapuram93b], fuzzy information retrieval [Kraft83], fuzzy data bases [Buckles82, Umano82, Bosc88, Zemankova-Leech84], fuzzy data analysis [Bandemer92, Bandemer93, Kruse87b, Tanaka87, Tanaka88, Näther90], and decision support systems [Bellman70, Zimmermann84, Zimmermann87, Chen92].

Frequently it turns out that it is advantageous to employ fuzzy methods in combination with other techniques such as neural networks, probabilistic reasoning, genetic algorithms, etc. rather than exclusively. A case in point is the growing number of so-called neurofuzzy consumer products [Asakawa94]. The line of argumentation for the industry is expressed in [Zadeh94]:

'In traditional — hard — computing the prime desiderata are precision, certainty and rigor. By contrast the point of departure in 'soft computing' is the thesis that precision and certainty carry a cost and that computation, reasoning, and decision making should exploit — wherever possible — the tolerance for imprecision and uncertainty.'

2.8.2 Fuzzy Sets and their Semantics

The fundamental treatise on the theory of fuzzy sets was written by L.A. Zadeh [Zadeh65]. Before him others had been interested in modelling vagueness, such as the philosopher M. Black, who characterized vague symbols by using consistency profiles [Black37], or the mathematician K. Menger, who first used the French term 'ensembles flou' and the English translation 'hazy set' for his idea of 'cloud-like points' [Menger51] in the context of probabilistic geometry. In [Goguen67] the notion of an L-fuzzy set was introduced to extend the class of possible membership functions. The idea to employ layered systems of sets to represent fuzzy sets was suggested first in [Gentilhomme68]. C.V. Negoita and D.A. Ralescu [Negoita75] showed by proving the so-called representation theorem that these representations are equivalent. The underlying algebraic point of view is described exhaustively and mathematically very clearly in [Goodman91a, Goodman91b]. The application of basic operations as a direct generalization of complement, intersection, and union for characteristic functions was proposed by L.A. Zadeh. The first axiomatic analysis in this domain is due to R.E. Bellman and M. Giertz. In [Bellman73] they show the uniqueness of (min, max) assuming that the mappings $\top$ and $\bot$ satisfy the axioms of continuity, distributivity, strict monotonicity of $\top(a,a)$ and $\bot(a,a)$ in a, and the inequality $\top(a,b) \leq \min\{a,b\}$ as well as $\bot(a,b) \geq \max\{a,b\}$.

As was proved in empirical studies, the conjunctive operator min and the disjunctive operator max often do not suffice, since they represent only a very coarse approximation of human assessment [Zimmermann80, Zimmermann91]. For this reason classes of operators that are adjusted to the problem under consideration, for example those suggested in [Yager80, Yager82, Hamacher78, Frank79, Dubois80b, Dombi82], and [Alsina83] have been in use for some time. The concepts of t-norm (triangular norm) and t-conorm (triangular conorm) serve as a mathematical foundation. Originally they were used to analyse metric spaces [Menger42] and were examined by B. Schweizer and A. Sklar [Schweizer61, Schweizer63, Schweizer83]. The basic ideas w.r.t. fuzzy sets are described in [Dubois82, Weber83] in a very comprehensible way.

From the point of view of fuzzy sets the importance of the extension principle was recognized in [Zadeh75]. It can be used to 'fuzzify' whole theories, as was done, for example, with fuzzy topological spaces [Lowen76], that we do not deal with in this book. We confined our attention in this chapter to an elementary application of the extension principle to generalize ordinary arithmetics to fuzzy sets of $\mathbf{R}$. Fuzzy arithmetics can be looked upon as a generalization of interval arithmetics [Moore66, Moore79]. The book [Kaufmann85] deals with this subject.

The notion of a fuzzy relation was first mentioned in [Zadeh71a]. In this publication, as a generalization of an ordinary equivalence relation, the notion of a 'similarity relation' was introduced, which is a reflexive, symmetrical, and transitive binary fuzzy relation, whose α-cuts are ordinary equivalence relations. The concepts of projection and cylindrical extension, which are important for approximate reasoning, were generalized to fuzzy sets by L.A. Zadeh [Zadeh75].

To apply fuzzy sets in a sensible way, the description of semantical aspects is of major importance. For this reason two of the numerous approaches to interpreting fuzzy sets are presented here. They are based on the works of the authors of this book. We recommend the papers [Hisdal88] and [Höhle91], in which some basic ideas are expressed.

The first approach regards fuzzy sets as a compressed representation of competing imprecise pieces of information and, by this, as an induced concept. Analogous points of view have proved to be satisfactory in other calculi; for example, the notions of *one-point coverage* for random sets [Nguyen78b, Nguyen84], *contour function* for belief functions [Shafer76], and *falling shadow* in set-valued statistics [Wang83, König92]. Our view on fuzzy sets as an induced concept is described in detail in [Gebhardt92b]. This publication deals with the semantic background, its mathematical realization, and especially with the consequences for operations on fuzzy sets, if they are to concur with this interpretation. Among other things it becomes apparent in which way the extension principle and some of the pairs of t-norm and t-conorm, that are used frequently in practice, cannot only be motivated, but also be proved to be unavoidable with the help of semantical conditions that are specifiable in a clear way. In addition we like to mention that the underlying modelling permits us to tackle a deeper foundation and — in correspondence to this — extension of possibility theory [Dubois87b], whose introduction in Chapter 3 is essentially based on approximate reasoning in knowledge-based systems as an important application field. Approaches similar to these were described in [Goodman82, Orlov78]; for an overview see [Dubois89a].

The second approach to interpreting fuzzy sets that we have discussed is based on the concept of an equality relation according to [Höhle91] or earlier works in which the names similarity relation [Zadeh71a, Ruspini90] and indistinguishability operator [Trillas84] were used. The principal aim is to manage crisp data even in imperfect contexts. The equality relations serve as a characterization of vague environments and induce fuzzy sets as points in this vague environments. This view on fuzzy sets proves to be helpful for the methods of fuzzy control, which are the subject of Chapter 4. The corresponding theoretical studies can be found in [Klawonn93a, Klawonn93b].

For reasons of space some essential concepts for fuzzy sets cannot be dealt with in detail here. Linguistic modelling, for example, requires fuzzy sets of type 2, that have fuzzy sets as degrees of membership [Mizumoto76].

For example, the type 2 fuzzy set *intelligent* can be formed out of the
(type 1) fuzzy sets *poor, average, superior*, and *outstanding*. Other concepts
used in modelling linguistic expressions are *fuzzy quantifiers* (*many, nearly
all, some*), *fuzzy truth values* (*very true, more or less wrong*), *fuzzy
probability values* (*very likely*), and *linguistic hedges* (*very, considerable*)
[Klir88, Zimmermann91].

Fuzzy sets and information theory are strongly related. The question of
how the degree of vagueness of a fuzzy set should be defined was inspected
comparatively early in an axiomatic way by the analysis of *measures of
fuzziness* [Knopfmacher75] and the application of alternative notions of
entropy [de Luca72, Kruse83].

A good survey on information measures that are used in several calculi of
uncertainty and other concepts, that are of some interest for the use of fuzzy
sets, can to be found in [Klir88, Klir87, Zadeh83a].

2.8.3   The Acquisition of Degrees of Membership

A main problem for the development of fuzzy systems is the choice of an
appropriate membership function. The analysis of the membership degrees
that are used in existing applications shows that the exact numerical value
in the interval $[0, 1]$ does not matter, as is the case in the frequentistic
interpretation of probabilities, but only its order of magnitude. This is affirmed
by psychological studies showing that a human is normally able to distinguish
between about five to nine degrees of membership only.

From a measure theoretical point of view [Krantz71a, Krantz71b, Luce90]
the interval $[0, 1]$ is only used as an ordinal scale. If L is a generalized set
and $\sqsubseteq$ a linear ordering of L, where the smallest and the greatest element
exist, then this ordinal scale can be used as a domain for L-fuzzy sets directly.
A preference relation $\sqsubseteq$ on a set X is a transitive relation, that satisfies the
condition $x \sqsubseteq y \vee y \sqsubseteq x$. If $\sqsubseteq$ is a preference relation on the finite non-empty
set X with the intuitive meaning 'The degree of membership of x (w.r.t. a
fuzzy set μ) is less or equal to the degree of membership of x' (w.r.t. μ)', then
it can be checked easily that there is a function $\mu_{\sqsubseteq} : X \to [0, 1]$ which has the
property

$$x \sqsubseteq x' \iff \mu_{\sqsubseteq}(x) \leq \mu_{\sqsubseteq}(x')$$

for all $x \in X$. If the function $\mu_{\sqsubseteq}$ is unambiguous with the exception of a
monotonic increasing transformation $f : [0, 1] \to [0, 1]$, i.e. if every function
$f \circ \mu_{\sqsubseteq}$ satisfies the condition for compatibility

$$x \sqsubseteq x' \iff f \circ \mu_{\sqsubseteq}(x) \leq f \circ \mu_{\sqsubseteq}(x'),$$

then this is called an ordinal scale level. The missing unambiguity is to be
taken into account if fuzzy sets are combined [French88, Turksen91].

If it is possible to represent not only preferences of the form $x \sqsubseteq x'$, but also those of second order by

'$[x \sqsubseteq x']$ is accepted with a degree less or equal to the one of $[x'' \sqsubseteq x''']$',
denoted by $[x \sqsubseteq x'] \lesssim [x'' \sqsubseteq x''']$,

and if $\lesssim$ satisfies the axioms of an algebraic differential structure [Krantz71a, Krantz71b, Luce90], then a function can be found that satisfies

$$x \sqsubseteq x' \iff \mu(x) \leq \mu(x')$$

and

$$[x \sqsubseteq x'] \lesssim [x'' \sqsubseteq x'''] \iff \mu(x') - \mu(x) \leq \mu(x''') - \mu(x'').$$

Here μ is the affine transformation

$$f : [0,1] \to [0,1], \qquad x \mapsto \alpha x + \beta, \qquad \alpha > 0.$$

The corresponding scale is commonly called the interval scale.

In probability theory the preferences between the elements of a σ-algebra have to be more restrictive to assure unambiguity of the corresponding probability measure [Savage72, Dubois89b, Cox46]. For fuzzy applications these metric scales are often too restrictive.

To acquire and to construct membership functions, several procedures have been suggested in the literature. Surveys can be found in [Dubois80a, Turksen91, Norwich84]. For the most part these procedures are based on statistical methods and on the analysis of preferences [Saaty74, Saaty78].

The choice of membership functions as well as the choice of operators is closely related to aspects of decision theory. In [French88, French84] the problems which have to be taken into consideration are examined.

In applications in engineering the choice of a certain precise membership function is less significant, since often only the qualitative properties of functions (e.g. monotony in a certain area) are needed and real numbers are used merely for reasons of representation, as can be proved by a sensitivity analysis in many cases. Nevertheless this fact should not seduce users to implement fuzzy systems with imprecise mathematical methods.

2.8.4 Fuzzy Logic

Fuzzy logic has to be seen in the context of multi-valued logic. Although the foundation of fuzzy sets is ascribed to L.A. Zadeh, he was not the first to allow the whole unit interval instead of the two truth values 'true' and 'false'. As early as 1922 J. Lukasiewicz discussed a logic with truth values in the unit interval as a generalization of his three-valued logic, in which he had besides the values 0 ('false') and 1 ('true') a value $\frac{1}{2}$ ('indefinite'). A collection of his most important writings can be found in [Lukasiewicz30]. The idea to deviate from the concept of two-valued logic was proposed as early as 1878 by H. MacColl [MacColl78]. Since then several types of multi-valued logic have been developed, but only a few of them are based on the unit interval. An introduction to the subject of multi-valued logic can be found in [Rosser52], in [Rescher69], which includes an ample historical overview till 1969, and in [Dolc02].

The use of more than two truth values can be motivated in the first place by the necessity to have additional truth values with a concrete meaning (e.g. 'indefinite') besides 'true' and 'false'. In the second place a syntactical procedure of deduction generally induces a Lindenbaum algebra, which can be used as the range of truth values. The Lindenbaum algebra is obtained as the quotient set of all logical expressions with regard to the equivalence relation that considers two logical expressions φ and ψ as being equal, if φ can be deduced from ψ as well as ψ from φ with the help of the syntactical procedure of deduction. The procedure of deduction induces an ordering relation on the equivalence classes, where the equivalence class of φ is less than or equal to the equivalence class of ψ, if and only if ψ can be proved from φ syntactically. In the case of classical logic the Lindenbaum algebra is a Boolean algebra. For intuitionistic logic a Heyting algebra [Dummett77] results, whereas the Lukasiewicz logic induces a MV algebra [Chang58]. With regard to logic the unit interval of the MV-algebra plays the same central role as the two-element Boolean algebra among the Boolean algebras. Therefore fuzzy logic, since it is based on the unit interval, has some parallels to the Lukasiewicz logic.

Whereas the approach to logic on the basis of a Lindenbaum algebra is mostly algebraic and lattice-theoretic [Rasiowa70], logic can also be motivated from the point of view of category theory like the topos theory [Goldblatt79]. In topos theory fuzzy sets can be interpreted as characterizing morphisms (functions) of subobjects (subsets). But topos theory is related rather to intuitionistic logic than to the Lukasiewicz logic. An overview of the relations between fuzzy sets and category theory is given in [Höhle91, Stout91] and [Rodabaugh92].

In the approaches to category theory, besides the non-sharp element relationship, non-sharp equality and local existence play an important role. Local existence means that a degree of existence is assigned to an object. If this degree of existence is smaller than 1, this object is said to exist locally. In the domain of fuzzy logic non-sharp equality and local existence are seldom dealt with, although the concepts of equality and existence are fundamental for every predicate calculus. On the other hand, in literature on fuzzy systems, besides the ordinary existence and universal quantifiers, *linguistic quantifiers* like 'many', 'the most', 'some', etc. can be found [Zadeh83a, Bouchon-Meunier92, Dubois90].

A fuzzy logic where fuzziness comes from indiscernibility between possible worlds modelled by a fuzzy similarity relation is described in [Ruspini91].

2.8.5   The SOLD System — An Implementation

A very important application field of fuzzy set theory is the modelling and statistical analysis of data that are afflicted with vagueness. In this connection we distinguish the two views of vagueness mentioned above. The first one regards a vague datum as an existing object, for example a physical grey scale picture. Therefore this view is called a *physical* interpretation of vague data. The second view, the *epistemic* interpretation, uses vague data only to describe a value, that is existing and precise, but not measurable with exactitude. Thus the first view does not examine real-valued data, but objects that are more complex. In the most simple case these are sets, as they turn up in the context of random sets [Matheron75, Kendall74, Stoyan87].

The key to a generalized probability theory of non-standard data is always based on a 'law of large numbers', whose validity in the case of set-valued data was proved in [Artstein75]. With regard to fuzzy data an analogous theorem can be proved [Ralescu82, Puri86, Kruse82c], that allows us to develop a fuzzy probability theory and by this to lay down a basis for mathematical statistics on fuzzy sets [Kruse82c, Kruse84]. In the books [Bandemer92, Kruse87d] the methods of fuzzy statistics and their application is described in detail. Comparable approaches were discussed in [Gil88, Czogala86, Hirota81].

Epistemic interpretation assumes the existence of an (unkown) original value, so that fuzzy sets serve to describe the vagueness and uncertainty of the exact location of this original in a set of possible values. This interpretation of fuzzy sets is treated in depth in Chapter 3. Usually the expression 'possibility distribution' is used instead of 'fuzzy set with an epistemic interpretation.' The idea to analyse such data statistically was first described in [Kwakernaak78a, Kwakernaak78b]. In [Kruse87d, Kruse92] some aspects of statistical inference of possibilistic data are dealt with. The methods analysed there have been incorporated into the software tool SOLD (Statistics On Linguistic Data) [Kruse87a, Kruse89], which consists of more than 20,000 lines of PASCAL-code. In [Bandemer92] similar methods are discussed. Further studies on the semantics of vague data can be found in [Gebhardt92a].

As an example for the application of many of the concepts, methods, and results discussed in this chapter, we briefly present the software tool SOLD (Statistics On Linguistic Data), which supports the modelling and statistical analysis of linguistic data, which are representable by fuzzy sets [Kruse89]. Commercial versions of SOLD, developed in the programming language PASCAL-XT within the framework of a cooperation contract between the

Computer Science Department of the University of Braunschweig and Siemens AG München, can be run under the operation systems BS2000 and SINIX.

An application of the SOLD system consists of two steps, which have to be considered separately with regard to their underlying concepts.

In the first step (*specification phase*) SOLD enables its user to create an application environment (e.g. to analyse weather data), that consists of a finite set of *attributes* (e.g. *clouding, temperature, precipitation*) with their *domains* (intervals of real numbers, e.g. $[0, 100]$ for the clouding of the sky in %). For each attribute A the user states several (possibly parameterized) *elementary linguistic values* (e.g. *cloudy* or *approximately*(75) as vague degrees of the clouding of the sky) and defines for all of these values w the fuzzy sets μ_w, that shall be associated with them. For this reason SOLD provides 15 different classes of parameterized fuzzy sets of $\mathbf{R}$ (e.g. triangular, rectangular, trapezoidal, Gaussian, and exponential functions) as well as 16 logical and arithmetical operators (*and, or, not*, $+, -, *, /, **$) and functions (e.g. *exp, log, min, max*), that are generalized to fuzzy sets using the extension principle.

The application of context-free generic grammars G_A permits the combination of elementary linguistic values by logic operators (*and, or, not*) and so-called linguistic hedges (*very, considerable*) to increase or decrease the specifity of vague concepts. By this, formal languages $L(G_A)$ are obtained, which consist of the linguistic expressions that are permitted to describe the values of the attributes A (e.g. *cloudless or fair* as a linguistic expression with respect to the attribute *clouding*).

In the second step (*analysis phase*) the application environments created in the specification phase can be applied to describe realizations of random samples by tuples of linguistic expressions. Since the random samples consist of existing numeric values, that generally cannot be observed exactly, the fuzzy sets, which are related to the particular linguistic expressions, are interpreted epistemically as vague data.

The SOLD system allows us to determine convex fuzzy estimators for several characteristic parameters of the generic random variables for the considered attributes (e.g. for the expected value, variance, p-quantile, and range). In addition SOLD calculates fuzzy estimates for the unknown parameters of several classes of given distributions and also determines fuzzy tests for one- or two-valued hypotheses with regard to the parameters of normally distributed random variables.

The algorithms incorporated in this tool are based on the original results about fuzzy statistics that were presented in the monograph [Kruse87d]. In this connection it should be mentioned that the estimation functions known from ordinary statistics can, of course, be generalized without any problems from real numbers to fuzzy sets with the help of the extension principle, but that the membership degrees of the resulting fuzzy estimator are hard to calculate in practice with the exception of trivial cases. In SOLD therefore, only fuzzy sets of the classes $F_{D_k}(\mathbf{R})$ (cf. Definition 2.39) are employed. In this case the operations to be performd can be reduced to the α-cuts of the involved fuzzy sets according to Theorem 2.40. Nevertheless the simplification achieved by this restriction does not guarantee that we gain an efficient implementation, since operations on α-cuts are not equivalent to elementary interval arithmetics. The difficulties that arise can be recognized already in the following example of determining a fuzzy estimator for the variance.

Let $U : \Omega \to \mathbf{R}$ be the random variable defined with respect to a probability space $(\Omega, \mathfrak{S}, P)$ and F_U its distribution function. By a realization $(u_1, \ldots, u_n) \in \mathbf{R}^n$ of a random sample $(U_1, \ldots, U_n)$ with random variables $U_1, \ldots, U_n : \Omega \to \mathbf{R}$, $n \geq 2$, that are completely independent and equally distributed according to F_U, the parameter $\text{Var } U$ can be estimated with the help of the variance of the random sample, defined as

$$S_n(U_1, \ldots, U_n) \overset{\text{def}}{=} \frac{1}{n-1} \left(\sum_{i=1}^{n} \left(U_i - \frac{1}{n} \sum_{j=1}^{n} U_j \right)^2 \right).$$

$S_n(U_1, \ldots, U_n)$ is an unbiased, consistent estimator for $\text{Var } U$.

If $(\mu_1, \ldots, \mu_n) \in [F_{D_k}(\mathbf{R})]^n$ is the description of a vague observation $(u_1, \ldots, u_n)$ for a given $k \in \mathbf{N}$, then by applying the extension principle we obtain the following fuzzy estimator for $\text{Var } U$:

$$\hat{S}_n : [F_{D_k}(\mathbf{R})]^n \ \to \ [F_{D_k}(\mathbf{R})],$$

$$\hat{S}_n(\mu_1, \ldots, \mu_n)(y) \ = \ \sup \left\{ \min\{\mu_1(x_1), \ldots, \mu_n(x_n)\} \ | \right.$$

$$\left. (x_1, \ldots, x_n) \in \mathbf{R}^n \wedge S_n(x_1, \ldots, x_n) = y \right\}.$$

For $\alpha \in (0, 1]$ Theorem 2.40 leads to the α-cuts

$$\left[\hat{S}_n(\mu_1, \ldots, \mu_n) \right]_\alpha$$

$$= \ S_n \left([\mu_1]_\alpha, \ldots, [\mu_n]_\alpha \right)$$

$$= \ \left\{ y \mid \exists (x_1, \ldots, x_n) \in \underset{i=1}{\overset{n}{\times}} [\mu_i]_\alpha : S_n(x_1, \ldots, x_n) = y \right\}$$

$$= \ \left\{ y \mid \exists (x_1, \ldots, x_n) \in \underset{i=1}{\overset{n}{\times}} [\mu_i]_\alpha : \frac{1}{n-1} \sum_{i=1}^{n} \left(x_i - \frac{1}{n} \sum_{j=1}^{n} x_j \right)^2 = y \right\}.$$

Since

$$S_n \left([\mu_1]_\alpha, \ldots, [\mu_n]_\alpha \right) \ \subseteq \ \frac{1}{n-1} \sum_{i=1}^{n} \left([\mu_i]_\alpha - \frac{1}{n} \sum_{j=1}^{n} [\mu_j]_\alpha \right)^2,$$

but equality does not hold in general, then $S_n([\mu_1]_\alpha, \ldots, [\mu_n]_\alpha)$ cannot be determined by elementary interval arithmetic.

Therefore the creation of SOLD had to be preceded by further mathematical considerations, that were helpful to the development of efficient algorithms for the calculation of fuzzy estimators. Some results can be found in [Kruse87d, Gebhardt90].

The fuzzy set ν calculated during the analysis phase by statistical inference with regard to an attribute A (e.g. fuzzy estimation for the variance of the *temperature*) is not transformed back to a linguistic expression by SOLD, as might be expected at first glance. The fundamental problem consists in the fact that in general no $w \in L(G_A)$ can be found, for which $\nu \equiv \mu_w$ holds. Consequently one is left to a *linguistic approximation* of ν, i.e. to find those linguistic expressions w of $L(G_A)$, whose interpretations μ_w approximate the fuzzy set ν under consideration as accurately as possible. The *distance* between two fuzzy sets is measured with the help of the *generalized Hausdorff pseudometric* d_∞, which is defined as

$$d_\infty(\mu, \mu') \;=\; \sup_{\alpha \in (0,1]} \{d_H([\mu]_\alpha, [\mu']_\alpha)\}, \quad (\mu, \mu') \in [F_C(\mathbf{R})]^2,$$

$$d_H(A, B) \;=\; \max\left\{\sup_{a \in A} \inf_{b \in B} |a - b|, \sup_{b \in B} \inf_{a \in A} |a - b|\right\}.$$

d_H is called the *Hausdorff pseudometric* w.r.t. $\mathbf{R}$.

The aim of this linguistic approximation is to determine a $w_{\text{opt}} \in L(G_A)$ that satisfies

$$\forall w \in L(G_A): \quad d_\infty(\mu_{w_{\text{opt}}}, \nu) \leq d_\infty(\mu_w, \nu).$$

Since this optimization problem in general is very diffcult and can lead to unsatisfactory approximations, if $L(G_A)$ is chosen unfavourably, (Hausdorff distance too large or linguistic expressions too complicated), SOLD uses the language $L(G_A)$ only to name the vague data appearing in the random samples related to A in an expressive way. SOLD calculates the Hausdorff distance $d_\infty(\mu_w, \nu)$ between ν and a fuzzy set μ_w, provided by the user as a linguistic expression $w \in L(G_A)$, that appears to be suitable, but does not carry out a linguistic approximation by itself, since the resulting linguistic expression would not be very useful in order to *make a decision* in consequence of the statistical inference.

To illustrate the application of the SOLD system we finally present a picture of a session that is taken from a prototype version of this software tool under the operating system GEMDOS (see Figure 18).

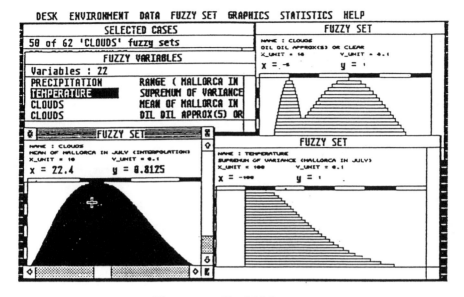

Figure 18   The SOLD system

2.8.6   Exercises

Exercise 2.1 Let $(A_\alpha)_{\alpha \in [0,1]}$ be the family of sets defined by

$$A_\alpha = \begin{cases} \left[1 - \sqrt{\ln(\tfrac{1}{\alpha})}, \ 1 + \sqrt{\ln(\tfrac{1}{\alpha})} \right], & \text{if } \alpha > 0 \\ \mathbf{R}, & \text{if } \alpha = 0. \end{cases}$$

Show that this system satisfies the conditions of Definition 2.15, and determine the corresponding fuzzy set $\mu \in F(\mathbf{R})$.

Exercise 2.2 Prove that the Yager family of t-norms and t-conorms has the following properties:

$$\lim_{p \to 0} \mathsf{T}''_p(a,b) = \mathsf{T}_{-1}(a,b),$$

$$\lim_{p \to 0} \perp''_p(a,b) = \perp_{-1}(a,b),$$

$$\lim_{p \to \infty} \mathsf{T}''_p(a,b) = \mathsf{T}_{\min}(a,b),$$

$$\lim_{p \to \infty} \perp''_p(a,b) = \perp_{\min}(a,b).$$

Let $C''_p(a) \overset{\text{def}}{=} (1 - a^p)^{\frac{1}{p}}$ be the Yager complement.

Show that $\perp_p''(a, C_p''(a)) = 1$ holds.

Exercise 2.3 Calculate the sum of two fuzzy sets of the type

$$\mu_{r,s}(t) = e^{-(\frac{t-r}{s})^2}, \quad r, s \in \mathbf{R}, \quad s > 0.$$

Exercise 2.4 Determine the set representation of

$$\mu : \mathbf{R} \to [0,1], \quad \mu(x) = \begin{cases} \frac{1}{2}x + \frac{1}{2}, & \text{if } -1 \leq x \leq 1 \\ 2 - x, & \text{if } 1 \leq x \leq 2 \\ 0, & \text{otherwise,} \end{cases}$$

and calculate the reciprocal value of μ.

Exercise 2.5 Let $\phi : \mathbf{N} \times \mathbf{N} \to \mathbf{N}; \quad \phi(a,b) = \max\{a, b\}.$

$$\mu_1 : \mathbf{N} \to [0,1]; \qquad \begin{aligned} &\mu_1(0) = 1, \quad \mu_1(1) = 0.5, \quad \mu_1(2) = 0.2, \\ &\mu_1(3) = 0.1, \quad \mu_1(n) = 0, \quad \text{for } n \geq 4, \end{aligned}$$

$$\mu_2 : \mathbf{N} \to [0,1]; \qquad \begin{aligned} &\mu_2(0) = 0.1, \quad \mu_2(1) = 0.4, \quad \mu_2(2) = 0.9, \\ &\mu_2(n) = 0, \quad \text{for } n \geq 3. \end{aligned}$$

Determine $\hat{\phi}(\mu_1, \mu_2)$.

Exercise 2.6 Let (X, δ) be a metric space.
Show that E_δ, defined as $E_\delta(x, x') = 1 - \min\{\delta(x, x'), 1\}$, is an equality relation w.r.t. T_{Luka}.

Exercise 2.7 We consider a reference set $X = \{m, w\}$ and fuzzy sets $\mu_1 : X \to [0,1], \mu_2 : X \to [0,1]$ with $\mu_1(m) = 0.8$, $\mu_1(w) = 0.2$, $\mu_2(m) = 0.4$, $\mu_2(w) = 0.6$. In addition let $C = \{c_1, \ldots, c_{10}\}$ be a set of contexts with the weights $P(\{c_i\}) = 0.1$, $i = 1, 2, \ldots, 10$. Determine random sets $\Gamma_1', \Gamma_1'', \Gamma_2', \Gamma_2''$ with the properties $\mu_{\Gamma_1'} = \mu_{\Gamma_1''} = \mu_1$, $\mu_{\Gamma_2'} = \mu_{\Gamma_2''} = \mu_2$,

$$\mu_{\Gamma_1' \cap \Gamma_2'}(x) = \max\{0, \mu_{\Gamma_1'}(x) + \mu_{\Gamma_2'}(x) - 1\}$$

and

$$\mu_{\Gamma_1'' \cap \Gamma_2''}(x) = \min\{\mu_{\Gamma_1''}(x), \mu_{\Gamma_2''}(x)\}.$$

Exercise 2.8 Show that the inequality of Example 2.50 holds.

Exercise 2.9 Let T and T' be two t-norms that satisfy $\mathsf{T}' \leq \mathsf{T}$. Show that if $E : X \times X \to [0,1]$ is an equality relation w.r.t. T, then E is an equality relation w.r.t. T'.

Exercise 2.10 Show that equation (2.41) defines an equality relation on $F(X)$ w.r.t. the t-norm T_{min}.

Exercise 2.11 Determine the residuals of the t-norm of the Weber family.

Exercise 2.12 Formulate and prove an analogue to Theorem 2.70 by replacing the Lukasiewicz implication by the Gödel implication.

Exercise 2.13 Let E and F be two equality relations on the sets X and Y, respectively, w.r.t. t-norm T.

Show that $G\left((x,y),(x',y')\right) = \mathsf{T}\left(E(x,x'),F(y,y')\right)$ defines an equality relation on $X \times Y$ w.r.t. T.

3

Possibilistic Reasoning

It has recently grown more and more accepted that there is a need for incorporating aspects of vagueness and uncertainty into knowledge-based systems, considering appropriate semantic foundations. Subsequent to earlier systems that worked rather heuristically, like MYCIN [Buchanan84] for medical analysis, some professional tools based on probability theory have been developed, which allow the execution of reasoning mechanisms on the basis of imperfect knowledge. Among these we especially mention HUGIN [Andersen89] and PATHFINDER [Heckerman90].

The topic of this chapter is to investigate an alternative approach to approximate reasoning in expert systems, which rather refers to *possibilistic* than probabilistic reasoning. We introduce a special concept of possibilistic focusing that differs from most other approaches in the way that we interpret possibility distributions as *information-compressed* representations of (not necessarily nested) random sets (cf. Section 2.6.1), and we prefer a *constraint-related* view of possibilistic reasoning which is *not* based on a specific conditioning concept for uncertain knowledge.

In Section 3.1 we inspect the important distinction between a *vague concept* and an *uncertain datum*. The epistemic interpretation of fuzzy sets, which is related to the characterization of uncertain data, leads to the notion of a *possibility distribution* and hence to the foundations of possibility theory. The assessment of uncertain data with respect to decision making suggests the introduction of alternative *measures of uncertainty*, which can be compared with the probability measures known from measure theory.

In Section 3.2 we present an application in the field of artificial intelligence by giving an outline of an expert system that performs approximate reasoning, if the whole knowledge about a certain application can be represented with the aid of possibility distributions on a multidimensional universe of discourse. We will distinguish between *general knowledge*, which is provided by experts (e.g. in form of possibilistic rules), and situation-dependent *evidential knowledge*, gained by observations, and used for *focusing* operations. How the underlying possibility distributions can be obtained as an interpretation of (possibilistic) inference rules, is discussed in Section 3.3.

In contrast to the above, in Section 3.4 we do not aim to examine *quantitative* dependencies within the knowledge base, but take advantage of representing *qualitative* dependencies by hypergraphs and thus achieve efficient propagation algorithms.

As an alternative approach to possibilistic reasoning, Section 3.5 provides an introduction to logic-based inference mechanisms. In the supplementary remarks of Section 3.6, besides numerous references, a short presentation of other approaches to possibility theory, which have been proposed in the literature, can be found.

3.1   Possibility Distributions and Uncertainty Measures

In this chapter we deal with some methods of knowledge representation and reasoning in systems that are based on the application of fuzzy set theory and that are equipped with a knowledge base which consists of conjunctively combined *inference rules* like

$$R: \textbf{If } X \textbf{ is } \mu_X \textbf{ then } Y \textbf{ is } \mu_Y,$$

where X, Y are two attributes (partially) characterizing a certain object type, and μ_X, μ_Y are two fuzzy sets on the domains of the datatypes related to these attributes.

As a simple example we inspect the vague statement

Fast cars are expensive.

It vaguely connects two attributes that characterize a car — namely its (maximum) *speed* and its *price*, described by the vague expressions *fast* and *expensive*, respectively, which can be interpreted by fuzzy sets. The inference rule that corresponds to this statement reads

$$R: \textbf{if } speed \textbf{ is } fast \textbf{ then } price \textbf{ is } expensive,$$

where we may assume the attached domains $\text{Dom}(speed) = [0, 300]$ (in km/h) and $\text{Dom}(price) = [0, 500.000]$ (in \$).

When inspecting vague rules, two problems arise on principle, namely:

(a) how to interpret and how to represent vague rules with the help of appropriate fuzzy sets, and

(b) how to find an inference mechanism that is founded on clear semantics and that permits approximate reasoning by means of a conjunctive general system of vague rules and case-specific vague facts.

Let us apply (a) and (b) to our introductory example. In (a) we state, for instance, a *fuzzy relation*

$$\mu_R : \text{Dom}(speed) \times \text{Dom}(price) \rightarrow [0, 1]$$

that represents R.

The approximate reasoning mechanism mentioned in (b) could be motivated by the aim to infer something about a certain car based on a given set of rules. If we know, for example, that a car A's maximum speed can be called *moderately fast* and that this expression can be described by a fuzzy set μ_F, then the question arises, what we may infer from

> the vague fact F :   *speed* is *moderately fast*
>
> and the vague rule R :   **if** *speed* is *fast*
>   **then** *price* is *expensive*

with respect to the price of A.

One would expect the applied inference mechanism to yield a vague conclusion, which can be described by a fuzzy set $\mu : \text{Dom}(price) \rightarrow [0, 1]$ and which depends on μ_F and μ_R only. But if we consider approximate reasoning in the framework of multi-valued logic, then R can be interpreted as an *implication rule*, and thus the inference mechanism, that is applied to F and R, has to correspond to a generalization of the *modus ponens*, well-known from two-valued logics. That a generalization is needed can easily be seen from the fact that the ordinary *modus ponens* would require the premise F to coincide with the antecedence of R (*speed* is *fast*) and would lead to the conclusion of R (*price* is *expensive*). But this does not agree with our example, since *moderately fast* and *fast* would certainly have to be represented by different fuzzy sets.

As we already mentioned in the previous chapter, Zadeh's view of fuzzy sets originated from the idea of extending the notion of a characteristic function of ordinary sets, such that it becomes possible not only to distinguish between membership and non-membership to a given reference set, but also to include gradual membership. An important motivation for this extension emerged from the aim to represent and interpret vague statements, as they are common in daily life and hence also appear as expert opinions in the domain of knowledge-based systems, within a suitable formal structure.

'It is cloudy,' for example, is a statement that describes an (abstract) object (namely the weather) with the help of a vague concept (*cloudy*) related to an attribute (clouding) characterizing the weather. The stated assertion very clearly reflects the common and reasonable behaviour of reducing information to the amount that is essential in a given situation in order to facilitate communication. In this sense the application of the vague concept *cloudy* tends to be more helpful than speaking of a more precise clouding degree of

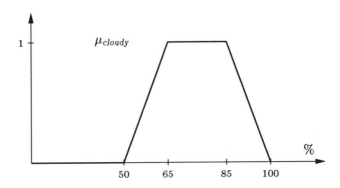

Figure 1   Fuzzy set interpretation of the vague concept *cloudy*

71.26 %, especially because it would be very difficult for an observer to make justified statements of that precision without extraordinary technical effort. If we regard *cloudy* as a linguistic description of a vague concept that can be interpreted by a fuzzy set, then the fuzzy set μ_{cloudy} shown in Figure 1 is such a possible interpretation.

We view $\mu_{cloudy} : X \rightarrow [0,1]$, where $X = [0,100]$ is the reference set of all possible degrees of clouding, as the fuzzy set of those degrees of clouding that reflect the vague concept *cloudy*, and $\mu_{cloudy}(x)$ as the *degree of membership*, with which $x \in X$ satisfies this concept *cloudy*, i.e. as the degree, with which x has to be called *cloudy* according to the chosen fuzzy set interpretation μ_{cloudy}.

But this *concept-oriented interpretation* of μ_{cloudy} as a fuzzy set does not help if we want to specify, for instance, our *uncertainty* about the degree of clouding observable at the Hannover airport weather station on March 8th 1992 at 2 p.m. Then *cloudy* rather serves as a linguistic description of an *uncertain datum*, since it is of less significance what we mean by *cloudy* conceptionally, but rather is more important to assume that at the time and place stated above there was a crisp degree of clouding $x_0 \in [0,100]$, which we want to describe on the basis of the information available by the uncertain datum alone. This is the so-called *epistemic view* of the function μ_{cloudy} as a *possibility distribution*, by which $\mu_{cloudy}(x)$ is regarded as the *degree of possibility* that $x \in X$ coincides with the actually existing, but inaccessible degree of clouding x_0:

$\mu_{cloudy}(x) = 0$ means that $x = x_0$ is impossible,

$\mu_{cloudy}(x) = 1$ means that $x = x_0$ is possible without any restrictions

$\mu_{cloudy}(x) \in (0,1)$ means that $x = x_0$ has to be regarded as being possible with the degree $\mu_{cloudy}(x)$ according to the given vague datum μ_{cloudy}.

With respect to their representation, fuzzy sets and possibility distributions have the same mathematical description but different interpretations. On the other hand these two notions are so closely related to one another that nearly all of the results about fuzzy sets elaborated in Chapter 2 — especially the extension principle — can be transferred to possibility distributions without change. That is, if the function μ_{cloudy} stated above is regarded as a possibility distribution and hence as an uncertain characterization of an actually existing degree of clouding x_0, then this fact can also be interpreted in the way that x_0 satisfies the vague concept *cloudy* by presupposition, and that the corresponding fuzzy set $\mu_{cloudy} \in F([0,100])$ is the set of all possible values for x_0. Therefore $\mu_{cloudy}(x)$, for an arbitrary $x \in [0,100]$, states the degree with which x belongs to μ_{cloudy}, but also the degree of possibility with which $x = x_0$ holds, i.e. that x is the original clouding x_0 we are searching for.

All possibility distributions can be treated as fuzzy sets this way, and in reverse, every fuzzy set $\mu \in F(X)$ satisfying the *normalization condition* ($\exists x \in X : \mu(x) = 1$) can be regarded as a possibility distribution on X. This is called the *epistemic interpretation of a fuzzy set*. Postulating the normalization condition for possibility distributions arises from the consideration that a possibility distribution which characterizes an actually existing, but only imperfectly observed state $x_0 \in X$ of a given object, has to contain at least one element x of the reference set, for which $x = x_0$ is regarded as being possible without restriction. Otherwise the chosen characterization would not be consistent. The precondition $x_0 \in X$ is sometimes called the *closed world assumption*, i.e. the unknown state of the object x_0 under consideration certainly is an element of the considered set X of all possible states.

To distinguish possibility distributions and fuzzy sets on the symbolic level, from now on we will denote possibility distributions by π, while the underlying reference set, which especially in the field of knowledge-based systems is often called *universe* (universe of discourse), will be denoted by Ω. The unknown object state that we are interested in, and which we describe by a possibility distribution π_0, will be denoted by ω_0.

Definition 3.1 *A* **possibility distribution** π *on* Ω *is a mapping from the reference set or the* **universe** Ω *into the unit interval, i.e.*

$$\pi : \Omega \to [0,1],$$

for which ($\exists \omega \in \Omega : \pi(\omega) = 1$) **(normalization condition)** *holds.* POSS(Ω) *denotes the set of all possibility distributions on* Ω.

The name 'possibility distribution' was chosen by analogy to probability theory and the notion of a probability distribution. Let us consider a probability space $(\Omega, \mathfrak{S}, P)$, where Ω is a set of elementary events, $\mathfrak{S}$ a σ-algebra of events, and P a probability measure. One of the numerous application domains of $(\Omega, \mathfrak{S}, P)$ refers to the characterization of a random experiment, whose possible results are elements of Ω. That is, if $A \in \mathfrak{S}$ is an arbitrary event, then $P(A)$ is the (objective) *probability* that the result of the experiment, which cannot be foreseen with the exception of trivial cases (e.g. $|\Omega| = 1$), is an element of A. With possibility distributions one can ask analogously what the *possibility* might be that $\omega_0 \in A$ holds. Note that the semantics of the two approaches deviates, since it makes, as an illustrative example, a great difference whether it is *probable* or *possible* for a person to eat four eggs at breakfast.

Instead of using a probability measure it is now the task to define alternative *uncertainty measures* that no longer focus on additivity, but coincide with the principles of *possibilistic* rather than *probabilistic* reasoning.

If we are given a possibility distribution $\pi_0 \in \mathrm{Poss}(\Omega)$ for an unknown object state $\omega_0 \in \Omega$ of interest, then, for all subsets $A \subseteq \Omega$ of possible candidates for ω_0, $\mathrm{Poss}_{\pi_0}(A) \overset{\text{def}}{=} \sup\{\pi_0(\omega) \mid \omega \in A\}$ with $\sup \emptyset \overset{\text{def}}{=} 0$ quantifies the possibility that $\omega_0 \in A$ is true.

Note that a high value of $\mathrm{Poss}_{\pi_0}(A)$ is by no means sufficient to make us somewhat certain that ω_0 indeed belongs to A. For this reason one should also consider $\mathrm{Poss}_{\pi_0}(\Omega \backslash A)$ and calculate $\mathrm{Nec}_{\pi_0}(A) \overset{\text{def}}{=} 1 - \mathrm{Poss}_{\pi_0}(\Omega \backslash A)$. The higher the value of $\mathrm{Nec}_{\pi_0}(A)$, the more certain we are that $\omega_0 \in A$ holds. This motivates the definition of the following two set functions.

Definition 3.2 *Let* $\pi \in \mathrm{POSS}(\Omega)$ *be a possibility distribution.*

(a) $\mathrm{Poss}_\pi : \mathfrak{P}(\Omega) \quad \longrightarrow \quad [0, 1]$,

$$\mathrm{Poss}_\pi(A) \quad \overset{\text{def}}{=} \quad \sup\{\pi(\omega) \mid \omega \in A\}, \qquad \sup \emptyset \overset{\text{def}}{=} 0$$

is called the **possibility measure of** π.

(b) $\mathrm{Nec}_\pi : \mathfrak{P}(\Omega) \quad \longrightarrow \quad [0, 1]$,

$$\mathrm{Nec}_\pi(A) \quad \overset{\text{def}}{=} \quad \inf\{1 - \pi(\omega) \mid \omega \in \Omega \backslash A\}, \qquad \inf \emptyset \overset{\text{def}}{=} 1$$

is called the **necessity measure of** π.

Example 3.3 Consider Figure 1 and assume that $\pi_0 \equiv \mu_{\text{cloudy}} \in \mathrm{POSS}([0, 100])$ is a possibility distribution for the degree of clouding at Hannover airport, March 8th, 1992, 2 p.m.

Choosing $A = [65, 70]$, we obtain $\mathrm{Poss}_{\pi_0}(A) = 1$ and $\mathrm{Nec}_{\pi_0}(A) = 0$, which means that $\omega_0 \in [65, 70]$ is completely possible w.r.t. π_0, but not certain (necessary) at all. It should be pointed out that π_0 must not be interpreted as a probability density function, since $\int_{[0,100]} \pi_0(\omega)d\omega = 35$. But also normalizing it to $\pi_0' \equiv \frac{1}{35}\pi_0$ and denoting P as the induced probability measure of π_0', yields $P(A) = \frac{1}{7}$, a result which is different from $\text{Poss}_{\pi_0}(A)$ and $\text{Nec}_{\pi_0}(A)$, respectively. □

Example 3.3 illustrates that *probability theory* and *possibility theory* (as a theory of possibility distributions, possibility measures, and their dual necessity measures) model quite different uncertainty phenomena. Both approaches are at least related in the sense that the probability of an event also entails its possibility. Whereas probability measures are additive, possibility measures and necessity measures are subadditive and superadditive, respectively.

The following theorem summarizes some more of their properties.

Theorem 3.4 *For all $\pi \in \text{POSS}(\Omega)$ and all $A, B \subseteq \Omega$:*

(a) $\text{Poss}_\pi(\emptyset) = \text{Nec}_\pi(\emptyset) = 0$,

(b) $\text{Poss}_\pi(\Omega) = \text{Nec}_\pi(\Omega) = 1$,

(c) $\text{Poss}_\pi(A \cup B) = \max\{\text{Poss}_\pi(A), \text{Poss}_\pi(B)\}$,

(d) $\text{Nec}_\pi(A \cap B) = \min\{\text{Nec}_\pi(A), \text{Nec}_\pi(B)\}$,

(e) $\text{Poss}_\pi(A \cap B) \leq \min\{\text{Poss}_\pi(A), \text{Poss}_\pi(B)\}$,

(f) $\text{Nec}_\pi(A \cup B) \geq \max\{\text{Nec}_\pi(A), \text{Nec}_\pi(B)\}$,

(g) $\text{Nec}_\pi(A) \leq \text{Poss}_\pi(A)$.

Proof:
(a) and (b) are trivial.

$$
\begin{aligned}
\text{(c)} \quad \text{Poss}_\pi(A \cup B) &= \sup\{\pi(\omega) \mid \omega \in A \cup B\} \\
&= \sup\{\pi(\omega) \mid \omega \in A \vee \omega \in B\} \\
&= \max\{\sup\{\pi(\omega) \mid \omega \in A\}, \sup\{\pi(\omega) \mid \omega \in B\}\} \\
&= \max\{\text{Poss}_\pi(A), \text{Poss}_\pi(B)\}.
\end{aligned}
$$

$$
\begin{aligned}
\text{(d)} \quad \text{Nec}_\pi(A \cap B) &= 1 - \text{Poss}_\pi(\Omega \backslash (A \cap B)) \\
&= 1 - \sup\{\pi(\omega) \mid \omega \in \Omega \backslash A \vee \omega \in \Omega \backslash B\} \\
&= 1 - \max\{\sup\{\pi(\omega) \mid \omega \in \Omega \backslash A\}, \\
&\qquad\qquad \sup\{\pi(\omega) \mid \omega \in \Omega \backslash B\}\} \\
&= \min\{1 - \text{Poss}_\pi(\Omega \backslash A), 1 - \text{Poss}_\pi(\Omega \backslash B)\} \\
&= \min\{\text{Nec}_\pi(A), \text{Nec}_\pi(B)\}.
\end{aligned}
$$

(e) and (f) are consequences of (c) and (d), if one notices that

$$
C \subseteq D \Longrightarrow \text{Nec}_\pi(C) \leq \text{Nec}_\pi(D)
$$

and

$$C \subseteq D \implies \text{Poss}_\pi(C) \leq \text{Poss}_\pi(D)$$

hold for arbitrary $C, D \subseteq \Omega$.

(g) For $\text{Poss}_\pi(A) < 1$, considering the normalization condition of possibility distributions, we get $\text{Poss}_\pi(\Omega \backslash A) = 1$ and therefore $\text{Nec}_\pi(A) = 0 \leq \text{Poss}_\pi(A)$.

If $\text{Poss}_\pi(A) = 1$, trivially it is $\text{Nec}_\pi(A) \leq \text{Poss}_\pi(A)$. $\qquad\qquad\square$

From a practical point of view, imprecise and uncertain knowledge about ω_0 is not always directly formalized with the aid of the characterizing possibility distribution π_0, but often through a set of *constraints* that qualify subsets A of Ω and induce some restrictions with respect to π_0. In this connection two types of specification should be distinguished:

- *Certainty qualification* is related to statements of the form 'it is certain at least at the degree α that $\omega_0 \in A$' (for short: 'A is α-certain for ω_0').
- *Possibility qualification* refers to statements of the form 'A is a (completely) possible range for ω_0 at least at the degree α' (for short: 'A is α-possible for ω_0').

In more detail, 'A is α-certain for ω_0' can be interpreted as the constraint $\text{Nec}_{\pi_0}(A) \geq \alpha$, whereas '$A$ is α-possible for ω_0' corresponds to $\Delta_{\pi_0}(A) \geq \alpha$, with

$$\begin{aligned} \Delta_{\pi_0} : \mathfrak{P}(\Omega) \quad &\rightarrow \quad [0,1], \\ \Delta_{\pi_0}(A) \quad &\overset{\text{def}}{=} \quad \inf\{\pi_0(\omega) \mid \omega \in A\} \end{aligned} \tag{3.1}$$

being the uncertainty measure of *guaranteed possibility* and

$$\begin{aligned} \nabla_{\pi_0} : \mathfrak{P}(\Omega) \quad &\rightarrow \quad [0,1], \\ \nabla_{\pi_0}(A) \quad &\overset{\text{def}}{=} \quad 1 - \Delta_{\pi_0}(\Omega \backslash A) \end{aligned} \tag{3.2}$$

its dual measure of *potential certainty*. The equality $\text{Nec}_{\pi_0}(A) = \inf\{1 - \pi_0(\omega) \mid \omega \in \Omega \backslash A\}$ entails the equivalence of the constraints $\text{Nec}_{\pi_0}(A) \geq \alpha$ and $(\forall \omega \in \Omega \backslash A : \pi_0(\omega) \leq 1 - \alpha)$. Incorporating the fact that there are no restrictions on $\omega \in A$, 'A is α-certain for ω_0' can be translated into $(\forall \omega \in \Omega : \pi_0(\omega) \leq \max\{\mathbb{I}_A(\omega), 1 - \alpha\})$.

As an alternative representation of this constraint we may use a *constraint function*

$$c : \Omega \rightarrow \mathfrak{P}([0,1]),$$

where $c(\omega)$ is the set of all non-rejected candidates for $\pi_0(\omega)$. In this sense, 'A is α-certain for ω_0' can be represented with the aid of the constraint function that is defined by

$$c(\omega) = [0, \max\{\mathbb{I}_A(\omega), 1 - \alpha\}], \quad \omega \in \Omega. \tag{3.3}$$

On the other hand, referring to possibility qualification and considering that $\Delta_{\pi_0}(A) = \inf\{\pi_0(\omega) \mid \omega \in A\}$, we obtain the equivalence of $\Delta_{\pi_0} \geq \alpha$ and $(\forall \omega \in A : \pi_0(\omega) \geq \alpha)$. Again, since there are no restrictions on values $\omega \in \Omega \backslash A$, '$A$ is α-possible for ω_0' can be translated into $(\forall \omega \in \Omega : \pi_0(\omega) \geq \min\{\mathbb{I}_A(\omega), \alpha\})$, represented by the constraint function

$$c : \Omega \rightarrow \mathfrak{P}([0,1]),$$
$$c(\omega) = [\min\{\mathbb{I}_A(\omega), \alpha\}, 1]. \tag{3.4}$$

When k pieces of (completely reliable) information constraining the possible values of ω_0 are available, formalized by $c_j : \Omega \rightarrow \mathfrak{P}([0,1])$, $j = 1, \ldots, k$, then the conjunction of these constraints leads to the constraint function

$$c : \Omega \rightarrow \mathfrak{P}([0,1]),$$
$$c(\omega) \overset{\text{def}}{=} \bigcap_{j=1}^{k} c_j(\omega). \tag{3.5}$$

Complete reliability of information sources ensures that π_0 is a possibility distribution. Otherwise the sources turn out to be conflicting and therefore partially inconsistent.

Assuming that all given constraints have been combined, we get a single constraint function $c_0 : \Omega \rightarrow \mathfrak{P}([0,1])$ that reflects our imperfect knowledge about ω_0. The absence of any constraint corresponds to the case of total ignorance, formalized by $c_0(\omega) = [0,1]$ for all $\omega \in \Omega$, while complete knowledge of ω_0 is modelled by $c_0(\omega_0) = \{1\}$ and $c_0(\omega) = \{0\}$ for all $\omega \in \Omega \backslash \{\omega_0\}$.

Since any possibility distribution $\pi \in \text{POSS}(\Omega)$ with $(\forall \omega \in \Omega : \pi(\omega) \in c_0(\omega))$ satisfies the underlying constraints, we need a metaconcept of possibilistic reasoning in order to realize a unique choice of π_0. Such a reasonable metaconcept is called the *principle of minimum specificity*. Playing a similar role in possibility theory as the application of the maximum entropy principle in probability theory, it embodies the commonsense claim that only those things which are impossible should be ruled out. With respect to this principle, for all $\omega \in \Omega$, $\pi_0(\omega)$ has to be defined as the largest degree of possibility that fits the available constraints, which means that $\pi_0(\omega) \overset{\text{def}}{=} \sup(c_0(\omega))$. Note that a possibility distribution $\pi \in \text{POSS}(\Omega)$ is called *at least as specific as* another possibility distribution $\pi' \in \text{POSS}(\Omega)$, if $(\forall \omega \in \Omega : \pi(\omega) \leq \pi'(\omega))$ holds. Hence, π_0 in fact is of minimum specificity among all possibility distributions $\pi \in \text{POSS}(\Omega)$ that satisfy c_0 in the way that $(\forall \omega \in \Omega : \pi(\omega) \in c_0(\omega))$ holds.

It should be pointed out that the postulate of the minimum specificity principle is reasonable, when certainty qualification is intended to be realized. In the case of possibility qualification we need a different information principle, expressing that anything which is not established as being possible can be neglected. This motivates a *principle of maximum specificity* and $\pi_0(\omega) \overset{\text{def}}{=} \inf(c_0(\omega))$.

Applying the considered approaches to the representation of uncertain and imprecise information on an unknown value $\omega_0 \in \Omega$ in terms of a possibility distribution $\pi_0 \in \text{POSS}(\Omega)$, we now turn to the question of how to operate on such possibility distributions. Suppose that $\omega_0 = (\omega_0^{(1)}, \ldots, \omega_0^{(n)}) \in \Omega$ is a tuple of single attribute values $\omega_0^{(i)} \in \Omega^{(i)}$, $i = 1, \ldots, n$, each of them imperfectly characterized by a constraint function $c^{(i)} : \Omega^{(i)} \to \mathfrak{P}([0,1])$ with induced possibility distribution $\pi_0^{(i)} \in \text{POSS}(\Omega^{(i)})$, obtained by postulation of the minimum specificity principle.

Given a mapping $\varphi : \times_{i=1}^{n} \Omega^{(i)} \to \Omega'$, we are interested in calculating the corresponding constraint function $c' : \Omega' \to \mathfrak{P}([0,1])$ for $\varphi(\omega_0)$, and its induced possibility distribution, denoted by $\hat{\varphi}(\pi_0^{(1)}, \ldots, \pi_0^{(n)})$. As $c^{(i)}$ does not contain any information on $\omega_0^{(j)}$, $j \neq i$, it can be extended to a constraint function for ω_0, which is

$$
\begin{aligned}
c_i : \quad &\Omega \quad \to \quad \mathfrak{P}([0,1]), \\
c_i(\omega^{(1)}, \ldots, \omega^{(n)}) \quad &\overset{\text{def}}{=} \quad c^{(i)}(\omega^{(i)}).
\end{aligned}
\tag{3.6}
$$

Conjunction of the constraints yields

$$
\begin{aligned}
c : \quad &\Omega \quad \to \quad \mathfrak{P}([0,1]), \\
c(\omega^{(1)}, \ldots, \omega^{(n)}) \quad &\overset{\text{def}}{=} \quad \bigcap_{i=1}^{n} c^{(i)}(\omega^{(i)}),
\end{aligned}
\tag{3.7}
$$

expressing that $\pi_0(\omega) \in c(\omega)$ has to be satisfied. Since $\pi_0(\omega) \in c(\omega)$ should entail $\hat{\varphi}(\pi_0^{(1)}, \ldots, \pi_0^{(n)}) \in c'(\varphi(\omega_0^{(1)}, \ldots, \omega_0^{(n)}))$ and nothing else,

$$
\begin{aligned}
c' : \Omega' \quad \to \quad &\mathfrak{P}([0,1]) \quad \text{is given by} \\
c'(\omega') \quad = \quad &\bigcup \{c(\omega) \mid \omega \in \Omega \text{ and } \omega' = \varphi(\omega)\}.
\end{aligned}
\tag{3.8}
$$

With respect to the principle of minimum specificity we obtain

$$
\begin{aligned}
\hat{\varphi}(\pi_0^{(1)}, &\ldots, \pi_0^{(n)})(\omega') \\
= \quad &\sup(c'(\omega')) \\
= \quad &\sup\{\min\{\pi_0^{(1)}(\omega^{(1)}), \ldots, \pi_0^{(n)}(\omega^{(n)})\}\} \mid \\
&(\omega^{(1)}, \ldots, \omega^{(n)}) \in \Omega \text{ and } \varphi(\omega^{(1)}, \ldots, \omega^{(n)}) = \omega'\},
\end{aligned}
$$

$$(3.9)$$
$$(3.10)$$

which is the well-known *extension principle*, already introduced for fuzzy sets in Section 2.4.

Note that the epistemic view of a normal fuzzy set $\pi_0 \in F_N(\Omega)$ as a possibility distribution for ω_0 means that the acceptance degrees $\pi_0(\omega) = \text{acc}(\omega \text{ belongs to } \pi_0)$ have to be interpreted as possibility degrees $\pi_0(\omega) = \text{possibility}(\omega = \omega_0)$. In fact,

$$\hat{\varphi}(\pi_0^{(1)}, \ldots, \pi_0^{(n)})(\omega') = \text{possibility}(\omega' = \varphi(\omega_0^{(1)}, \ldots, \omega_0^{(n)}))$$
$$= \text{acc}(\omega' \text{ belongs to } \hat{\varphi}(\pi_0^{(1)}, \ldots, \pi_0^{(n)}))$$

coincides with the motivation of Section 2.4. Nevertheless it has to be emphasized that the extension principle for fuzzy sets can be modified using another pair of underlying t-norm and t-conorm, while $(\min, \max)$ has turned out to be the only choice when starting with the semantics of a possibility distribution and the principle of minimum specificity.

3.2 Concept of a Focusing Expert System for Possibilistic Data

An important application of the concepts presented in the previous section refers to the problem of *possibilistic reasoning* in knowledge-based systems. Let $\omega_0 \in \Omega$ denote the unknown current state of an object under consideration. Ω may be viewed as the Cartesian product $\Omega = \times_{i=1}^{n} \Omega^{(i)}$ of domains $\Omega^{(i)}$, attached to attributes $A^{(i)}$ that are chosen to characterize ω_0. Hence, we suppose that ω_0 is an n-tuple $(\omega_0^{(1)}, \ldots, \omega_0^{(n)})$ of attribute values $\omega_0^{(i)} \in \Omega^{(i)}$, $i = 1, \ldots, n$.

The theory of *approximate reasoning*, as proposed by L.A. Zadeh, is a methodology for representing imperfect knowledge about ω_0 in terms of possibility distributions, and using it in order to get most specific characterizations of the single attribute values $\omega_0^{(i)}$.

The main aim of the subsequent sections will be to present in this sense a mathematical model for a special class of expert systems, in which general expert knowledge about a part of the world can be represented with the help of possibility distributions. The user of such a system is able to integrate application-dependent observations of the actual state of an object under consideration, representable by possibility distributions, and to determine the conclusions about this state resulting from the given knowledge base by applying an appropriate inference mechanism.

If $\omega_0 \in \Omega$ is the object state we are interested in, $\varrho \in \text{POSS}(\Omega)$ the representation of the given general expert knowledge about all object states in the chosen part of the world, and $\varepsilon \in \text{POSS}(\Omega)$ is the representation of a special observation of ω_0, then the a priori knowledge ϱ about ω_0 can be made more specific on account of ε by a conjunctive combination of ϱ and ε resulting in $\varrho_\varepsilon \overset{\text{def}}{=} \min\{\varrho, \varepsilon\}$. In this case ϱ is sometimes called the underlying *knowledge base* and ε is called *evidence* with respect to ω_0. If ϱ and ε do not contradict one another, ϱ_ε will be a possibility distribution on Ω again. Then we have $(\forall \omega \in \Omega : \varrho_\varepsilon(\omega) \leq \varrho(\omega))$; i.e. ϱ_ε is *at least as specific as* ϱ and hence at least as informative as an uncertain characterization of ω_0.

Definition 3.5 *Let* $\pi_1, \pi_2 \in \mathrm{POSS}(\Omega)$ *be two possibility distributions on an arbitrary universe* Ω. π_1 *is called* **at least as specific as** π_2 *(in symbols:* $\pi_1 \sqsubseteq \pi_2$), *if for all* $\omega \in \Omega$ *the inequality* $\pi_1(\omega) \leq \pi_2(\omega)$ *holds.*

If in addition $\pi_1 \not\equiv \pi_2$ *holds, then* π_1 *is called* **more specific than** π_2 *(in symbols:* $\pi_1 \sqsubset \pi_2$).

For the sake of simplicity the expert systems considered here do not include concepts for modifying the structure of the expert knowledge (so-called *general belief change concepts*), but serve only to achieve the specialization of the (static) expert knowledge given *a priori* about the object state ω_0 under consideration with the help of some pieces of evidence. In such a case one sometimes speaks of *focusing* the knowledge about ω_0 by means of the evidence gathered. Therefore we will call the expert systems we conceptualize in this section *possibilistic F-expert systems*.

For the present we will reduce our examinations to the special case of those F-expert systems that can manage imprecise data. That is, these systems allow us to characterize imprecisely a state $\omega_0 \in \Omega$ on the basis of incomplete information about ω_0 with the help of a subset $E \subseteq \Omega$, for which $\omega_0 \in E$ is assumed. Obviously this is a special case, since E can also be interpreted as a possibility distribution $\mathbb{I}_E \in \mathrm{POSS}(\Omega)$.

By the next motivating example we make clear which structural properties the inspected expert systems in general should have.

Example 3.6 We consider a knowledge base containing information about authorities and their competences with respect to oil contamination of waters by trading vessels. The part of the world to be modelled is formed by the three attributes *place of contamination*, *competence*, and *action* with their domains

$$\Omega^{(1)} \overset{\mathrm{def}}{=} \mathrm{Dom}(place\ of\ contamination) \overset{\mathrm{def}}{=} \{z_3, z_2, z_1, ca, rd, ld\},$$
$$\Omega^{(2)} \overset{\mathrm{def}}{=} \mathrm{Dom}(competence) \overset{\mathrm{def}}{=} \{fd, hc, wp, ep\},$$
$$\Omega^{(3)} \overset{\mathrm{def}}{=} \mathrm{Dom}(action) \overset{\mathrm{def}}{=} \{ad, ac, fi, im\}.$$

As possible places of contamination we distinguish the *open sea* (z_3), the *12-mile zone* (z_2), the *3-mile zone* (z_1), the *canal* (ca), the *refuelling dock* (rd), and the *loading dock* (ld). In addition we assume that the competences in case of contamination occurring reside with the *fire department* (fd), the *harbour control* (hc), the *water police* (wp), and the *environmental protection agency* (ep). Finally let the possible actions be pronouncing an *admonition* (ad), drawing up a *bill* (*bl*), imposing a *fine* (*fi*), and carrying out an *imprisonment* (*im*).

The universe $\Omega = \Omega^{(1)} \times \Omega^{(2)} \times \Omega^{(3)}$ consists of all *states* (*elementary events*) that are permitted *a priori* within the part of the world under consideration. For example, the element $(z_2, ep, ad) \in \Omega$ expresses that the pronouncement of an admonition by the environmental protection agency for a contamination that happened in the 12-mile zone is a possible elementary event and thus can occur as an actual case of oil contamination. But not all of the combinations contained in Ω are elementary events that are compatible to our background knowledge. The fire department, for instance, will never arrest anyone, i.e. (z_3, fd, im) does not belong to the accepted states.

Having chosen $\Omega = \Omega^{(1)} \times \Omega^{(2)} \times \Omega^{(3)}$ and thus having carried out a coarse formal structuring of the part of the world under consideration, the second phase in building our simple expert system consists in including background knowledge about this part of the world by specifying the set $R \subseteq \Omega$ of all states that are possible or allowed *a posteriori*. If a large number of attributes has to be considered, then R is not stated directly by experts in practice, but results from statements given by experts (rules) that define relations on the domains of a selection of all possible attributes. In our example an expert may have stated the following rules with respect to the *place of contamination* and the *competence*:

(*R₁*) Only the environmental protection agency is competent for the open sea and the 12-mile zone.

(*R₂*) The competence of the harbour control is restricted to the docks.

(*R₃*) The water police and the environmental protection agency have no competence for the docks.

(*R₄*) The environmental protection agency is not competent for the canal.

(*R₅*) Only the harbour control is competent for the refuelling dock.

(*R₆*) The fire department is not equipped to carry out operations in the open sea.

Each of the rules stated above defines a relation $R_i \subseteq \Omega^{(1)} \times \Omega^{(2)}$. The rule (*R₃*), for example, leads to

$$R_3 \stackrel{\text{def}}{=} \Omega^{(1)} \times \Omega^{(2)} \setminus \{(rd, wp), (ld, wp), (rd, ep), (ld, ep)\}.$$

We assume that all statements are true and that they can therefore be combined conjunctively. This becomes apparent, if we form the set-theoretic intersection of the corresponding relations, so that

$$R^* \stackrel{\text{def}}{=} R_1 \cap R_2 \cap \ldots \cap R_6$$

R^*	fd	hc	wp	ep
z_3	o	o	o	•
z_2	o	o	o	•
z_1	•	o	•	•
ca	•	o	•	o
rd	o	•	o	o
ld	•	•	o	o

Table 1   A relation between places of contamination and competences

R^{**}	ad	bl	fi	im
fd	o	•	•	o
hc	o	•	o	o
wp	o	•	•	•
ep	•	o	o	o

Table 2   A relation between competences and actions

is the relation between the domains of the attributes *place of contamination* and *competence* induced by our six rules. Table 1 shows the description of R^*.

If an entry is marked with o, then the corresponding pair of attribute values is not compatible with the expert knowledge specified by R^*, whereas an entry marked with • is meant to express that the corresponding pair cannot be excluded on account of the expert knowledge and therefore has to be regarded as being possible until it contradicts a rule, which was modified or newly entered into the knowledge base by a belief revision process.

By means of R^* we obtain the set

$$\left\{ (\omega^{(1)}, \omega^{(2)}, \omega^{(3)}) \in \Omega \mid (\omega^{(1)}, \omega^{(2)}) \in R^* \right\} \subseteq \Omega$$

of all states within our part of the world that are compatible with the stated rules.

We now add further expert knowledge that allows assertions about competences and possible actions and that may be representable as the relation $R^{**} \subseteq \Omega^{(2)} \times \Omega^{(3)}$ shown in Table 2.

After including R^{**} we finally obtain

$$R \stackrel{\text{def}}{=} \left\{ (\omega^{(1)}, \omega^{(2)}, \omega^{(3)}) \in \Omega \mid (\omega^{(1)}, \omega^{(2)}) \in R^* \wedge (\omega^{(2)}, \omega^{(3)}) \in R^{**} \right\}$$

as the *knowledge base* of our expert system. It can be used to infer more precise conclusions from the (imprecise) data that are available with respect to an actual oil contamination.

Let the following *facts* (*evidence* gathered by observers) be known about an accident:

(E_1) The water police were not involved.

(E_2) The accident happened either in the canal or at the refuelling dock.

(E_1) and (E_2) can be formalized as follows:

$$E_1 \stackrel{\text{def}}{=} \Omega^{(2)} \backslash \{wp\},$$

$$E_2 \stackrel{\text{def}}{=} \{ca, rd\}.$$

The set of all states compatible with (E_1) and (E_2) is

$$E \stackrel{\text{def}}{=} \left\{ (\omega^{(1)}, \omega^{(2)}, \omega^{(3)}) \in \Omega \mid \omega^{(2)} \in E_1 \wedge \omega^{(1)} \in E_2 \right\}.$$

To draw conclusions from R and E, we assume the validity of the *closed world assumption*, which ensures that there is an original state $\omega_0 = (\omega_0^{(1)}, \omega_0^{(2)}, \omega_0^{(3)}) \in \Omega$ characterizing the considered accident in our part of the world. For this original state it is $\omega_0 \in R \cap E$ and therefore

$$\omega_0^{(1)} \in \{ca, rd\},$$

$$\omega_0^{(2)} \in \{fd, hc\},$$

$$\omega_0^{(3)} \in \{bl, fi\}.$$

The oil contamination happened in the canal or at the refuelling dock (this information corresponds to E_2). The harbour control or the fire department was involved; the taken action consisted in charging an account or in imposing a fine (this information exceeds E_1 and is due to the inclusion of the knowledge base). □

From Example 3.6 it becomes obvious that we may use product spaces and operations on them in order to represent expert knowledge and user evidence, but also to describe the inference mechanism which finally has to be applied. The subsequent definitions introduce the required notions and a simplified symbolism for them.

Definition 3.7 *A family* $\mathcal{U} = (\Omega^{(i)})_{i \in \mathbf{N}_n}$ *of non-empty domains* $\Omega^{(i)}$, $i = 1, \ldots, n$, $n \in \mathbf{N}$, *is called a* **universe** *of dimension n.* $\mathbf{N}_n \overset{\text{def}}{=} \{1, 2, \ldots, n\}$ *is the* **index set** *related to this universe.*

For any non-empty index set $I \subseteq \mathbf{N}_n$ *we define the* **product space**

$$\Omega^I \overset{\text{def}}{=} \times_{i \in I} \Omega^{(i)}.$$

For the product space $\Omega^{\mathbf{N}_n}$ *we write* Ω, *for short.*

Remark 3.8 Let $I = \{i_1, i_2, \ldots, i_m\}$ with $i_1 < i_2 < \ldots < i_m$ and $0 < m \leq n$ be an index subset of $\mathbf{N}_n$. Any element $\omega \in \Omega^I$ can be represented by an m-tuple $(\omega^{(i_1)}, \ldots, \omega^{(i_m)})$, for which $\omega^{(i)} \in \Omega^{(i)}$, $i \in I$, holds. $\square$

Definition 3.9 *Let* $\mathcal{U} = (\Omega^{(i)})_{i \in \mathbf{N}_n}$ *be a universe of dimension n. In addition let* C, S, *and* T *be index subsets of* $\mathbf{N}_n$, *which satisfy the conditions* $T = S \cup C$, $S \cap C = \emptyset$, *and* $S \neq \emptyset$.

(a) $\text{red}_S^T : \Omega^T \quad \to \quad \Omega^S,$

     $\text{red}_S^T(\omega^T) \quad = \quad \omega^S$ *with* $(\forall i \in S : (\omega^S)^{(i)} = (\omega^T)^{(i)})$

     is called **pointwise projection** *of* Ω^T *onto* Ω^S.

(b) $\Pi_S^T : \mathfrak{P}(\Omega^T) \quad \to \quad \mathfrak{P}(\Omega^S),$

     $\Pi_S^T(A) \quad = \quad \left\{ \omega^S \in \Omega^S \mid \exists \omega^T \in A : \text{red}_S^T(\omega^T) = \omega^S \right\}$

     is called the **projection** *of* Ω^T *onto* Ω^S.

(c) $\hat{\Pi}_S^T : \mathfrak{P}(\Omega^S) \quad \to \quad \mathfrak{P}(\Omega^T),$

     $\hat{\Pi}_S^T(B) \quad = \quad \left\{ \omega^T \in \Omega^T \mid \text{red}_S^T(\omega^T) \in B \right\}$

     is called the **cylindrical extension** *of* Ω^S *onto* Ω^T.

Now we are equipped with the formal aids to describe in general the mathematical model for expert systems of the type presented in Example 3.6, including the inference mechanism which has to be employed in its application. The definition and application of these expert systems consist of four phases, which can be distinguished conceptually as follows:

In the *first phase* the knowledge acquisition which is initiated by one or more experts leads to appointing the attributes $X^{(1)}, X^{(2)}, \ldots, X^{(n)}$, $n \in \mathbf{N}$, that are essential to model the part of the world under consideration, and their domains $\Omega^{(1)}, \Omega^{(2)}, \ldots, \Omega^{(n)}$. Fixing this universe $\mathcal{U} \overset{\text{def}}{=} (\Omega^{(i)})_{i \in \mathbf{N}_n}$ provides the *representation structure* for the expert knowledge. $\Omega = \Omega^{\mathbf{N}_n}$ forms the set of all states of the part of the world under consideration, that are *a priori* possible.

In the *second phase* rules are formulated that express *general dependencies* between the domains of the involved attributes $X^{(1)}, X^{(2)}, \ldots, X^{(n)}$ on account of some expert's knowledge. The single rules R_j, $j = 1, \ldots, l$, $l \in \mathbf{N}$, normally do not concern all attributes, but only a small number $X^{(i)}$, $i \in M_j$, which are identified by an index set $M_j \subseteq \mathbf{N}_n$ of low cardinality.

Therefore R_j can be represented as a subset of the low-dimensional product space Ω^{M_j}. Since R_j contains no information about those attributes not contained in M_j, $\hat{\Pi}_{M_j}^{\mathbf{N}_n}(R_j)$ consists of all states of the part of the world that are compatible with the rule R_j. Rules that refer to the same index set can be combined without any difficulties by intersecting the corresponding relations, such that only a single rule remains. For this reason we assume that we can choose a finite set of rules $R_1, \ldots, R_l$ with $R_j \subseteq \Omega^{M_j}$, $j = 1, \ldots, l$, whose index sets M_j are pairwise different (but of course not necessarily pairwise disjoint in general) and that satisfy the condition $\bigcup_{j=1}^{l} M_j = \mathbf{N}_n$. This condition ensures that the represented expert knowledge covers all attributes. To avoid partial dependencies, we further assume that no index set is contained in any other.

Definition 3.10 *Let $\mathbf{N}_n$ be the index set of a universe $\mathcal{U}$. A set $\mathcal{M} = \{M_1, M_2, \ldots, M_l\}$ with $\emptyset \neq M_j \subseteq \mathbf{N}_n$, $j = 1, \ldots, l$, $l \in \mathbf{N}$, is called a* **modularization of $\mathbf{N}_n$,** *if the following holds:*

(a) $\displaystyle\bigcup_{j=1}^{l} M_j = \mathbf{N}_l$

(b) $\forall i, j \in \mathbf{N}_n : i \neq j \Rightarrow M_i \not\subseteq M_j.$

We assume that all statements given by the experts are absolutely reliable and therefore can be combined conjunctively. This assumption suggests the next definition.

Definition 3.11 *Let $\mathcal{U} = (\Omega^{(i)})_{i \in \mathbf{N}_n}$ be a universe of dimension n and $\mathcal{M}$ a modularization of $\mathbf{N}_n$. $\mathcal{R}(\mathcal{U}, \mathcal{M}) = \{R^M \mid M \in \mathcal{M}\}$ with $R^M \subseteq \Omega^M$ is called a* **rule base** *with respect to $\mathcal{U}$ and $\mathcal{M}$, and*

$$R \overset{\text{def}}{=} \bigcap_{M \in \mathcal{M}} \hat{\Pi}_M^{\mathbf{N}_n}(R^M)$$

is called the **knowledge base** *induced by $\mathcal{R}(\mathcal{U}, \mathcal{M})$.*

Having acquired and represented the given knowledge, in the *third phase* it follows the application of the gained knowledge base to actual problems. According to the *closed world assumption* we assume that in a given application there is an actually existing object state that can be characterized within the modelled part of the world by an n-tuple $(\omega_0^{(1)}, \ldots, \omega_0^{(n)}) \in \Omega$. In addition we assume that the user of the knowledge base is able to represent his observations of the attribute values $\omega_0^{(i)}$, $i = 1, 2, \ldots, n$, as *evidence (facts)*

$E^{(i)} \subseteq \Omega^{(i)}$. Just as the experts, the user is supposed to give only reliable information, so that $\omega_0^{(i)} \in E^{(i)}$, $i = 1, \ldots, n$, can be supposed to be true. At best, $E^{(i)}$ contains only one element, which would be a precise observation of $\omega_0^{(i)}$. At worst, $E^{(i)} = \Omega^{(i)}$ does not allow a restriction of $\omega_0^{(i)}$ and therefore reflects complete ignorance of this attribute value. In the second phase the set of all object states R, that were regarded possible w.r.t. the expert knowledge, was induced by the stated rules $R_1, \ldots, R_l$. Analogously the evidence $E^{(i)}$ determines the set E of all object states that are possible in the application under consideration from a user's point of view.

As far as the evidence contributed by the user is concerned, we will introduce the generalization that he may not only state one-dimensional observations $E^{(i)} \subseteq \Omega^{(i)}$, $i \in \mathbf{N}_n$, but may also use relations between attributes as facts. However, this extension is restricted by the requirement that this evidence must not establish causal dependencies between attributes that have not already been set up by the modularization $\mathcal{M}$ resulting from the rule base.

In a user-specified evidence system E_k, $k = 1, \ldots, m$, $m \in \mathbf{N}$, with $E_k \subseteq \Omega^{N_k}$, $\emptyset \neq N_k \subseteq \mathbf{N}_n$, it has therefore to be ensured for all $M \in \mathcal{M}$ and all $k \in \{1, \ldots, m\}$ that either $N_k \subseteq M$ or $N_k \cap M = \emptyset$ holds. The index sets assigned to the pieces of evidence E_k form a partition of $\mathbf{N}_n$ that is compatible with $\mathcal{M}$.

Definition 3.12 *Let $\mathcal{U}$ be an n-dimensional universe and $\mathcal{M} = \{M_1, \ldots, M_l\}$ a modularization of $\mathbf{N}_n$.*

*A partition $\mathcal{N} = \{N_1, \ldots, N_m\}$, $\emptyset \neq N_k \subseteq \mathbf{N}_n$, $k = 1, \ldots, m$, of $\mathbf{N}_n$ is called **compatible with** $\mathcal{M}$, if for all $M \in \mathcal{M}$ there is a subset $\mathcal{N}^* \subseteq \mathcal{N}$ of partitions, such that $M = \bigcup \{N \mid N \in \mathcal{N}^*\}$ holds.*

Definition 3.13 *Let $\mathcal{U} = (\Omega^{(i)})_{i \in \mathbf{N}_n}$ be an n-dimensional universe, let $\mathcal{M}$ be a modularization of $\mathbf{N}_n$ and $\mathcal{N}$ a partition of $\mathbf{N}_n$ compatible with $\mathcal{M}$.*

*$\mathcal{E}(\mathcal{U}, \mathcal{N}) = \{E^N \mid N \in \mathcal{N}\}$ with $E^N \subseteq \Omega^N$ is called the **evidence system** w.r.t. $\mathcal{U}$ and $\mathcal{N}$, and*

$$E \stackrel{\text{def}}{=} \bigcap_{N \in \mathcal{N}} \hat{\Pi}_N^{\mathbf{N}_n} \left(E^N \right)$$

*is called the **total evidence** induced by $\mathcal{E}(\mathcal{U}, \mathcal{N})$.*

We now turn to the final *fourth phase*, in which a user of the expert system employs the knowledge base R and an application-induced total evidence E to carry out an *inference* (i.e. to draw conclusions) in order to identify as exactly as possible the object state $(\omega_0^{(1)}, \ldots, \omega_0^{(n)})$ that is described imprecisely by R and E. The user will be especially interested in specific characterizations of the single attributes $\omega_0^{(i)}$, i.e. in the sets $\kappa^{(i)} = \Pi_{\{i\}}^{\mathbf{N}_n}(R \cap E)$. Summarizing, we gain the following definition of an F-expert system:

Definition 3.14 *An* **F-expert system** $\mathcal{X}$ *is a 4-tuple*
$\mathcal{X} = (\mathcal{U}, \mathcal{M}, \mathcal{N}, \mathcal{R}(\mathcal{U}, \mathcal{M}))$ *with an n-dimensional universe* $\mathcal{U} = (\Omega^{(i)})_{i \in N_n}$, *a modularization* $\mathcal{M}$ *of* N_n, *and a partition* $\mathcal{N}$ *of* N_n, *compatible with* $\mathcal{M}$, *and a rule base* $\mathcal{R}(\mathcal{U}, \mathcal{M})$. *If* R *is the knowledge base induced by* $\mathcal{R}(\mathcal{U}, \mathcal{M})$, *and* E *the total evidence induced by an evidence system* $\mathcal{E}(\mathcal{U}, \mathcal{N})$, *then we call*

$$\sigma(\mathcal{X}, \mathcal{E}(\mathcal{U}, \mathcal{N})) \stackrel{\text{def}}{=} R \cap E$$

the **state** *of* $\mathcal{X}$ *w.r.t.* $\mathcal{E}(\mathcal{U}, \mathcal{N})$ *and*

$$\kappa^{(i)} \stackrel{\text{def}}{=} \Pi^{N_n}_{\{i\}}(\sigma(\mathcal{X}, \mathcal{E}(\mathcal{U}, \mathcal{N}))), \quad i = 1, \ldots, n,$$

the **i-th restrictions** *of* $\sigma(\mathcal{X}, \mathcal{E}(\mathcal{U}, \mathcal{N}))$.

Example 3.6 (continuation) By the given definitions we can describe more precisely the expert system, which was introduced earlier in this section. It refers to the inspection of competences and possible actions if oil contamination of a body of water has occured.

1. Phase: Fixing the structure of representation

A universe $(\Omega^{(i)})_{i \in N_3}$ with domains $\Omega^{(1)}, \Omega^{(2)}, \Omega^{(3)}$ belonging to the attributes *place of contamination*, *competence*, and *action*, respectively, is given.

2. Phase: Building the knowledge base

The rules $R_1, \ldots, R_6$, which are provided by the first expert, refer to the index set $M_1 = \{1, 2\}$ and hence are combined to a single rule $R^{M_1} \stackrel{\text{def}}{=} R^* \subseteq \Omega^{M_1}$. The statements of the second expert lead to the rule $R^{M_2} \stackrel{\text{def}}{=} R^{**} \subseteq \Omega^{M_2}$ with $M_2 = \{2, 3\}$. The knowledge base induced by R^{M_1} and R^{M_2} is therefore:

$$R = \hat{\Pi}^{\{1,2,3\}}_{\{1,2\}}(R^{M_1}) \cap \hat{\Pi}^{\{1,2,3\}}_{\{2,3\}}(R^{M_2}).$$

3. Phase: Application-dependent consideration of total evidence

By means of the facts given by the user with respect to an oil contamination that actually occured, we infer the following restrictions:

$$E^{(1)} = \{ca, rd\} \quad \text{(results from } (E_2)),$$
$$E^{(2)} = \{fd, hc, ep\} \quad \text{(results from } (E_1)),$$
$$E^{(3)} = \Omega^{(3)} \quad \text{(no information about the action taken)}.$$

The total evidence described by $E^{(1)}$, $E^{(2)}$, and $E^{(3)}$ therefore is

$$E = \bigcap_{i \in \{1,2,3\}} \hat{\Pi}_{\{i\}}^{\{1,2,3\}}(E^{(i)}).$$

4. Phase: Carrying out inference

For the i-th restrictions it follows that $\kappa^{(i)} = \Pi_{\{i\}}^{\{1,2,3\}}(R \cap E)$, $i = 1, 2, 3$, and hence

$$
\begin{aligned}
\kappa^{(1)} &= \{ca, rd\}, \\
\kappa^{(2)} &= \{fd, hc\}, \\
\kappa^{(3)} &= \{bl, fi\}.
\end{aligned}
$$

Our expert system can be stated applying the symbols used in this example as follows:

$$
\begin{array}{ll}
\mathcal{X} = (\mathcal{U}, \mathcal{M}, \mathcal{N}, \mathcal{R}(\mathcal{U}, \mathcal{M})) & \text{(F-expert system),} \\
\mathcal{U} = (\Omega^{(i)})_{i \in \mathbb{N}_3} & \text{(universe),} \\
\mathcal{M} = \{\{1, 2\}, \{2, 3\}\} & \text{(modularization),} \\
\mathcal{N} = \{\{1\}, \{2\}, \{3\}\} & \text{(partition compatible to } \mathcal{M}), \\
\mathcal{R}(\mathcal{U}, \mathcal{M}) = \{R^{\{1,2\}}, R^{\{2,3\}}\} & \text{(rule base).}
\end{array}
$$

In this application the i-th restrictions $\kappa^{(i)}$, $i = 1, 2, 3$, were determined using the evidence system $\mathcal{E}(\mathcal{U}, \mathcal{N}) = \{E^{(1)}, E^{(2)}, E^{(3)}\}$.   □

F-expert systems are developed in order to focus on imprecise knowledge. As a generalization, the concept of a possibilistic F-expert system $\mathcal{X} = (\mathcal{U}, \mathcal{M}, \mathcal{N}, \mathcal{R}(\mathcal{U}, \mathcal{M}))$ can be introduced without difficulties. Their rule base $\mathcal{R}(\mathcal{U}, \mathcal{M})$ does not consist of ordinary relations $R^M \subseteq \Omega^M$, $M \in \mathcal{M}$, but of possibility distributions $\varrho^M \in \text{POSS}(\Omega^M)$, each of them describing an uncertain relation (a possibilistic relation) between the values of the attributes selected by the index set M. In an analogous way evidence systems $\mathcal{E}(\mathcal{U}, \mathcal{N})$ with possibilistic pieces of evidence $\varepsilon^N \in \text{POSS}(\Omega^N)$, $N \in \mathcal{N}$, are incorporated into our considerations. ε^N is a possibility distribution for the attributes (referenced by N) of the object state $\omega_0 \in \Omega$ that has to be characterized. Just as in the case of F-expert systems the knowledge that is generally available about ω_0 from $\mathcal{R}(\mathcal{U}, \mathcal{M})$ and additional, situation-dependent pieces of evidence w.r.t. ω_0, formalized by $\mathcal{E}(\mathcal{U}, \mathcal{N})$, are employed to identify the most specific possibility distributions describing ω_0 that are compatible with the entire knowledge. The gained possibility distribution $\sigma(\mathcal{X}, \mathcal{E}(\mathcal{U}, \mathcal{N}))$ is called a *state of* $\mathcal{X}$ with respect to $\mathcal{E}(\mathcal{U}, \mathcal{N})$. To find $\sigma(\mathcal{X}, \mathcal{E}(\mathcal{U}, \mathcal{N}))$, some considerations have to be made first.

Let $M \in \mathcal{M}$ be an index subset and ϱ^M the corresponding possibility distribution taken from the rule base $\mathcal{R}(\mathcal{U}, \mathcal{M})$. ϱ^M does not tell us anything about the attributes induced by $\mathbf{N}_n \setminus M$. If the object state ω_0 is to be characterized by a possibility distribution ϱ on Ω based on ϱ^M, then for all $\omega, \omega' \in \Omega$ with $\omega \neq \omega'$ and $\mathrm{red}_M^{\mathbf{N}_n}(\omega) = \mathrm{red}_M^{\mathbf{N}_n}(\omega')$, we have to require $\varrho(\omega) = \varrho(\omega') = \varrho^M\left(\mathrm{red}_M^{\mathbf{N}_n}(\omega)\right)$, since otherwise ϱ^M would have a non-intended influence on the degrees of possibility of attributes values indexed by $\mathbf{N}_n \setminus M$. The uniquely determined possibility distribution ϱ is called the cylindrical extension of ϱ^M to Ω.

Definition 3.15 *Let $\mathcal{U} = (\Omega^{(i)})_{i \in \mathbf{N}_n}$ be an n-dimensional universe, M an index set with $\emptyset \neq M \subseteq \mathbf{N}_n$ and $\mathrm{ext}_M^{\mathbf{N}_n}$ the mapping defined by*

$$\mathrm{ext}_M^{\mathbf{N}_n} : \mathrm{POSS}(\Omega^M) \quad \rightarrow \quad \mathrm{POSS}(\Omega),$$
$$\mathrm{ext}_M^{\mathbf{N}_n}(\pi)(\omega) \quad \overset{\mathrm{def}}{=} \quad \pi\left(\mathrm{red}_M^{\mathbf{N}_n}(\omega)\right).$$

$\mathrm{ext}_M^{\mathbf{N}_n}(\pi)$ *is called the* **cylindrical extension of π to Ω** *for $\pi \in \mathrm{POSS}(\Omega^M)$.*

We use cylindrical extensions to state the knowledge base ϱ induced by the rule base $\mathcal{R}(\mathcal{U}, \mathcal{M})$. For $\mathcal{R}(\mathcal{U}, \mathcal{M}) = \{\varrho^M \mid M \in \mathcal{M}\}$ we obtain

$$\varrho = \min\left\{\mathrm{ext}_M^{\mathbf{N}_n}(\varrho^M) \mid M \in \mathcal{M}\right\}, \tag{3.11}$$

where the min-operator is chosen to conjunctively combine the occuring possibility distributions $\mathrm{ext}_M^{\mathbf{N}_n}(\varrho^M)$. The application of an appropriate t-norm for such a conjunctive combination is out of the question, since possibility distributions $\pi \in \mathrm{POSS}(\Omega)$ can be regarded as epistemic interpretations of fuzzy sets satisfying the normalization condition ($\exists \omega \in \Omega : \pi(\omega) = 1$). The t-norm min ensures the compatibility with the application of the extension principle to possibility distributions. For this we will also use it in the sequel of this chapter.

Whereas $\mathrm{ext}_M^{\mathbf{N}_n}(\varrho^M) \in \mathrm{POSS}(\Omega)$ is always true, the condition for normalization need not necessarily hold for the resulting knowledge base ϱ. But if ϱ is a possibility distribution, then we call this knowledge base *non-contradictory*.

Definition 3.16 *Let $\mathcal{U} = (\Omega^{(i)})_{i \in \mathbf{N}_n}$ be a universe of dimension n and $\mathcal{M}$ a modularization of $\mathbf{N}_n$. We call $\mathcal{R}(\mathcal{U}, \mathcal{M}) = \{\varrho^M \mid M \in \mathcal{M}\}$ with $\varrho^M \in \mathrm{POSS}(\Omega^M)$ a* **possibilistic rule base** *w.r.t. $\mathcal{U}$ and $\mathcal{M}$, and the function*

$$\varrho : \Omega \quad \rightarrow \quad [0, 1],$$
$$\varrho(\omega) \quad \overset{\mathrm{def}}{=} \quad \min\left\{\mathrm{ext}_M^{\mathbf{N}_n}(\varrho^M)(\omega) \mid M \in \mathcal{M}\right\}$$

the **possibilistic knowledge base** *induced by* $\mathcal{R}(\mathcal{U}, \mathcal{M})$.

$\mathcal{R}(\mathcal{U}, \mathcal{M})$ *is called* **non-contradictory,** *if* $\varrho \in \mathrm{POSS}(\Omega)$ *holds.*

In an analogous way the notion of total evidence can be generalized to possibility distributions in the following way:

Definition 3.17 *Let* $\mathcal{U} = (\Omega^{(i)})_{i \in \mathbf{N}_n}$ *be an n-dimensional universe,* $\mathcal{M}$ *a modularization of* $\mathbf{N}_n$ *and* $\mathcal{N}$ *a partition of* $\mathbf{N}_n$ *compatible with* $\mathcal{M}$.

Then $\mathcal{E}(\mathcal{U}, \mathcal{N}) = \{\varepsilon^N \mid N \in \mathcal{N}\}$ *with* $\varepsilon^N \in \mathrm{POSS}(\Omega^N)$ *is called a* **possibilistic evidence system** *w.r.t.* $\mathcal{U}$ *and* $\mathcal{N}$. *The function*

$$\varepsilon : \Omega \quad \rightarrow \quad [0, 1],$$
$$\varepsilon(\omega) \quad \stackrel{\mathrm{def}}{=} \quad \min\left\{\mathrm{ext}_N^{\mathbf{N}_n}(\varepsilon^N)(\omega) \mid N \in \mathcal{N}\right\}$$

is called the **total evidence** *induced by* $\mathcal{E}(\mathcal{U}, \mathcal{N})$.

$\mathcal{E}(\mathcal{U}, \mathcal{N})$ *is called* **non-contradictory,** *if and only if* $\varepsilon \in \mathrm{POSS}(\Omega)$ *holds.*

The state $\sigma(\mathcal{X}, \mathcal{E}(\mathcal{U}, \mathcal{N}))$ of a possibilistic F-expert system $\mathcal{X}$, that depends on general knowledge (i.e. $\mathcal{R}(\mathcal{U}, \mathcal{M})$) and situation-dependent additional knowledge (i.e. $\mathcal{E}(\mathcal{U}, \mathcal{N})$), is determined by conjunctively combining the knowledge base ϱ and the total evidence ε, that is

$$\sigma(\mathcal{X}, \mathcal{E}(\mathcal{U}, \mathcal{N})) \equiv \min\{\varrho, \varepsilon\}. \tag{3.12}$$

If $\sigma(\mathcal{X}, \mathcal{E}(\mathcal{U}, \mathcal{N})) \notin \mathrm{POSS}(\Omega)$ holds, this indicates that $\mathcal{R}(\mathcal{U}, \mathcal{M})$ and $\mathcal{E}(\mathcal{U}, \mathcal{N})$ are inconsistent (incompatible). Otherwise $\sigma(\mathcal{X}, \mathcal{E}(\mathcal{U}, \mathcal{N}))$ can be calculated in order to determine the i-th restrictions $\kappa^{(i)}$, which are the projections of the possibility distribution $\sigma(\mathcal{X}, \mathcal{E}(\mathcal{U}, \mathcal{N}))$ onto the particular domains $\Omega^{(i)}$, $i = 1, \ldots, n$. The corresponding operation can be realized by a simple application of the extension principle to the projections $\mathrm{red}_{\{i\}}^{\mathbf{N}_n}$.

Definition 3.18 *Let* $\mathcal{U} = (\Omega^{(i)})_{i \in \mathbf{N}_n}$ *be an n-dimensional universe,* M *an index set with* $\emptyset \neq M \subseteq \mathbf{N}_n$, *and* $\mathrm{proj}_M^{\mathbf{N}_n}$ *the mapping which is defined by*

$$\mathrm{proj}_M^{\mathbf{N}_n} : \mathrm{POSS}(\Omega) \quad \rightarrow \quad \mathrm{POSS}(\Omega^M),$$
$$\mathrm{proj}_M^{\mathbf{N}_n}(\pi)(\omega) \quad \stackrel{\mathrm{def}}{=} \quad \sup\left\{\pi(\omega') \mid \omega' \in \Omega \;\wedge\; \omega = \mathrm{red}_M^{\mathbf{N}_n}(\omega')\right\}.$$

If $\pi \in \mathrm{POSS}(\Omega)$, *then* $\mathrm{proj}_M^{\mathbf{N}_n}(\pi)$ *is called the* **projection of** π **onto** Ω^M.

The i-th restriction of $\sigma(\mathcal{X}, \mathcal{E}(\mathcal{U}, \mathcal{N}))$ is therefore defined as

$$\kappa^{(i)} \equiv \mathrm{proj}_{\{i\}}^{\mathbf{N}_n}(\sigma(\mathcal{X}, \mathcal{E}(\mathcal{U}, \mathcal{N}))). \tag{3.13}$$

If $\omega_0 \in \Omega$ is an object state which is characterized by general knowledge $\mathcal{R}(\mathcal{U}, \mathcal{M})$ and evidential knowledge $\mathcal{E}(\mathcal{U}, \mathcal{N})$, then $\kappa^{(i)}$ is the most specific of all possibility distributions that are compatible with $\mathcal{R}(\mathcal{U}, \mathcal{M})$ and $\mathcal{E}(\mathcal{U}, \mathcal{N})$ and that characterize the attribute value $\text{red}_{\{i\}}^{N_n}(\omega_0)$. From this we achieve the following definition of a possibilistic F-expert system.

Definition 3.19 *A possibilistic F-expert system $\mathcal{X}$ is a 4-tuple $\mathcal{X} = (\mathcal{U}, \mathcal{M}, \mathcal{N}, \mathcal{R}(\mathcal{U}, \mathcal{M}))$, consisting of an n-dimensional universe $\mathcal{U} = (\Omega^{(i)})_{i \in N_n}$, a modularization $\mathcal{M}$ of N_n, a partition $\mathcal{N}$ of N_n compatible with $\mathcal{M}$, and a non-contradictory possibilistic rule base $\mathcal{R}(\mathcal{U}, \mathcal{M})$ w.r.t. $\mathcal{U}$ and $\mathcal{M}$. If ϱ is the possibilistic knowledge base induced by $\mathcal{R}(\mathcal{U}, \mathcal{M})$, and ε the total evidence induced by a non-contradictory possibilistic evidence system $\mathcal{E}(\mathcal{U}, \mathcal{N})$ w.r.t. $\mathcal{U}$ and $\mathcal{N}$, then the function*

$$\sigma(\mathcal{X}, \mathcal{E}(\mathcal{U}, \mathcal{N})) : \Omega \quad \rightarrow \quad [0, 1],$$
$$\sigma(\mathcal{X}, \mathcal{E}(\mathcal{U}, \mathcal{N}))(\omega) \quad \overset{\text{def}}{=} \quad \min\{\varrho(\omega), \varepsilon(\omega)\}$$

*is called the **state** of $\mathcal{X}$ w.r.t. $\mathcal{E}(\mathcal{U}, \mathcal{N})$. $\sigma(\mathcal{X}, \mathcal{E}(\mathcal{U}, \mathcal{N}))$ is called **consistent**, if and only if it is a possibility distribution on Ω. Finally, given a consistent $\sigma(\mathcal{X}, \mathcal{E}(\mathcal{U}, \mathcal{N}))$, we call the possibility distributions $\kappa^{(i)} \in \text{POSS}(\Omega^{(i)})$ defined by*

$$\kappa^{(i)}(\omega) \quad \overset{\text{def}}{=} \quad \text{proj}_{\{i\}}^{N_n}(\sigma(\mathcal{X}, \mathcal{E}(\mathcal{U}, \mathcal{N})))(\omega), \quad i = 1, 2, \ldots, n,$$

*the **i-th restrictions** of $\sigma(\mathcal{X}, \mathcal{E}(\mathcal{U}, \mathcal{N}))$.*

Example 3.6 (continuation) We modify $\mathcal{X} = (\mathcal{U}, \mathcal{M}, \mathcal{N}, \mathcal{R}(\mathcal{U}, \mathcal{M}))$ to a very simple possibilistic F-expert system, applying the rule base $\mathcal{R}(\mathcal{U}, \mathcal{M}) \overset{\text{def}}{=} \{\varrho^{\{1,2\}}, \varrho^{\{2,3\}}\}$, where the two possibility distributions are given as shown in Table 3.

This is a non-contradictory rule base, since we may, for example, calculate

$$\varrho(z_2, wp, bl) = \min\left\{ \left(\text{ext}_{\{1,2\}}^{\{1,2,3\}}\left(\varrho^{\{1,2\}}\right)\right)(z_2, wp, bl), \right.$$
$$\left. \left(\text{ext}_{\{2,3\}}^{\{1,2,3\}}\left(\varrho^{\{2,3\}}\right)\right)(z_2, wp, bl)\right\}$$
$$= \min\left\{\varrho^{\{1,2\}}(z_2, wp), \varrho^{\{2,3\}}(wp, bl)\right\}$$
$$= 1.$$

As the evidence system we consider $\mathcal{E}(\mathcal{U}, \mathcal{N}) \overset{\text{def}}{=} \{\varepsilon^{\{1\}}, \varepsilon^{\{2\}}, \varepsilon^{\{3\}}\}$, where $\varepsilon^{\{2\}} \equiv \mathbb{I}_{\Omega^{(2)}}, \varepsilon^{\{3\}} \equiv \mathbb{I}_{\Omega^{(3)}}$, and $\varepsilon^{\{1\}}$ is defined in Table 4.

It is a trivial fact that the total evidence ε induced by $\mathcal{E}(\mathcal{U}, \mathcal{N})$ is non-contradictory. Since, for example, $\varrho(z_2, wp, bl) = \varepsilon(z_2, wp, bl) = 1$ holds, $\mathcal{X}$ is in a consistent state $\sigma(\mathcal{X}, \mathcal{E}(\mathcal{U}, \mathcal{N}))$, which assigns the degrees of possibility shown in Table 5 to the object states $(\omega_1, \omega_2, \omega_3) \in \Omega$.

$$\omega_2$$

		fd	hc	wp	ep
	ld	1	1	1	1
	rd	1	1	1	1
ω_1	ca	0.4	0	1	1
	z_1	0.4	0	0.8	1
	z_2	0.4	0	1	1
	z_3	1	0	1	1

$\varrho^{\{1,2\}}(\omega_1, \omega_2)$

$$\omega_3$$

		ad	bl	fi	im
	fd	0	1	1	0
ω_2	hc	0	1	0	0
	wp	0	1	1	1
	ep	1	0	0	0

$\varrho^{\{2,3\}}(\omega_2, \omega_3)$

Table 3   Representation of $\varrho^{\{1,2\}}$ and $\varrho^{\{2,3\}}$

ω_1	ld	rd	ca	z_1	z_2	z_3
$\varepsilon^{\{1\}}(\omega_1)$	0	0	0.3	0.7	1	0

Table 4   Representation of $\varepsilon^{\{1\}}$

		ω_3			
		ad	bl	fi	im
ω_2	fd	0	0.3	0.3	0
	hc	0	0	0	0
	wp	0	0.3	0.3	0.3
	ėp	0.3	0	0	0

$\sigma(\mathcal{X}, \mathcal{E}(\mathcal{U}, \mathcal{N}))(ca, \omega_2, \omega_3)$

		ω_3			
		ad	bl	fi	im
ω_2	fd	0	0.4	0.4	0
	hc	0	0	0	0
	wp	0	0.7	0.7	0.7
	ep	0.7	0	0	0

$\sigma(\mathcal{X}, \mathcal{E}(\mathcal{U}, \mathcal{N}))(z_1, \omega_2, \omega_3)$

		ω_3			
		ad	bl	fi	im
ω_2	fd	0	0.4	0.4	0
	hc	0	0	0	0
	wp	0	1	1	1
	ep	1	0	0	0

$\sigma(\mathcal{X}, \mathcal{E}(\mathcal{U}, \mathcal{N}))(z_2, \omega_2, \omega_3)$

Table 5   Determining the state $\sigma(\mathcal{X}, \mathcal{E}(\mathcal{U}, \mathcal{N}))$

ω_1	ld	rd	ca	z_1	z_2	z_3
$\kappa^{(1)}(\omega_1)$	0	0	0.3	0.7	1	0

ω_2	fd	hc	wp	ep
$\kappa^{(2)}(\omega_2)$	0.4	0	1	1

ω_3	ad	bl	fi	im
$\kappa^{(3)}(\omega_3)$	1	1	1	1

Table 6 i-th restrictions of $\sigma\,(\mathcal{X}, \mathcal{E}(\mathcal{U}, \mathcal{N}))$

If the object state $\omega_0 \in \Omega$ is characterized by $\mathcal{R}(\mathcal{U}, \mathcal{M})$ and $\mathcal{E}(\mathcal{U}, \mathcal{N})$, then for its attribute values $\omega_i = \text{red}_{\{i\}}^{\{1,2,3\}}(\omega_0)$, $i = 1, 2, 3$, we obtain the possibility distributions shown in Table 6.

These i-th restrictions $\kappa^{(i)}$ of $\sigma\,(\mathcal{X}, \mathcal{E}(\mathcal{U}, \mathcal{N}))$ are the most specific possibility distributions that can be determined from $\mathcal{R}(\mathcal{U}, \mathcal{M})$ and $\mathcal{E}(\mathcal{U}, \mathcal{N})$ for the attribute values of ω_0. The application of even more specific possibility distributions would therefore be a specialization of the knowledge given with respect to ω_0 that surpasses $\mathcal{R}(\mathcal{U}, \mathcal{M})$ and $\mathcal{E}(\mathcal{U}, \mathcal{N})$ and hence cannot be justified. $\Box$

3.3   Interpretation of Possibilistic Inference Rules

In our consideration of (possibilistic) F-expert systems $(\mathcal{U}, \mathcal{M}, \mathcal{N}, \mathcal{R}(\mathcal{U}, \mathcal{M}))$ with n-dimensional universe $\mathcal{U} = (\Omega^{(i)})_{i \in \mathbf{N}_n}$ we have, up to now, always directly referred to the underlying rule base $\mathcal{R}(\mathcal{U}, \mathcal{M})$ in describing the knowledge base, and we have implicitly assumed that an expert provides the ordinary relations $R^M \subseteq \Omega^M$ or the possibility distributions $\varrho^M \in \text{POSS}(\Omega^M)$, respectively, contained in it. However, in many domains of application a different mode of knowledge representation is preferred, which is to use a system of conjunctively combined rules that have to be interpreted in the sense of logical implications. In the case of an F-expert system the involved imprecise inference rules have the form

$$R_j : \textbf{if } \xi^{S_j} \textbf{ in } A_j \textbf{ then } \xi^{T_j} \textbf{ in } B_j, \quad j = 1, \ldots, r,$$

where $A_j \subseteq \Omega^{S_j}$, $B_j \subseteq \Omega^{T_j}$, $\emptyset \neq S_j \subseteq \mathbf{N}_n$, and $\emptyset \neq T_j \subseteq \mathbf{N}_n$, $S_j \cap T_j = \emptyset$.

The rule R_j establishes a relation between the values of those attributes that are selected by the index sets S_j and T_j. ξ is meant to denote a variable with domain Ω, and $\xi^{S_j} \stackrel{\text{def}}{=} \text{red}_{S_j}^{N_n}(\xi)$ and $\xi^{T_j} \stackrel{\text{def}}{=} \text{red}_{T_j}^{N_n}(\xi)$ are its projections onto Ω^{S_j} or Ω^{T_j}, respectively.

R_j expresses that for all object states $\xi \in \Omega$ compatible with this imprecise inference rule, the validity of $\xi^{S_j} \in A_j$ entails the validity of $\xi^{T_j} \in B_j$. But if $\xi^{S_j} \in \Omega^{S_j} \backslash A_j$, the rule R_j does not allow us to make a statement about the value of ξ^{T_j} more specific than the simple fact that $\xi^{T_j} \in \Omega^{T_j}$ holds.

When R_j is represented as a relation on $\Omega^{S_j \cup T_j}$, it follows that

$$
\begin{aligned}
R_j &= \left(\hat{\Pi}_{S_j}^{S_j \cup T_j}(A_j) \cap \hat{\Pi}_{T_j}^{S_j \cup T_j}(B_j) \right) \cup \hat{\Pi}_{S_j}^{S_j \cup T_j}(\Omega^{S_j} \backslash A_j) \\
&= \hat{\Pi}_{S_j}^{S_j \cup T_j}(\Omega^{S_j} \backslash A_j) \cup \hat{\Pi}_{T_j}^{S_j \cup T_j}(B_j).
\end{aligned}
\tag{3.14}
$$

If in an application the given expert knowledge can be formalized with the help of imprecise inference rules and a system of relations $(R_j)_{j=1}^r$ induced by them, then for the relations R^M, $M \in \mathcal{M}$, of the rule base $\mathcal{R}(\mathcal{U}, \mathcal{M})$ of the corresponding F-expert system we obtain:

$$
R^M = \Pi_M^{N_n} \left(\bigcap_{j=1}^{r} \hat{\Pi}_{S_j \cup T_j}^{N_n}(R_j) \right).
\tag{3.15}
$$

Example 3.6 (continuation) To state the relations between places of contamination and competences, that have to be taken into account in case of an occurring oil contamination, we used six rules, presented in an informal way. As an example we repeat rule (R_1):

(R_1) Only the environmental protection agency is competent for the open sea and the 12-mile zone.

Transferred to the notation of imprecise inference rules, (R_1) can be represented by

$$
R_1 : \quad \textbf{if } \textit{place of contamination in } \{z_3, z_2\} \\
\textbf{then } \textit{competence in } \{ep\}
$$

or by

$$
R_1 : \quad \textbf{if } \xi^{S_1} \textbf{ in } A_1 \textbf{ then } \xi^{T_1} \textbf{ in } B_1,
$$

where $\xi \in \Omega$, $S_1 = \{1\}$, $T_1 = \{2\}$, $A_1 = \{z_3, z_2\}$, and $B_1 = \{ep\}$.

Analogously we obtain the other rules:

R_2 : **if** *competence* **in** $\{hc\}$
    then *place of contamination* **in** $\{rd, ld\}$,

R_3 : **if** *place of contamination* **in** $\{rd, ld\}$
    then *competence* **in** $\text{Dom}(competence)\backslash\{wp, ep\}$,

R_4 : **if** *place of contamination* **in** $\{ca\}$
    then *competence* **in** $\text{Dom}(competence)\backslash\{fd\}$,

R_5 : **if** *place of contamination* **in** $\{rd\}$
    then *competence* **in** $\{hc\}$,

R_6 : **if** *competence* **in** $\{fd\}$
    then *place of contamination* **in** $\text{Dom}(place\ of\ contamination)\backslash\{z_3\}$.

The transition to a relational description of the imprecise inference rules yields, for example,

$$R_3 = \hat{\Pi}_{\{1\}}^{\{1,2\}}(\{z_3, z_2, z_1, ca\}) \cup \hat{\Pi}_{\{2\}}^{\{1,2\}}(\{fd, hc\})$$

$$= \Omega^{\{1,2\}}\backslash\{(wp, rd), (wp, ld), (fd, rd), (fd, ld)\}.$$

All listed rules refer to the index set M_1 of the established modularization $\mathcal{M} = \{M_1, M_2\}$ with $M_1 = \{1, 2\}$ and $M_2 = \{2, 3\}$.

Finally we calculate $R^{M_1} = \bigcap_{i=1}^{6} R_i$. This relation was already shown in Table 1. □

Based on the simple concept of imprecise inference rules, we proceed to the more general case of a system $(R_j)_{j=1}^{r}$ of conjunctively combined possibilistic inference rules, stated in the form

$$R_j : \textbf{if } \xi^{S_j} \textbf{ is } \mu_j \textbf{ then } \xi^{T_j} \textbf{ is } \nu_j, \quad j = 1, \ldots, r,$$

with $\mu_j \in \text{POSS}(\Omega^{S_j})$, $\nu_j \in \text{POSS}(\Omega^{T_j})$, and $\emptyset \neq S_j \subseteq \mathbf{N}_n$, $\emptyset \neq T_j \subseteq \mathbf{N}_n$, $S_j \cap T_j = \emptyset$. Just as above, the rule R_j establishes a relation between the values of the attributes indexed by the sets S_j and T_j.

ξ is a variable whose values can be arbitrary possibility distributions on Ω. Analogous to the imprecise inference rules we will designate the projections of ξ onto Ω^{S_j} or Ω^{T_j}, as ξ^{S_j} and ξ^{T_j}, respectively, i.e. $\xi^{S_j} \equiv \text{proj}_{S_j}^{\mathbf{N}_n}(\xi)$ and $\xi^{T_j} \equiv \text{proj}_{T_j}^{\mathbf{N}_n}(\xi)$.

The symbol **is**, appearing in possibilistic inference rules, serves as a linguistic description of the operator $\sqsubseteq$ and is therefore to be read as 'is at least as specific as'. The application of **is** instead of $\sqsubseteq$ has the advantage that in this way each rule R_j can very clearly reflect a part of the expert knowledge, provided that ξ^{S_j}, μ_j, ξ^{T_j}, and ν_j are replaced by appropriate linguistic terms and are regarded merely as an interpretation of these terms.

Viewed as a logical implication, a possibilistic inference rule R_j expresses that for all object states $\omega_0 \in \Omega$, compatible with it, the following holds: If $\xi \in \text{POSS}(\Omega)$ is a possibility distribution that characterizes ω_0, then the validity of $\xi^{S_j} \sqsubseteq \mu_j$ entails the validity of $\xi^{T_j} \sqsubseteq \nu_j$.

In the case of $\xi^{S_j} \not\sqsubseteq \mu_j$, the rule does not impose any restriction on ξ^{T_j}. On the other hand, considering the implication mentioned above, a specialization of ξ^{T_j}, surpassing $\xi^{T_j} \sqsubseteq \mathbb{I}_{\Omega^{T_j}}$, can be evoked for $\min\{\xi^{S_j}, \mu_j\} \not\equiv 0$. In the light of this interpretation we now determine an appropriate representation of possibilistic inference rules R_j by possibility distributions $\pi_j \in \text{POSS}(\Omega^{S_j \cup T_j})$. For this reason we first formalize the conditions that R_j imposes on the inference mechanism connected with R_j.

If we regard the desired possibility distribution $\pi_j \in \text{POSS}(\Omega^{S_j \cup T_j})$ as a possibilistic relation to characterize the projection $\text{red}_{S_j \cup T_j}^{N_n}(\omega_0)$ of an observable object state $\omega_0 \in \Omega$, and if we describe the observation of $\text{red}_{S_j}^{N_n}(\omega_0)$ added to this general expert knowledge w.r.t. ω_0 using a possibility distribution $\xi^{S_j} \in \text{POSS}(\Omega^{S_j})$, then the inference mechanism is expected to produce a possibility distribution $\xi^{T_j} \in \text{POSS}(\Omega^{T_j})$, which is as specific as possible for $\text{red}_{T_j}^{N_n}(\omega_0)$, resulting from a conjunctive combination of ξ^{S_j} and π_j. Note that $\xi^{T_j} \in \text{POSS}(\Omega^{T_j})$ has to be valid only in those cases in which ξ^{T_j} and π_j lead to a consistent state of the corresponding possibilistic F-expert system.

The inference mechanism is based on the partially defined function

$$\text{infer}_j : \Omega^{S_j} \times \Omega^{S_j \cup T_j} \to \Omega^{T_j},$$

$$\text{infer}_j(\omega, \omega') \overset{\text{def}}{=} \begin{cases} \text{red}_{T_j}^{S_j \cup T_j}(\omega'), & \text{if } \text{red}_{S_j}^{S_j \cup T_j}(\omega') = \omega \\ \text{undefined}, & \text{otherwise} \end{cases} \qquad (3.16)$$

and produces, applying the extension principle, $\xi^{T_j} \equiv \widehat{\text{infer}}_j(\xi^{S_j}, \pi_j)$.

In order to determine π_j, we reconsider the required condition

$$\xi^{S_j} \sqsubseteq \mu_j \implies \xi^{T_j} \sqsubseteq \nu_j. \qquad (3.17)$$

With the exception of some irrelevent trivial cases, π_j is not uniquely determined by this condition. But since $\xi^{T_j} \sqsubseteq \nu_j$ does not tell in which way ξ^{T_j} is expected to be more specific than ν_j, it is reasonable to search for the least specific possibility distribution π_j that satisfies the condition (3.17) in connection with the discussed inference mechanism. Related to the principle of minimum specificity we therefore require for π_j that

$$\widehat{\text{infer}}_j(\mu_j, \pi_j) \equiv \nu_j, \qquad (3.18)$$

which corresponds to the condition (3.17) for $\xi^{S_j} \equiv \mu_j$.

From Theorem 2.36(a), which can be directly transferred from normal fuzzy sets to possibility distributions, we get the following description of the strict α-cuts of the possibility distribution $\widehat{\mathrm{infer}}_j(\mu_j, \pi_j)$:

$$\left[\widehat{\mathrm{infer}}_j(\mu_j, \pi_j)\right]_{\underline{\alpha}} = \mathrm{infer}_j\left([\mu_j]_{\underline{\alpha}}, [\pi_j]_{\underline{\alpha}}\right). \tag{3.19}$$

For $\alpha \in [0, 1)$ we calculate:

$$\mathrm{infer}_j\left([\mu_j]_{\underline{\alpha}}, [\pi_j]_{\underline{\alpha}}\right)$$
$$= \left\{\omega \in \Omega^{T_j} \mid (\exists \omega' \in [\mu_j]_{\underline{\alpha}})\,(\exists \omega'' \in [\pi_j]_{\underline{\alpha}}) : \mathrm{infer}_j(\omega', \omega'') = \omega\right\}$$
$$= \left\{\omega \in \Omega^{T_j} \mid (\exists \omega' \in [\mu_j]_{\underline{\alpha}})\,(\exists \omega'' \in [\pi_j]_{\underline{\alpha}}) : \right.$$
$$\left. \mathrm{red}_{S_j}^{S_j \cup T_j}(\omega'') = \omega' \wedge \mathrm{red}_{T_j}^{S_j \cup T_j}(\omega'') = \omega\right\}$$
$$= \Pi_{T_j}^{S_j \cup T_j}\left(\hat{\Pi}_{S_j}^{S_j \cup T_j}\left([\mu_j]_{\underline{\alpha}}\right) \cap [\pi_j]_{\underline{\alpha}}\right).$$

From this it follows that (3.18) can only be true for those possibility distributions π_j, for which

$$[\pi_j]_{\underline{\alpha}} \subseteq \hat{\Pi}_{T_j}^{S_j \cup T_j}\left([\nu_j]_{\underline{\alpha}}\right) \cup \hat{\Pi}_{S_j}^{S_j \cup T_j}\left(\Omega^{S_j} \backslash [\mu_j]_{\underline{\alpha}}\right) \tag{3.20}$$

holds. Hence, for $\omega \in \Omega^{S_j \cup T_j}$, $\omega^{S_j} \overset{\text{def}}{=} \mathrm{red}_{S_j}^{S_j \cup T_j}(\omega)$ and $\omega^{T_j} \overset{\text{def}}{=} \mathrm{red}_{T_j}^{S_j \cup T_j}(\omega)$, we obtain the condition

$$\omega \in [\pi_j]_{\underline{\alpha}} \implies \left(\omega^{S_j} \in [\mu_j]_{\underline{\alpha}} \wedge \omega^{T_j} \in [\nu_j]_{\underline{\alpha}}\right) \vee \omega^{S_j} \notin [\mu_j]_{\underline{\alpha}}$$
$$\iff \left(\mu_j(\omega^{S_j}) > \alpha \wedge \nu_j(\omega^{T_j}) > \alpha\right) \vee \mu_j(\omega^{S_j}) \leq \alpha,$$

which means that the uniquely determined possibility distribution π_j of minimal specificity is the *Gödel relation* $\varrho^{\text{Gödel}}[\mu_j, \nu_j] \in \mathrm{POSS}\left(\Omega^{S_j \cup T_j}\right)$ induced by μ_j and ν_j, defined as

$$\varrho^{\text{Gödel}}[\mu_j, \nu_j](\omega) \overset{\text{def}}{=} \begin{cases} \nu_j(\omega^{T_j}), & \text{if } \mu_j(\omega^{S_j}) > \nu_j(\omega^{T_j}) \\ 1, & \text{if } \mu_j(\omega^{S_j}) \leq \nu_j(\omega^{T_j}). \end{cases} \tag{3.21}$$

Note that the Gödel relation not only satisfies (3.18), but, of course, also condition (3.17), since

$$\xi^{S_j} \sqsubseteq \mu_j \implies \xi^{T_j} \equiv \widehat{\mathrm{infer}}_j\left(\xi^{S_j}, \varrho^{\text{Gödel}}[\mu_j, \nu_j]\right) \equiv \nu_j$$

holds, and in addition, conforming with the incorporation of the principle of minimal specificity, it ensures

$$\min\{\xi^{S_j}, \mu_j\} \equiv 0 \implies \widehat{\mathrm{infer}}_j\left(\xi^{S_j}, \varrho^{\text{Gödel}}[\mu_j, \nu_j]\right) \equiv \mathbb{1}_{\Omega^{T_j}}.$$

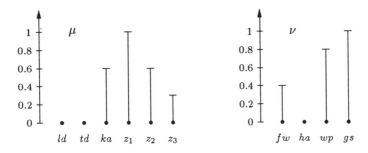

Figure 2   Two possibility distributions μ and ν

		ω_2			
		fd	hc	wp	ep
	ld	1	1	1	1
	rd	1	1	1	1
ω_1	ca	0.4	0	1	1
	z_1	0.4	0	0.8	1
	z_2	0.4	0	1	1
	z_3	1	0	1	1

$\varrho^{\mathrm{G\ddot{o}del}}[\mu,\nu](\omega_1,\omega_2)$

Table 7 . Tabular presentation of the Gödel relation induced by μ and ν

Example 3.6 (continuation) An expert states the following possibilistic inference rule as a representation of the uncertain relation between places of contamination and authorities:

$$R: \quad \textbf{if } \xi^{\{1\}} \textbf{ is } \mu \textbf{ then } \xi^{\{2\}} \textbf{ is } \nu.$$

The variable ξ takes its values in $\mathrm{POSS}(\Omega)$. The constant possibility distributions $\mu \in \mathrm{POSS}(\Omega^{(1)})$ and $\nu \in \mathrm{POSS}(\Omega^{(2)})$ may be defined as shown in Figure 2.

They induce the Gödel relation $\varrho^{\mathrm{G\ddot{o}del}}[\mu,\nu]$ presented in Table 7, that coincides with the possibility distribution already shown in Table 4.   □

With the foregoing explanations we finish Example 3.6 that has accompanied us through this chapter. For a further illustration of Gödel relations we inspect another simple example which employs a continuum instead of the finite universe we have used so far.

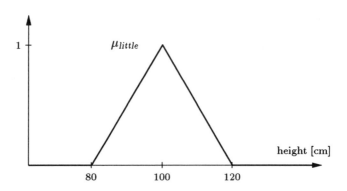

Figure 3   Possibility distribution of the linguistic value 'little'

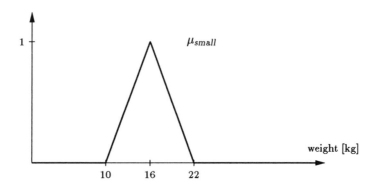

Figure 4   Possibility distribution of the linguistic value 'small'

Example 3.20 An expert states an uncertain relation between body-height and weight of human beings with the help of the rule

$$R_1 : \textbf{if } height \textbf{ is } little \textbf{ then } weight \textbf{ is } small.$$

For a more detailed specification in terms of a possibilistic infer-
ence rule he uses the two-dimensional universe $\mathcal{U} = (\Omega^{(1)}, \Omega^{(2)})$
with the domains $\Omega^{(1)} \overset{\text{def}}{=} \mathrm{Dom}(height) \overset{\text{def}}{=} [40, 250] \subset \mathbf{R}$ (in cm) and
$\Omega^{(2)} \overset{\text{def}}{=} \mathrm{Dom}(weight) \overset{\text{def}}{=} [1.5, 300] \subset \mathbf{R}$ (in kg). He interprets the linguistic values 'little' and 'small' with the help of the possibility distributions
$\mu_1 \in \mathrm{POSS}(\Omega^{(1)})$ and $\nu_1 \in \mathrm{POSS}(\Omega^{(2)})$, graphically presented in Figures 3
and 4, respectively.

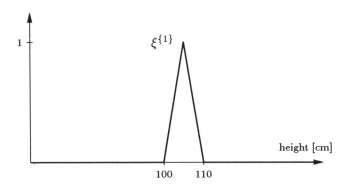

Figure 5   Possibility distribution of the body-height of Marcel

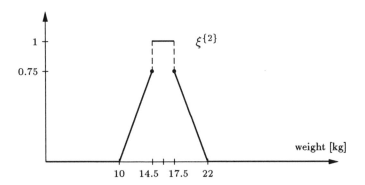

Figure 6   Resulting possibility distribution of the weight of Marcel (in kg)

The induced Gödel relation $\varrho^{\text{Gödel}}[\mu_1, \nu_1] \in \text{POSS}(\Omega)$ is used to identify the most specific possibility distribution $\xi^{\{2\}} \in \text{POSS}(\Omega^{(2)})$, compatible with R_1, and satisfying the principle of minimum specificity for the weight of a boy named Marcel, whose body-height is characterized with the help of the possibility distribution $\xi^{\{1\}} \in \text{POSS}(\Omega^{(1)})$ shown in Figure 5.

For $\xi^{\{2\}} \equiv \widehat{\text{infer}}_1\left(\xi^{\{1\}}, \varrho^{\text{Gödel}}[\mu_1, \nu_1]\right)$ we obtain the graphical representation in Figure 6.

By adding further possibilistic inference rules of the type

$$R_j : \textbf{if } height \textbf{ is } \mu_j \textbf{ then } weight \textbf{ is } \nu_j, \quad j = 1, \ldots, r,$$

and their corresponding Gödel relations $\varrho^{\text{Gödel}}[\mu_j, \nu_j]$, the whole expert knowledge about the uncertain relation between *height* and *weight* could be specified. In a possibilistic F-expert system $(\mathcal{U}, \mathcal{M}, \mathcal{N}, \mathcal{R}(\mathcal{U}, \mathcal{M}))$, maybe extended, if necessary, by some attributes (e.g. *age*, *profession*, etc.), with n-dimensional universe $\mathcal{U} = (\Omega^{(i)})_{i \in N_n}$, $n \geq 2$, the rules $R_1, R_2, \ldots, R_r$ would be represented by $\varrho^{\{1,2\}} \overset{Df}{\equiv} \min_{j=1,\ldots,r} \{\varrho^{\text{Gödel}}[\mu_j, \nu_j]\}$ and thus be entered with respect to the index set $\{1, 2\} \in \mathcal{M}$ into the rule base $\mathcal{R}(\mathcal{U}, \mathcal{M}) = \{\varrho^M \mid M \in \mathcal{M}\}$. Note that $\varrho^{\{1,2\}}$ should turn out to be a possibility distribution in order to contribute for a non-contradictory rule base $\mathcal{R}(\mathcal{U}, \mathcal{M})$.   □

3.4 Knowledge Representation and Propagation Using Hypergraphs

It was the aim of the two foregoing sections to introduce the concept of a (possibilistic) F-expert system and to consider how if-then rules, well-known in the field of knowledge-based systems, are to be interpreted within the suggested formal framework.

We now turn to the development of an appropriate *propagation algorithm*, which enables us to determine the states $\sigma(\mathcal{X}, \mathcal{E}(\mathcal{U}, \mathcal{N}))$ of (possibilistic) F-expert systems $\mathcal{X}$ with respect to evidence systems $\mathcal{E}(\mathcal{U}, \mathcal{N})$ and to efficiently compute their i-th restrictions $\kappa^{(i)}$.

At first we will focus our investigations on the special case of ordinary F-expert systems which were introduced in Definition 3.14. As we already know, $\kappa^{(i)}$ can be stated as

$$\kappa^{(i)} = \Pi_{\{i\}}^{N_n} \left(\bigcap_{M \in \mathcal{M}} \hat{\Pi}_M^{N_n}(R^M) \cap \bigcap_{N \in \mathcal{N}} \hat{\Pi}_N^{N_n}(E^N) \right)$$

$$= \Pi_{\{i\}}^{N_n} \left(\sigma(\mathcal{X}, \mathcal{E}(\mathcal{U}, \mathcal{N})) \right). \tag{3.22}$$

From this representation we find a trivial propagation algorithm that computes $\kappa^{(i)}$ by calculating the cylindrical extensions, intersections, and projections stated above. But as simple as this algorithm appears to be, it is little applicable, since it determines $\kappa^{(i)}$ by $\sigma(\mathcal{X}, \mathcal{E}(\mathcal{U}, \mathcal{N}))$ and therefore operates on n-ary relations, whose cardinality cause several problems.

As a clarification we inspect an example that is rather elementary compared to actual applications. Let $\mathcal{X} = (\mathcal{U}, \mathcal{M}, \mathcal{N}, \mathcal{R}(\mathcal{U}, \mathcal{M}))$ be an F-expert system with a 10-dimensional universe $\mathcal{U} = (\Omega^{(i)})_{i \in N_{10}}$ and domains $\Omega^{(i)}$ with $|\Omega^{(i)}| = 6$, $i = 1, 2, \ldots, 10$. Let $M = \{1, 2\} \in \mathcal{M}$ be an index set of the modularization $\mathcal{M}$, and $R^M \in \mathcal{R}(\mathcal{U}, \mathcal{M})$, $|R^M| = 12$, a relation of the rule base $\mathcal{R}(\mathcal{U}, \mathcal{M})$. For the calculation of $\kappa^{(i)}$, the mentioned propagation algorithm uses, for instance, the relation $\hat{\Pi}_M^{N_{10}}(R^M)$, which is impossible to manage because of its extreme number of tuples $|\hat{\Pi}_M^{N_{10}}(R^M)| = 12 \cdot 6^8 = 20.155.392$.

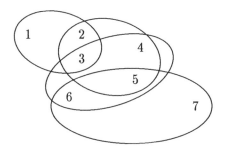

Figure 7 The hypergraph $H_\mathcal{M}$ induced by $\mathcal{M}$

Carrying out the propagation, one should therefore try to shift the intersection operations executed in Ω to low-dimensional subspaces in order to transform the mentioned *global* algorithm (i.e. referring to Ω) to a *locally* operating one (i.e. referring to Ω^M with $M \in \mathcal{M}$). Since the corresponding subspaces are predetermined by the modularization $\mathcal{M}$, which expresses the existing qualitative dependencies between all involved attributes, it suggests representing $\mathcal{M}$ not only as a set of index sets, but also with the aid of a hypergraph.

Definition 3.21 *A* **hypergraph** *is a pair* $(V, \mathfrak{E})$ *of a finite set V of* **vertices** *and a set* $\mathfrak{E} \subseteq \mathfrak{P}(V)$ *of* **hyperedges,** *that satisfy the following conditions:*

$$(a) \qquad \forall E \in \mathfrak{E}:\ E \neq \emptyset$$

$$(b) \qquad \bigcup_{E:E\in\mathfrak{E}} E = V$$

It is obvious that any modularization $\mathcal{M}$ of $\mathbf{N}_n$ induces the hypergraph $H_\mathcal{M} = (\mathbf{N}_n, \mathcal{M})$.

Example 3.22 Let $\mathcal{M} = \{\{1,2,3\}, \{2,3,4,5\}, \{3,4,5,6\}, \{5,6,7\}\}$ be a modularization of $\mathbf{N}_7$. $H_\mathcal{M} = (\mathbf{N}_7, \mathcal{M})$ can be represented graphically as shown in Figure 7.   □

Before we consider the desired local propagation algorithm for F-expert systems, we introduce the following property of hyperedges.

Definition 3.23 *Let* $E_1, E_2 \in \mathfrak{E}$ *be two different hyperedges of the hypergraph* $(V, \mathfrak{E})$. E_1 *and* E_2 *are called* **overlapping** *if*

$$E_1 \not\subseteq E_2\ \wedge\ E_2 \not\subseteq E_1\ \wedge\ E_1 \cap E_2 \neq \emptyset.$$

holds.

Overlapping hyperedges $M_1, M_2 \in \mathcal{M}$ of a hypergraph $H_{\mathcal{M}}$, induced by a modularization $\mathcal{M}$ of an F-expert system $\mathcal{X} = (\mathcal{U}, \mathcal{M}, \mathcal{N}, \mathcal{R}(\mathcal{U}, \mathcal{M}))$, are that important, because they point out the qualitative dependencies that exist between the attributes adressed by M_1 and M_2. We therefore call $H_{\mathcal{M}}$ the *dependency hypergraph* of the underlying F-expert system $\mathcal{X}$. In this connection we observe, with respect to Definition 3.10, that any pair $M_1, M_2 \in \mathcal{M}$ of index sets of a modularization $\mathcal{M}$ satisfies the conditions $M_1 \not\subseteq M_2$ and $M_2 \not\subseteq M_1$, so that in this case the property 'overlapping' is ensured already by the existence of an attribute of $\mathcal{X}$ adressed by M_1 as well as M_2.

The local propagation algorithm for F-expert systems presented below is written in a self-explanatory notation similar to PASCAL. All type declarations have been omitted. By calling the procedure *focus*, which uses the global variables $\mathcal{M}, \mathcal{N}, \mathcal{R}(\mathcal{U}, \mathcal{M})$, and $\mathcal{E}(\mathcal{U}, \mathcal{N})$, whose names refer to the notation used up to now, and on the basis of the evidence system $\mathcal{E}(\mathcal{U}, \mathcal{N}) = \{E^N \mid N \in \mathcal{N}\}$, the modified relations system $\mathcal{R}(\mathcal{U}, \mathcal{M}) = \{R^M \mid M \in \mathcal{M}\}$ with $\left(\forall M \in \mathcal{M} : R^M := \Pi_M^{N_n}(\sigma(\mathcal{X}, \mathcal{E}(\mathcal{U}, \mathcal{N})))\right)$ can be determined. The symbol ':=' is to be read as an assignment operator.

The underlying idea of the algorithm is that information about those attribute values that are no longer possible in the light of the included evidence has to be distributed within the hypergraph via overlapping hyperedges. This information is used to exclude tuples from the relations R^M, which may lead to an exclusion of further attribute values. The most specific relations are determined by this procedure only if the rule base is *non-redundant*, i.e. none of the relations R^M in $R(\mathcal{U}, \mathcal{M})$ contains a tuple that cannot result from a projection of the induced knowledge base R onto Ω^M. In the case of a non-redundant rule base the i-th restrictions of $\sigma(\mathcal{X}, \mathcal{E}(\mathcal{U}, \mathcal{N}))$ can be computed after an execution of the propagation algorithm by

$$\kappa^{(i)} = \Pi_{\{i\}}^M(R^M) \text{ for all } M \in \mathcal{M} \text{ with } i \in M.$$

The modified rule base is non-redundant, too. On the other hand, if the rule base is redundant, the projections stated above lead to supersets of $\kappa^{(i)}$, $i = 1, \ldots, n$.

Definition 3.24 *A rule base* $\mathcal{R}(\mathcal{U}, \mathcal{M}) = \{R^M \mid M \in \mathcal{M}\}$ *w.r.t. an n-dimensional universe* $\mathcal{U}$ *and a modularization* $\mathcal{M}$ *of* N_n *is called* **non-redundant**, *if*

$$\forall M \in \mathcal{M} : R^M = \Pi_M^{N_n}\left(\bigcap_{M^* \in \mathcal{M}} \hat{\Pi}_{M^*}^{N_n}\left(R^{M^*}\right)\right).$$

Note that using a hypergraph representation for all existing dependencies among the values of the considered attributes is connected with an implicit *decomposability assumption*, which reads

$$R = \bigcap_{M \in \mathcal{M}} \hat{\Pi}_M^{N_n} \left(\Pi_M^{N_n}(R) \right).$$

From a more illustrative point of view, non-redundancy therefore means that all tuples contained in a relation R^M in $\mathcal{R}(\mathcal{U}, \mathcal{M})$ result from the projection of the induced (decomposable) knowledge base onto Ω^M.

Algorithm 3.25

procedure focus;

  var $\mathcal{N}^*; N; M; R^*$;

  begin
    $\mathcal{N}^* := \mathcal{N}$;
    while $\mathcal{N}^* \neq \emptyset$ **do**
    begin
      choose an $N \in \mathcal{N}^*$;
      $\mathcal{N}^* := \mathcal{N}^* \backslash \{N\}$;
      choose an $M \in \mathcal{M}$ with $N \subseteq M$;
      $R^* := \hat{\Pi}_N^M(E^N) \cap R^M$;
      if $R^* \neq R^M$
      then begin
        $R^M := R^*$;
        propagate(M)
      end
    end
  end {focus};

procedure propagate(M);

  var $\mathcal{M}^*; R^*$;

  begin
    $\mathcal{M}^* := \mathcal{M} \backslash \{M\}$;
    while $\mathcal{M}^* \neq \emptyset$ **do**
    begin
      choose an $M^* \in \mathcal{M}^*$;
      $\mathcal{M}^* := \mathcal{M}^* \backslash \{M^*\}$;
      if M^* and M overlap
      then begin
        $R^* := R^{M^*} \cap \hat{\Pi}_{M^* \cap M}^{M^*} \left(\Pi_{M^* \cap M}^M (R^M) \right)$;
        if $R^* \neq R^{M^*}$

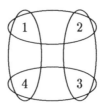

Figure 8   Dependency-hypergraph of $\mathcal{X}_r$

then begin
    $R^{M^*} := R^*;$
    propagate(M^*)
  end
    end
  end
end {propagate};   □

In general the local propagation algorithm does not operate on n-ary relations, but at most on $\max\{|M| \mid M \in \mathcal{M}\}$-ary relations. But the costs that are connected with computing the $\kappa^{(i)}$ do not only depend on the cardinality of the relations incorporated in $\mathcal{R}(\mathcal{U}, \mathcal{M})$, but also on the structure of the hypergraph $H_\mathcal{M}$. By the next example we show that hypergraphs containing at least one *cycle* can lead, worst case assumed, to unacceptable inefficiency of the presented propagation algorithm.

The notion of a cycle will be formalized for hypergraphs subsequently. Its intuitive meaning should be clear from the example.

Example 3.26 Let $\mathcal{X}_r = (\mathcal{U}_r, \mathcal{M}, \mathcal{N}, \mathcal{R}_r(\mathcal{U}_r, \mathcal{M}))$, $r \in \mathbf{N}$,
be an F-expert system with a four-dimensional universe
$\mathcal{U}_r = \left(\Omega_r^{(1)}, \Omega_r^{(2)}, \Omega_r^{(3)}, \Omega_r^{(4)}\right)$, $\Omega_r^{(i)} = \{0, 1, \ldots, r\}$, $i = 1, \ldots, 4$,
a modularization $\mathcal{M} = \{M_1, M_2, M_3, M_4\}$ with
$M_1 = \{1, 2\}$, $M_2 = \{2, 3\}$, $M_3 = \{3, 4\}$, $M_4 = \{1, 4\}$,
a partition $\mathcal{N} = \{\{1\}, \{2\}, \{3\}, \{4\}\}$ of $\mathbf{N}_4$ compatible with $\mathcal{M}$,
and a rule base $\mathcal{R}_r(\mathcal{U}_r, \mathcal{M}) = \{R_r^{M_1}, R_r^{M_2}, R_r^{M_3}, R_r^{M_4}\}$,
$R_r^{M_1} = R_r^{M_2} = R_r^{M_3} = \{(i, i) | i \in \{0, 1, \ldots, r\}\}$,
$R_r^{M_4} = \{((i + 1) \bmod r, i) \mid i \in \{0, \ldots, r - 1\}\} \cup \{(r, r)\}$.

If the propagation Algorithm 3.25 is applied to the evidence system
$\mathcal{E}_r(\mathcal{U}_r, \mathcal{N}) = \left\{E_r^{\{1\}}, E_r^{\{2\}}, E_r^{\{3\}}, E_r^{\{4\}}\right\}$ with
$E_r^{\{1\}} = \Omega_r^{(1)} \backslash \{0\}$, $E_r^{\{2\}} = \Omega_r^{(2)}$, $E_r^{\{3\}} = \Omega_r^{(3)}$, $E_r^{\{4\}} = \Omega_r^{(4)}$,

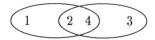

Figure 9   Dependency-hypergraph of $\mathcal{X}'_r$

the relations contained in the rule base $\mathcal{R}_r(\mathcal{U}_r, \mathcal{M})$ are specified as follows:

$$R_r^{M_j} := \{(r, r)\}, \quad j = 1, 2, 3, 4.$$

As i-th restrictions $\kappa^{(i)}$ of the state $\sigma(\mathcal{X}_r, \mathcal{E}_r(\mathcal{U}_r, \mathcal{N}))$ of $\mathcal{X}$, induced by $\mathcal{E}_r(\mathcal{U}_r, \mathcal{N})$, we then obtain

$$\kappa^{(i)} = \{r\}, \quad i = 1, 2, 3, 4.$$

The propagation can, for example, be carried out over the sequence of hyperedges $(M_1, M_2, M_3, M_4, M_1, M_2, \ldots)$. Any call of the procedure *propagate* is due to an intersection that removes just one element from one of the relations R^M or R^{M^*}. Hence, to obtain the final result, $4r$ executions of the procedure *propagate* are needed, i.e. the hypergraph $H_{\mathcal{M}}$ is traversed r times.

The problems of efficiency, which become obvious here, are in the first place connected with the redundancy of the initial state of the rule base $\mathcal{R}_r(\mathcal{U}_r, \mathcal{M})$. In the second place they are a consequence of an inappropriate modularization $\mathcal{M}$, which reflects cyclic dependencies between all of the involved attributes. With respect to this example, this suggests removing the cycles by fusing vertices.

As an alternative we choose the F-expert system $\mathcal{X}'_r = (\mathcal{U}_r, \mathcal{M}', \mathcal{N}, \mathcal{R}_r(\mathcal{U}_r, \mathcal{M}'))$ with

$$\mathcal{M}' = \{M'_1, M'_2\}, \qquad M'_1 = M_1 \cup M_4, \quad M'_2 = M_2 \cup M_3,$$

$$\mathcal{R}_r(\mathcal{U}_r, \mathcal{M}') = \{R_r^{M'_1}, R_r^{M'_2}\}, \qquad R_r^{M'_1} = \hat{\Pi}_{M_1}^{M'_1}(R_r^{M_1}) \cap \hat{\Pi}_{M_4}^{M'_1}(R_r^{M_4}),$$

$$R_r^{M'_2} = \hat{\Pi}_{M_2}^{M'_2}(R_r^{M_2}) \cap \hat{\Pi}_{M_3}^{M'_2}(R_r^{M_3}),$$

and its induced hypergraph $H_{\mathcal{M}'}$ shown in Figure 9.

$\mathcal{X}_r$ and $\mathcal{X}'_r$ are called *equivalent*, since assuming equal total evidence, characterized by $\mathcal{E}(\mathcal{U}_r, \mathcal{N})$, they lead to the same state $\sigma(\mathcal{X}_r, \mathcal{E}(\mathcal{U}_r, \mathcal{N})) = \sigma(\mathcal{X}'_r, \mathcal{E}(\mathcal{U}_r, \mathcal{N}))$.

If the local propagation algorithm is applied to $\mathcal{X}'$ with respect to $\mathcal{E}(\mathcal{U}_r, \mathcal{N})$, then the initial relations

$$R_r^{M'_1} = \{(i, i, (i + r - 1) \bmod r) \mid i \in \{0, \ldots, r - 1\}\} \cup \{(r, r, r)\},$$

$$R_r^{M'_2} = \{(i, i, i) \mid i \in \{0, 1, \ldots, r\}\},$$

which remain redundant, are specialized to the relations

$$R_r^{M_1'} := R_r^{M_2'} := \{(r, r, r)\}.$$

The result is now already achieved after the second execution of the procedure *propagate*. $\square$

Definition 3.27
Let $\mathcal{X}_1 = (\mathcal{U}_1, \mathcal{M}_1, \mathcal{N}_1, \mathcal{R}(\mathcal{U}_1, \mathcal{M}_1))$ and $\mathcal{X}_2 = (\mathcal{U}_2, \mathcal{M}_2, \mathcal{N}_2, \mathcal{R}(\mathcal{U}_2, \mathcal{M}_2))$ be two *F-expert systems*. $\mathcal{X}_1$ and $\mathcal{X}_2$ are called **equivalent**, if the following conditions hold:

(a) $\mathcal{U}_1 = \mathcal{U}_2$,

(b) $\mathcal{N}_1 = \mathcal{N}_2$,

(c) for all evidence systems $\mathcal{E}(\mathcal{U}_1, \mathcal{N}_1)$:
$\sigma(\mathcal{X}_1, \mathcal{E}(\mathcal{U}_1, \mathcal{N}_1)) = \sigma(\mathcal{X}_2, \mathcal{E}(\mathcal{U}_1, \mathcal{N}_1))$.

Without considering the problem that rule bases of F-expert systems can be redundant, Example 3.26 indicates that using an alternative F-expert system can lead to a considerable gain in efficiency with respect to propagation. Indeed it will become evident later that it might be helpful to restrict ourselves to systems $\mathcal{X} = (\mathcal{U}, \mathcal{M}, \mathcal{N}, \mathcal{R}(\mathcal{U}, \mathcal{M}))$, whose hypergraph $H_{\mathcal{M}}$, induced by the modularization $\mathcal{M}$, does not have any cycles. The following definitions introduce the notion of an *acyclic* hypergraph.

Definition 3.28 Let $H = (V, \mathfrak{E})$ be a hypergraph. A hyperedge $T \in \mathfrak{E}$ is called a **twig**, if $|\mathfrak{E}| = 1$ holds or if there is a hyperedge $E \in \mathfrak{E} \backslash \{T\}$, $E \cap T \neq \emptyset$, so that $(\forall E' \in \mathfrak{E} \backslash \{T\} : T \cap E' \subseteq E)$.

Hence, a twig is a hyperedge that is separated from the rest of the hypergraph by another hyperedge.

Definition 3.29 A hypergraph $H = (V, \mathfrak{E})$ is called a **hypertree**, if there is a numbering $E : \{1, 2, \ldots, |\mathfrak{E}|\} \rightarrow \mathfrak{E}$ of its hyperedges, such that $E(k)$ is a twig in the hypergraph $\left(\bigcup_{i=1}^{k} E(i), \{E(i) \mid i \in \{1, \ldots, k\}\} \right)$ for $k = 1, \ldots, |\mathfrak{E}|$. The finite sequence $(E(k))_{k=1}^{|\mathfrak{E}|}$ is called a **construction sequence** for H.

Example 3.26 (continuation) The hyperedges E_1 and E_4 of the hypergraph in Figure 10 are twigs, since E_1 is separated from the rest of the hypergraph by E_2, and E_4 is separated by E_3. $\square$

Normally the application of Definition 3.29 turns out to be rather impracticable, if the aim is to determine whether a given hypergraph is a

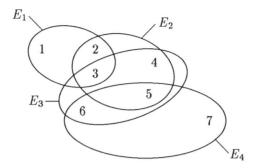

Figure 10   This hypergraph is a hypertree with the construction
sequences (E_1, E_2, E_3, E_4) and (E_4, E_3, E_2, E_1)

hypertree or not. A very simple possibility to test the hypertree property is
based on the idea of reversing the process, that is, to determine a *reduction
sequence* instead of a constructing one, which, if the given hypergraph is
a hypertree, resolves it by removing all of its vertices and hyperedges. A
procedure based on this idea is the so-called *Graham Reduction*, which can be
stated as follows:

Algorithm 3.30 [Graham Reduction] A hypergraph $H = (V, \mathfrak{E})$ is a
hypertree if and only if all vertices of H can be removed by iteratively applying
one of the following operations:

(i) Remove a vertex that occurs in only one hyperedge.

(ii) Remove a hyperedge that is contained in another hyperedge.   □

Example 3.26 (continuation) The application of the non-deterministic
Algorithm 3.30 leads, for example, to the following reduction:

(1) remove vertex 1,

(2) remove hyperedge $\{2, 3\}$,

(3) remove vertex 2,

(4) remove hyperedge $\{3, 4, 5\}$,

(5) remove vertex 3,

(6) remove vertex 4,

(7) remove hyperedge $\{5, 6\}$,

(8) remove vertex 5,

(9) remove vertex 6,

(10) remove vertex 7.

Note that the removal of a vertex not only changes the set of vertices, but also the set of hyperedges.   □

Hypertrees can be viewed as acyclic hypergraphs. If the dependency hypergraph $H_\mathcal{M}$, induced by the modularization $\mathcal{M}$ of an F-expert system $\mathcal{X} = (\mathcal{U}, \mathcal{M}, \mathcal{N}, \mathcal{R}(\mathcal{U}, \mathcal{M}))$, is a hypertree, it turns out to be quite simple to develop an efficient propagation procedure, which will be presented as Algorithm 3.34 later on.

A simple but nevertheless satisfactory estimation of the costs of the propagation Algorithm 3.25 can be carried out, if $H_\mathcal{M}$ belongs to the special class of *simple hypertrees*. But since these structures are generally too restrictive for dependency hypergraphs of F-expert systems, we pursue another approach that is based on the fact that every F-expert system can be transformed into an equivalent system with a non-redundant rule base. This is possible even without changing the dependency hypergraph, whereas alternative methods try to achieve an equivalent F-expert system with an induced simple hypertree as a dependency hypergraph by means of a modified modularization.

We now introduce the notion of a simple hypertree, and in Theorem 3.33 we relate the costs of propagation with respect to different preconditions to the number of calls of procedure 'propagate' needed in the worst case, when Algorithm 3.25 is applied. The considerable part of the execution costs that is due to the intersection operations carried out in the propagation algorithm, is consciously neglected in Theorem 3.33, since it is influenced not only by the dependency hypergraph $H_\mathcal{M}$, but also by the cardinality of the case-specific relations of the rule base $\mathcal{R}(\mathcal{U}, \mathcal{M})$.

Definition 3.31 *Let* $H = (V, \mathfrak{E})$ *be a hypertree.* $G(H) = (V', \mathfrak{E}')$ *with* $V' = V \cup \mathfrak{E}$, $\mathfrak{E}' = \{\{v, E\} \mid E \in \mathfrak{E},\ v \in E\}$ *is called the* **undirected graph induced by** H. *H is called a* **simple hypertree**, *if* $G(H)$ *is acyclic, i.e. if there is no sequence* $(E_i)_{i=0}^k$ *of hyperedges* $E_i \in \mathfrak{E}'$ *with* $E_i = \{v_i, v_{i+1}\}$ *and* $v_0 = v_k$.

Example 3.26 (continuation) The undirected graph induced by the hypertree shown in Figure 10 can be represented in the well-known form of ordinary graphs as shown in Figure 11.

We observe that ordinary graphs are special hypergraphs, namely those that contain only hyperedges of cardinality 2. The graph presented above obviously contains various cycles.   □

Example 3.32 Figures 12 and 13 show a simple hypertree and its induced acyclic undirected graph.   □

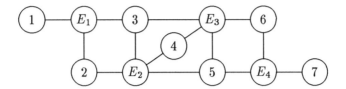

Figure 11   Induced undirected graph of the hypertree of figure 10

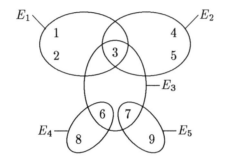

Figure 12   A hypertree H

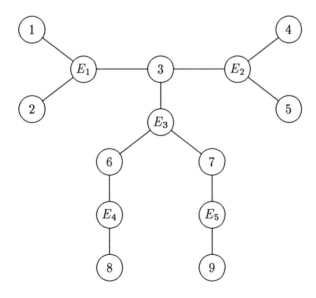

Figure 13   The undirected graph $G(H)$ induced by the hypertree H

Theorem 3.33 *If the propagation Algorithm 3.25 is applied to an evidence system $\mathcal{E}(\mathcal{U}, \mathcal{N})$ w.r.t. an F-expert system $\mathcal{X} = (\mathcal{U}, \mathcal{M}, \mathcal{N}, \mathcal{R}(\mathcal{U}, \mathcal{M}))$ with rule base $\mathcal{R}(\mathcal{U}, \mathcal{M}) = \{R^M \mid M \in \mathcal{M}\}$ and dependency hypergraph $H_\mathcal{M}$ induced by $\mathcal{M}$, then the number k of calls of the procedure* propagate *can be estimated as follows:*

(a) $k \leq \sum\limits_{M \in \mathcal{M}} |R^M|,$

(b) $H_\mathcal{M}$ *simple hypertree* $\implies$ $k \leq |\mathcal{N}| \cdot |\mathcal{M}|,$

(c) $\mathcal{R}(\mathcal{U}, \mathcal{M})$ *non-redundant* $\implies$ $k \leq |\mathcal{N}| \cdot |\mathcal{M}|.$

Proof:
We refer to the variables and statements used in Algorithm 3.25:

(a) A call *propagate(M)*, $M \in \mathcal{M}$, is executed in the procedure *focus* as well as (recursively) in the procedure *propagate* only if previously an assignment has been made to the global variable R^M changing its current value. Such a change is always connected with a removal of an element from R^M. For this reason not more than $\sum\limits_{M \in \mathcal{M}} |R^M|$ changes and therefore just as many calls to the procedure *propagate* can occur.

We prepare the proof of (b) and (c) by some considerations:

The iterative statement within the procedure *focus* is executed $|\mathcal{N}|$ times. Each execution leads to at most one call of procedure *propagate*. If *propagate* is called for an $M \in \mathcal{M}$, then the change of value of R^M causing this call can entail a change of the values of R^{M^*} for those $M^* \in \mathcal{M}$ that can be reached from M by a sequence of overlapping hyperedges in $H_\mathcal{M}$. If R^{M^*} is changed for some $M^* \in \mathcal{M}$ (which possibly is M again), this results in a recursive call of procedure *propagate* with current parameter M^*.

We will show that under the preconditions stated in (b) or (c), any relation R^{M^*}, $M^* \in \mathcal{M} \backslash \{M\}$, is modified at most once during the execution of *propagate(M)*, whereas R^M remains unchanged.

Since recursive calls of *propagate* are executed only in the sequel of a change of values, any call of *propagate(M)* from the procedure *focus* can entail at most $|\mathcal{M}| - 1$ additional (recursive) calls to procedure *propagate*. Given these conditions, we count at most $|\mathcal{N}| \cdot |\mathcal{M}|$ calls of this procedure.

(b) If $H_\mathcal{M}$ is a simple hypertree, then the undirected graph $G(H_\mathcal{M})$, induced by it, does not contain a cycle. Two or more pairwise overlapping hyperedges have therefore always exactly one vertex in common.

Hence, given an $M \in \mathcal{M}$, any $M^* \in \mathcal{M} \backslash \{M\}$ can be reached from M by exactly one sequence of overlapping hyperedges. Since an occuring procedure call *propagate(M*)* cannot lead to a modification of the value of R^{M^*} (the intersections of hyperedges of $H_\mathcal{M}$ containing at most one element), for each $M^* \in \mathcal{M}$, at most one change of value of R^{M^*} is to be counted.

(c) For an arbitrary call $propagate(M)$, $M \in \mathcal{M}$, let $\omega \in R^M$ be a tuple removed from R^M by the foregoing assignment. Let $R_{\text{old}}^{M^*}$ with $M^* \in \mathcal{M}$ denote the value of R^{M^*} prior to this assigment, R_{cur}^M denote the value of R^M after the assignment, and $R_{\text{new}}^{M^*}$ denote the value of R^{M^*} after the execution of the propagation algorithm, then we obtain:

(i) $R_{\text{cur}}^M = R_{\text{old}}^M \backslash \{\omega\}$,

(ii) $R_{\text{new}}^M = \Pi_M^{N_n} \left(\bigcap_{M^* \in \mathcal{M} \backslash \{M\}} \hat{\Pi}_{M^*}^{N_n} \left(R_{\text{old}}^{M^*} \right) \cap \hat{\Pi}_M^{N_n} \left(R_{\text{cur}}^M \right) \right)$.

Incorporating the presupposition of non-redundancy w.r.t. $\mathcal{R}(\mathcal{U}, \mathcal{M})$, we calculate:

(iii) $R_{\text{old}}^M = \Pi_M^{N_n} \left(\bigcap_{M^* \in \mathcal{M}} \hat{\Pi}_{M^*}^{N_n} \left(R_{\text{old}}^{M^*} \right) \right)$.

Combining these three conditions yields:

$$R_{\text{new}}^M \overset{\text{(i),(ii)}}{=} \Pi_M^{N_n} \left(\bigcap_{M^* \in \mathcal{M}} \hat{\Pi}_{M^*}^{N_n} \left(R_{\text{old}}^{M^*} \right) \cap \hat{\Pi}_M^{N_n} \left(R_{\text{cur}}^M \right) \right)$$

$$\overset{\text{(iii)}}{=} R_{\text{old}}^M \cap R_{\text{cur}}^M$$

$$\overset{\text{(i)}}{=} R_{\text{cur}}^M.$$

Hence, after the call of $propagate(M)$, R^M has not been changed by the propagation algorithm.

Together with the considerations mentioned above this completes the proof.

□

Example 3.26 (continuation) We have already mentioned that applying propagation Algorithm 3.25 to the specified evidence system $\mathcal{E}(\mathcal{U}_r, \mathcal{N})$ entails $4r$ calls of procedure $propagate$.

The dependency hypergraph assigned to the F-expert system $\mathcal{X}_r$ contains a cycle, and hence it is no hypertree. Since $r = |R_r^{M_i}|$, $i = 1, 2, 3, 4$, we calculate $4r = \sum_{M \in \mathcal{M}} |R^M|$. Therefore the upper bound stated in Theorem 3.33(a) is actually reached.

In order to avoid the inefficiency of the propagation algorithm when applied to $\mathcal{X}_r$, we introduced the F-expert systems $\mathcal{X}_r'$, which are equivalent to $\mathcal{X}_r$. They are based on the idea of fusing hyperedges of $H_\mathcal{M}$ such that the

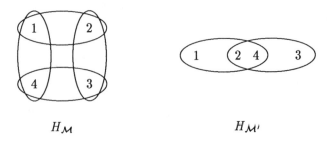

$H_\mathcal{M}$ $H_{\mathcal{M}'}$

Figure 14   Dependency hypergraph before and after fusion of edges

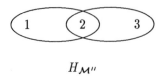

$H_{\mathcal{M}''}$

Figure 15   Dependency hypergraph resulting from a fusion of vertices

dependency hypergraph $H_{\mathcal{M}'}$ induced by the new modularization $\mathcal{M}'$ is a hypertree.

Alternatively $H_\mathcal{M}$ can be modified to a simple hypertree $H_{\mathcal{M}''}$ by fusing the vertices 2 and 4 into a single vertex (see Figure 14). But to do this, a structural change of the whole expert system (especially a redefinition of the universe $\mathcal{U}_r$) has to be carried out:

$$\mathcal{X}_r'' \stackrel{\text{def}}{=} \{\mathcal{U}_r'', \mathcal{M}'', \mathcal{N}'', \mathcal{R}_r(\mathcal{U}'', \mathcal{M}'')\},$$

$$\mathcal{U}_r'' \stackrel{\text{def}}{=} \left(\Omega_r^{(1)}, \Omega_r^{(2)} \times \Omega_r^{(4)}, \Omega_r^{(3)}\right),$$

$$\mathcal{M}'' \stackrel{\text{def}}{=} \{\mathcal{M}_1'', \mathcal{M}_2''\}, \; \mathcal{M}_1'' \stackrel{\text{def}}{=} \{1,2\}, \; \mathcal{M}_2'' \stackrel{\text{def}}{=} \{2,3\},$$

$$\mathcal{N}'' \stackrel{\text{def}}{=} \{\{1\}, \{2\}, \{3\}\},$$

$$R^{M_1''} \stackrel{\text{def}}{=} \{(i,(i,(i+r-1)\bmod r)) \mid i \in \{0,1,\ldots,r-1\}\} \cup \{(r,(r,r))\},$$

$$R^{M_2''} \stackrel{\text{def}}{=} \{(i,(i,i)) \mid i \in \{0,1,\ldots,r\}\}.$$

The induced hypergraph $H_{\mathcal{M}''}$ shown in Figure 15 is a simple hypertree. The vertices 2 and 4 of the hypertree $H_{\mathcal{M}'}$ are now represented by the vertex 2, with the *two*-dimensional domain $\Omega_r^{(2)} \times \Omega_r^{(4)}$.

Modifying the evidence systems in the corresponding way, we define
$\mathcal{E}(\mathcal{U}_r'', \mathcal{N}_r'') \stackrel{\text{def}}{=} \left\{ E_r^{\{1\}}, E_r^{\{2\}}, E_r^{\{3\}} \right\}$, where
$E_r^{\{1\}} = \Omega_r^{(1)} \backslash \{0\}$, $E_r^{\{2\}} = \Omega_r^{(2)} \times \Omega_r^{(4)}$ and $E_r^{\{3\}} = \Omega_r^{(3)}$.
According to Theorem 3.33(b) it follows that only $k \leq |\mathcal{M}''| = 2$ calls of
procedure *propagate* are necessary. It is easy to prove that exactly two calls
are executed.

If we want to determine the non-redundant rule base $\mathcal{R}_r^{\text{opt}}(\mathcal{M}) =$
$\left\{ R_{\text{opt}}^{M_1}, R_{\text{opt}}^{M_2}, R_{\text{opt}}^{M_3}, R_{\text{opt}}^{M_4} \right\}$ without changing the modularization $\mathcal{M}$, we have
to take the whole knowledge base R into account, since the cycle contained
in $H_\mathcal{M}$ affects all four vertices of the hypergraph.

With $R = \bigcap_{i=1}^{4} \hat{\Pi}_{M_i}^{\{1,2,3,4\}}(R_r^{M_i}) = \{(r,r,r,r)\}$ it follows that $R_{\text{opt}}^{M_1} = R_{\text{opt}}^{M_2} =$
$R_{\text{opt}}^{M_3} = R_{\text{opt}}^{M_4} = \{(r,r)\}$, where $R_{\text{opt}}^{M_i} \stackrel{\text{def}}{=} \Pi_{M_i}^{\{1,2,3,4\}}(R)$ is defined.

$\mathcal{X}_r^{\text{opt}} = (\mathcal{U}_r, \mathcal{M}, \mathcal{N}, \mathcal{R}_r^{\text{opt}}(\mathcal{U}_r, \mathcal{M}))$ is an F-expert system equivalent to $\mathcal{X}_r$
with unchanged dependency hypergraph $H_\mathcal{M}$, but non-redundant rule base
$\mathcal{R}_r^{\text{opt}}(\mathcal{U}_r, \mathcal{M})$.

Applying the propagation algorithm to the evidence system $\mathcal{E}(\mathcal{U}_r, \mathcal{N})$, we
expect $k \leq |\mathcal{M}| = 4$ executions of procedure *propagate*, considering the fact
that only one call of procedure *focus* occurs. It can be easily verified that the
upper bound is actually reached, that is, exactly four calls of this procedure
happen.   □

With the help of the presented method of *hyperedge fusion*, any F-expert
system with an arbitrary dependency hypergraph can be transformed into an
equivalent system with induced dependency *hypertree*. Accepting additional
elementary structural changes with respect to the underlying universe, even a
transformation into a simple hypertree can be achieved, where the presented
method of *vertex fusion* is to be employed. But in both cases we have to take
into account that the number of dimensions of the relations contained in the
rule base increases, and this leads to another kind of efficiency problem that is
due to considerably higher costs of the intersection operations carried out in
the propagation algorithm. These costs have been neglected in Theorem 3.33.

For this reason an alternative approach suggests itself, which ignores the
cycles possibly occuring in the dependency hypergraph, and which determines
a non-redundant rule base without a preceding change of the modularization.
Exploiting the structure of the dependency hypergraph, the high-dimensional
relation, representing the whole knowledge base, normally need not be
included in the calculations, but one can confine attention to an explicit
inspection of those low-dimensional subspaces that are identified by the
hyperedges of the hypertree emerging from edge fusion.

Even if the construction of a non-redundant rule base often requires considerable effort, we have to observe that this effort has only to be applied once with respect to the given expert knowledge. Thereafter the constructed rule base can be employed for arbitrary evidence systems (i.e. for arbitrary processes of focusing).

We conclude our examinations of efficient propagation in F-expert systems with the remark that the presented propagation Algorithm 3.25 can be improved even more, if certain additional conditions are satisfied. A possible approach is to avoid having to call the procedure *propagate* — in the worst case — for every $N \in \mathcal{N}$ from the procedure *focus* by *propagate*(M) with some $M \in \mathcal{M}$, for which $N \subseteq M$ holds, but to strive for a common propagation after having incorporated all pieces of evidence of the given evidence system into the relations R^M. Such a simplification can be easily achieved, if $H_{\mathcal{M}}$ is a hypertree.

The underlying idea is to traverse $H_{\mathcal{M}}$ forward according to one of its construction sequences, and then backward reversing the sequence of hyperedges, and, in doing so, to carry out the necessary changes to the relations R^M assigned to the hyperedges M.

After collecting all pieces of evidence, they can be distributed in a third pass. In contrast to Algorithm 3.25, the propagation continues even if no change of a relation R^M has occured. Since we consider only hypertrees, even rule bases that are redundant initially can thus be transformed to non-redundant ones by propagation.

We obtain the following propagation algorithm, which is modified in comparison with 3.25 to deal with (not necessarily simple) hypertrees, and which needs only $|\mathcal{N}| + 3|\mathcal{M}|$ intersection operations. The variable and procedure names are chosen just as in Algorithm 3.25. In addition let $(M(k))_{k=1}^{|\mathcal{M}|}$ be a construction sequence for $H_{\mathcal{M}}$.

Algorithm 3.34

procedure focus;
    var $\mathcal{N}^*$; M; N; j; k;
    begin
      {include evidence system}
      $\mathcal{N}^* := \mathcal{N}$;
      while $\mathcal{N}^* \neq \emptyset$ **do**
      begin
        Choose an $N \in \mathcal{N}^*$;
        $\mathcal{N}^* := \mathcal{N}^* \backslash \{N\}$;
        Choose an $M \in \mathcal{M}$ with $N \subseteq M$;
        $R^M := \hat{\Pi}_N^M(E^N) \cap R^M$
      end;

{propagate}
for $k := 2$ **to** $|\mathcal{M}|$ **do** {collect evidence: forward}
begin
  $M := M(j)$ mit $j = \max\{j \mid M(j) \text{ overlaps } M(k) \text{ and } j < k\}$;
  $R^{M(k)} := R^{M(k)} \cap \hat{\Pi}_{M \cap M(k)}^{M(k)} \left(\Pi_{M \cap M(k)}^M (R^M) \right)$
end;
for $k := |\mathcal{M}| - 1$ **to** 1 **do** {collect evidence: backward}
begin
  $M := M(j)$ with $j = \min\{j \mid M(j) \text{ overlaps } M(k) \text{ and } j > k\}$;
  $R^{M(k)} := R^{M(k)} \cap \hat{\Pi}_{M \cap M(k)}^{M(k)} \left(\Pi_{M \cap M(k)}^M (R^M) \right)$
end;
for $k := 2$ **to** $|\mathcal{M}|$ **do** {distribute evidence}
begin
  $M := M(j)$ with $j = \max\{j \mid M(j) \text{ overlaps } M(k) \text{ and } j < k\}$;
  $R^{M(k)} := R^{M(k)} \cap \hat{\Pi}_{M \cap M(k)}^{M(k)} \left(\Pi_{M \cap M(k)}^M (R^M) \right)$
end;
end; {focus}

$\square$

The results collected in this section can be directly transferred to the more general possibilistic F-expert systems, because the examined dependency hypergraphs are induced only by the modularization of the considerd expert system, but are not influenced by possibility distributions concerning the rule base and the evidence system.

Furthermore Algorithm 3.25, which is essential for our examinations, is only to be modified as far as the occuring intersection operations have to be replaced by the *min* operation, and all cylindrical extensions and projections have to be changed w.r.t. possibility distributions instead of ordinary relations. However, this does not have any effect on the number of calls of procedure *propagate*, which governed our examinations. But we have to take into account that the costs to execute the mentioned operations increase, since possibilistic rule bases or evidence systems naturally entail higher representational effort. Some aspects of the implementation are discussed in the supplementary remarks of Section 3.6.

3.5   Logic-based Inference Mechanisms

The methods presented up to now in this chapter assumed that all statements have to be made with respect to a fixed reference set or a given domain. Considering (measured) physical quantities this assumption holds due to the fact that generally a basic set of possible values can be stated (e.g. the interval

$[0, 300]$ for the body-height of human beings). But in the domain of databases only very seldom a fixed reference set is given. At a certain point in time the reference set can be determined, but it changes due to the removal of old and the addition of new records.

A database containing information about several software packages may hold, for example, the following information:

> Any software package that can be run under version 4
> of the operating system can also be run under version 5.

This rule can be easily described in the form of a logical implication $\forall s$: $(v4(s) \rightarrow v5(s))$, where the predicate $vn(s)$ is true if and only if s is a software package that can be run under version n of the operating system $(n = 4, 5)$. The reference set of all software packages need not be specified any further.

If in addition to certain rules, such as the one just considered, uncertain or vague knowledge is also to be included in the database, concepts for their management have to be provided, which are not based on set-theoretic, but on a logical grounds.

To do this, logical propositions are associated with numbers ranging from 0 to 1. If $d(\varphi)$ denotes the number associated with the proposition φ, then we cannot conclude anything about $d(\psi)$ from the information that $d(\varphi) = \alpha$ and $d(\varphi \rightarrow \psi) = \beta$, as long as we do not know what the numbers α and β mean. It is therefore necessary to fix semantics for these numbers to gain a reasonable inference mechanism in the form of a generalized *modus ponens* (from p and $p \rightarrow q$ deduce q), which permits given $d(\varphi \rightarrow \psi)$ and $d(\varphi)$, to determine the value of $d(\psi)$ or at least a lower bound for it from the lower bounds of $d(\varphi \rightarrow \psi)$ and $d(\varphi)$.

In classical logic, syntax and semantics are strictly distinguished. The syntactical or proof-theoretical aspect of logic deals with the derivation of propositions using formal rules of deduction in the form of the *modus ponens*. If a proposition φ can be proved syntactically from a set of propositions (axioms) A, we write $A \vdash \varphi$. The syntactical part of a logical system consists in the formal manipulation of certain symbols without assigning a precise meaning to them.

For semantics or the model-theoretical aspects of classical logic it is assumed that given a set A of statements, then 'worlds' are to be modelled, in which all propositions of A are true. In a 'world' or an interpretation, one of the truth values *true* or *false* can be assigned to a proposition. If in every model of A, i.e. an interpretation of A in which all propositions of A are *true*, the proposition φ also has the value *true*, then φ can be deduced semantically from A, in symbols: $A \models \varphi$.

Semantics state how the formulae of a logic are to be understood. But in general the assertion $A \models \varphi$ cannot be proved schematically, since to achieve a proof, all interpretations of A, and that may be an infinite number, have to be checked. Therefore syntactical procedures are employed that allow us to deduce propositions schematically. Of great importance is the *correctness* of such a deduction procedure, which means that only those statements can be deduced syntactically that are true in all interpretations. Formally this means that $A \vdash \varphi$ implies $A \models \varphi$. In reverse, *completeness* means that a statement that is true in all interpretations can be deduced schematically, i.e. $A \models \varphi$ implies $A \vdash \varphi$.

This strict separation of syntax and semantics is often not realized in those kinds of logic that assign a degree of uncertainty or a degree of truth $d(\varphi) \in [0, 1]$ to a statement φ. In many cases only a procedure is given, by which $d(\psi)$ can be determined from the knowledge of $d(\varphi)$ and $d(\varphi \to \psi)$. But a clean modelling requires a clear description of the phenomena that are to be represented. The phenomena (un-)certainty about the correctness of a statement and gradual truth value are in principle different concepts, but frequently they are not modelled separately. To demonstrate their difference, we take the following model as a basis:

A 'world' or a situation is to be modelled with the help of logical propositions. We inspect a proposition φ and a situation S to be modelled and distinguish four cases:

(i) crisp proposition (φ), crisp information (about S):

φ is a clearly defined proposition that is either true or false. S uniquely determines the situation to be modelled. φ could be, for example, the proposition 'John is 182 cm tall.' Since S uniquely determines the situation, the actual value of the height of John is known such that it can be determined without doubt, whether φ is true or false. This case can be modelled in terms of classical logic.

(ii) vague proposition, crisp information:

In this case only a gradual truth value can be assigned to the proposition φ. φ could be, for example, 'John is tall.' The height of John is known (crisp information), e.g. 185 cm. In this example it is not reasonable to classify φ uniquely as true or false. Only a gradual truth value can be assigned to φ. To describe this situation, truth-functional multi-valued logics are appropriate. (cf. Section 3.5.2).

(iii) crisp proposition, non-crisp information:

This case corresponds to an inversion of case (ii). φ would be the proposition 'John is 185 cm tall,' whereas the exact height of John is not known, but only that he is tall. In this example the proposition φ is either true or false, even if the truth value cannot be determined uniquely. Therefore only a number can be specified that expresses the certainty that φ is true, or how far it is regarded as being possible that φ is true. To model this case possibilistic logic could be employed (cf. Section 3.5.1).

(iv) vague proposition, non-crisp information:

This is the most complicated case, in which gradual truth values as well as uncertainty about the (gradual) truth value of the proposition appear in combination. For the proposition φ 'John is tall' no unique gradual truth value can be determined from the information that Joe is small and John is considerably taller than Joe. A possible approach to deal with this example consists in the application of fuzzy truth values, which means, that fuzzy sets on $[0,1]$ are permitted as truth values.

In case (iv) the single phenomenon of cases (ii) and (iii) appear combined, by which a clean modelling of this case can be carried out only at large cost. For this reason and to clarify the difference between degrees of (un-)certainty and gradual truth values, we only consider cases (ii) and (iii) in the following two sections. Additionally we restrict ourselves to the treatment of propositional logic. An extension to first order predicate logic is possible in principle, but this would require deeper knowledge about predicate logic and thus exceed the framework of logical calculi intended to be considered in this book.

3.5.1 Possibilistic Logic

The basic idea of a *possibilistic logic* is to assign to every logical proposition a number that expresses, to which degree the proposition is regarded as being necessarily true. In the following we inspect the set $\mathcal{P} = \{p_0, p_1, p_2, \ldots\}$ of *atoms* or *atomic propositions*. $\mathcal{L}$ denotes the set of all logical expressions that can be built out of the atoms of $\mathcal{P}$, the logical connectives $\wedge, \vee$, and $\neg$, and possibly some parentheses (and).

A *clause* is a logical expression of the form

$$\varphi_1 \vee \varphi_2 \vee \ldots \vee \varphi_n,$$

where φ_i $(i = 1, \ldots, n)$ is an atom or the negation of an atom.

Clauses are usually interpreted as rules, since the clause $\neg p_{i_1} \vee \ldots \vee \neg p_{i_k} \vee p_{i_{k+1}} \vee \ldots \vee p_{i_n}$ is equivalent to the implication $p_{i_1} \wedge \ldots \wedge p_{i_k} \longrightarrow p_{i_{k+1}} \vee \ldots \vee p_{i_n}$. For the set of all clauses we write $\mathcal{K}_0$. $\mathcal{K} = \mathcal{K}_0 \cup \{\top, \bot\}$ contains all clauses and the terms $\top$ and $\bot$, that are always true or false, respectively. Any proposition-logical expression can be written equivalently as a conjunction of clauses.

Definition 3.35 *A **necessity measure** N on $\mathcal{L}$ is a mapping $N : \mathcal{L} \to [0,1]$, that has the following properties:*

(i) $N(\top) = 1,$

(ii) $N(\bot) = 0,$

(iii) $N(\varphi \wedge \psi) = \min\{N(\varphi), N(\psi)\},$

(iv) if φ and ψ are equivalent logical expressions, it follows $N(\varphi) = N(\psi)$.

$N(\varphi)$ indicates to what extent φ is regarded as being necessarily true. From the properties (ii), (iii), and (iv) it follows that for all $\varphi \in \mathcal{L}$ at least one of the two values $N(\varphi)$ and $N(\neg\varphi)$ has to be zero. For a conjunction of two propositions φ and ψ, the value $N(\varphi \wedge \psi)$ can be determined from the knowledge of $N(\varphi)$ and $N(\psi)$ alone. But this does not hold for disjunction and negation. For this reason possibilistic logic is called *non-truth-functional*.

Example 3.36 Let

$$N : \mathcal{L} \to [0,1], \qquad \varphi \mapsto \left\{ \begin{array}{ll} 1, & \text{if } \varphi \text{ is a tautology} \\ 0, & \text{otherwise.} \end{array} \right.$$

Then $N(p_0) = N(\neg p_0) = N(p_1) = N(\neg p_1) = 0$, but $N(p_0 \vee \neg p_0) = N(\top) = 1 \neq 0 = N(p_0 \vee p_1)$. Analogously, for negation, $N(\bot) = N(p_0) = 0$, but $N(\neg\bot) = N(\top) = 1 \neq 0 = N(\neg p_0)$. $\square$

For disjunction, in general only $N(\varphi \vee \psi) \geq \max\{N(\varphi), N(\psi)\}$ holds, since $N(\varphi) = N(\varphi \wedge (\varphi \vee \psi)) = \min\{N(\varphi), N(\varphi \vee \psi)\}$ or $N(\psi) = N(\psi \wedge (\varphi \vee \psi)) = \min\{N(\psi), N(\varphi \vee \psi)\}$.

Definition 3.37 *A mapping* $\Pi : \mathcal{L} \to [0,1]$ *is called a* **possibility measure** *on $\mathcal{L}$ if there is a necessity measure N such that*

$$\Pi(\varphi) = 1 - N(\neg\varphi)$$

holds.

Possibility measures form the dual concept to necessity measures. They also have the properties (i), (ii), and (iv) of Definition 3.35. In addition it always holds that $\Pi(\varphi \vee \psi) = \max\{\Pi(\varphi), \Pi(\psi)\}$, whereas in general only $\Pi(\varphi \wedge \psi) \leq \min\{\Pi(\varphi), \Pi(\psi)\}$ is satisfied. The value $\Pi(\varphi)$ indicates to what extent the validity of the proposition φ is regarded as being possible. In the following we restrict ourselves to necessity measures and do not deal with possibility measures any further.

Definition 3.38 $\mathcal{W} \subseteq \mathcal{K} \times (0,1]$ *is called an* **uncertain knowledge base** *if* $\{(\top, \alpha) \mid \alpha \in (0,1]\} \subseteq \mathcal{W}$ *holds.*

An uncertain knowledge base consists of a set of formulae, to each of which a number is assigned. A pair of the form $(\varphi, \alpha) \in \mathcal{K} \times (0,1]$ is called an *uncertain clause*. α specifies a lower bound for an (unknown) necessity measure N, i.e. $(\varphi, \alpha) \in \mathcal{W}$ means $N(\varphi) \geq \alpha$. $\{(\top, \alpha) \mid \alpha \in (0,1]\} \subseteq \mathcal{W}$ is required only for formal purposes, to ensure that, if (φ, α) can be deduced from $\mathcal{W}$, (φ, β) with

$\beta \leq \alpha$ can be deduced from $\mathcal{W}$. An ordinary knowledge base $W \subseteq \mathcal{K}$ (set of axioms) corresponds to the uncertain knowledge base $\mathcal{W} = W \times \{1\}$. To deduce an uncertain clause (φ, α) from an uncertain knowledge base $\mathcal{W}$, we need a proof procedure. To achieve this we generalize the resolution principle. The *possibilistic resolution principle* can be formalized as follows:

$$\frac{(\psi \vee \varphi_1, \alpha_1) \, (\neg\psi \vee \varphi_2, \alpha_2)}{(\varphi_1 \vee \varphi_2, \min\{\alpha_1, \alpha_2\})}.$$

The possibilistic resolution principle signifies that from the knowledge of $(\psi \vee \varphi_1, \alpha_1)$ and $(\neg\psi \vee \varphi_2, \alpha_2)$, which corresponds to $N(\psi \vee \varphi_1) \geq \alpha_1$ and $N(\neg\psi \vee \varphi_2) \geq \alpha_2$, respectively, it can be concluded, that $(\varphi_1 \vee \varphi_2, \min\{\alpha_1, \alpha_2\})$, i.e. $N(\varphi_1 \vee \varphi_2) \geq \min\{\alpha_1, \alpha_2\}$. The justification can be achieved from:

$N(\psi \vee \varphi_1) \geq \alpha_1$ and $N(\neg\psi \vee \varphi_2) \geq \alpha_2$ implies

$$
\begin{aligned}
N(\varphi_1 \vee \varphi_2) &\geq \min\{N(\varphi_1 \vee \varphi_2), N((\psi \vee \varphi_1) \wedge (\neg\psi \vee \varphi_2))\} \\
&= N((\varphi_1 \vee \varphi_2) \wedge ((\psi \vee \varphi_1) \wedge (\neg\psi \vee \varphi_2))) \\
&= N((\psi \vee \varphi_1) \wedge (\neg\psi \vee \varphi_2)) \\
&= \min\{N(\psi \vee \varphi_1), N(\neg\psi \vee \varphi_2)\} \\
&\geq \min\{\alpha_1, \alpha_2\}.
\end{aligned}
$$

With the help of the possibilistic resolution principle new uncertain clauses can be deduced step by step from an uncertain knowledge base.

Definition 3.39 *Let $\mathcal{W}$ and $\mathcal{W}'$ be uncertain knowledge bases. $\mathcal{W}'$ can be* **deduced directly** *from $\mathcal{W}$, if there is an uncertain clause $(\varphi, \alpha) \in \mathcal{K} \times (0, 1]$ such that:*

(i) $\mathcal{W} = \mathcal{W}' \setminus \{(\varphi, \alpha)\}$

(ii) *There are clauses $\psi, \varphi_1, \varphi_2 \in \mathcal{K}$, $\alpha_1, \alpha_2 \in (0, 1]$ with*

    (a) $\alpha = \min\{\alpha_1, \alpha_2\}$

    (b) φ *is equivalent to $\varphi_1 \vee \varphi_2$*

    (c) $(\psi \vee \varphi_1, \alpha_1), (\neg\psi \vee \varphi_2, \alpha_2) \in \mathcal{W}$.

Definition 3.40 *Let $\mathcal{W}$ be an uncertain knowledge base and $(\varphi, \alpha) \in \mathcal{K} \times (0, 1]$ an uncertain clause. We say that (φ, α)* **can be proved** *from $\mathcal{W}$ (in symbols: $\mathcal{W} \vdash (\varphi, \alpha)$), if there is a sequence $\mathcal{W}_0, \mathcal{W}_1, \ldots, \mathcal{W}_n$ of uncertain knowledge bases with*

(i) $\mathcal{W}_0 = \mathcal{W}$

(ii) $\mathcal{W}_{i+1}$ *can be deduced directly from $\mathcal{W}_i$*

(iii) $(\varphi, \alpha) \in \mathcal{W}_n$.

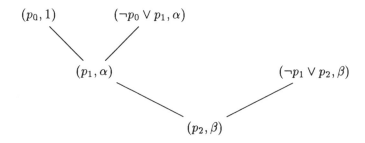

Figure 16   Possibilistic resolution for (p_2, β) from Example 3.41

Example 3.41 We consider the following (uncertain) knowledge base:

- If Anne is a student, then she is probably less than thirty years old.
- If Anne is less than thirty years old, then she is probably unmarried.
- If Anne is a student and has children, then she most probably is not unmarried.
- Anne is a student.

To deal with this knowledge base in the framework of possibilistic logic, we define the following atoms:

$$p_0 : \quad \text{Anne is a student.}$$
$$p_1 : \quad \text{Anne is less than thirty years old.}$$
$$p_2 : \quad \text{Anne is unmarried.}$$
$$p_3 : \quad \text{Anne has children.}$$

Formally the uncertain knowledge base can now be stated as:

$$\mathcal{W} = \{(\neg p_0 \vee p_1, \alpha), (\neg p_1 \vee p_2, \beta), (\neg p_0 \vee \neg p_3 \vee \neg p_2, \gamma),$$
$$(p_0, 1)\} \cup \{(\top, \delta) \mid \delta \in (0, 1]\}$$

with $0 < \beta < \alpha < \gamma < 1$.

Applying possibilistic resolution, it can be proved that $\mathcal{W} \vdash (p_2,)$, i.e. Anne is probably unmarried.

If we add to our uncertain knowledge base $\mathcal{W}$ the proposition that Anne has children, i.e. if we consider the knowledge base $\mathcal{W}' = \mathcal{W} \cup \{(p_3, 1)\}$, it turns out that $\mathcal{W}'$ is β-inconsistent, i.e. $\mathcal{W}' \vdash (\bot, \beta)$ can be proved. The resolution proofs are shown schematically in Figures 16 and 17. $\square$

Having fixed the syntax of possibilistic logic, we now turn to semantics. An (ordinary) interpretation $I : \mathcal{P} \to \{w, f\}$ assigns to every atom one of the two

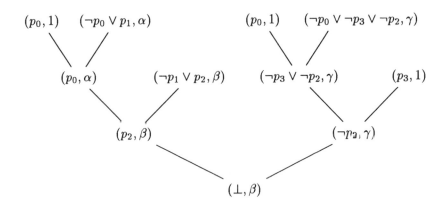

Figure 17   Possibilistic resolution for $(\bot, \beta)$ from Example 3.41

truth values *true* and *false*. Therefore a truth value induced by I according to the corresponding rules for $\wedge, \vee$, and $\neg$, can be assigned uniquely to any logical expression of $\mathcal{L}$, i.e. I can be extended canonically to the mapping $I^* : \mathcal{L} \rightarrow \{w, f\}$.

An uncertain knowledge base induces a fuzzy set on the set $\mathfrak{P}(\mathcal{P})$ of all interpretations.

Definition 3.42 *Let $\mathcal{W}$ be an uncertain knowledge base and $(\varphi, \alpha) \in \mathcal{K} \times (0, 1]$.*

(i)   *The fuzzy set $\mu_{(\varphi, \alpha)} \in F(\mathfrak{P}(\mathcal{P}))$ is given as*

$$\mu_{(\varphi, \alpha)}(I) = \begin{cases} 1, & \text{if } I^*(\varphi) = w \\ 1 - \alpha, & \text{otherwise.} \end{cases}$$

(ii)   *$\mu_{\mathcal{W}} \in F(\mathfrak{P}(\mathcal{P}))$ is being fixed by*

$$\mu_{\mathcal{W}}(I) = \inf \left\{ \mu_{(\varphi, \alpha)}(I) \mid (\varphi, \alpha) \in \mathcal{W} \right\}.$$

(iii)   *The value*

$$\text{cons}(\mathcal{W}) = \sup \left\{ \mu_{\mathcal{W}}(I) \mid I \in \mathfrak{P}(\mathcal{P}) \right\}$$

*is called a **degree of consistency** of $\mathcal{W}$.*

(iv)   *The value*

$$\text{inc}(\mathcal{W}) = 1 - \text{cons}(\mathcal{W})$$

*is called a **degree of inconsistency** of $\mathcal{W}$.*

The value $\mu_{(\varphi,\alpha)}(I) \in [0,1]$ indicates, to what extent the interpretation I is compatible with the uncertain clause (φ,α). If φ is true with respect to the interpretation I, then I is absolutely incompatible with (φ,α). If it is $I^*(\varphi) = false$, although a degree of necessity of at least α is assigned to φ, then I is compatible with (φ,α) only to the degree $1-\alpha$. In an analogous way the compatibility of an interpretation I with an uncertain knowledge base $\mathcal{W}$, i.e. a set of uncertain clauses, is defined as the infimum of the compatibilities of I with the uncertain clauses of $\mathcal{W}$.

The degree of consistency is the greatest degree with which an interpretation can be compatible with the uncertain knowledge base $\mathcal{W}$.

The subsequent definition describes what we mean by semantical deducibility in possibilistic logic.

Definition 3.43 *Let $\mathcal{W}$ be an uncertain knowledge base, $(\varphi,\alpha) \in \mathcal{K} \times (0,1]$. We write*

$$\mathcal{W} \models (\varphi,\alpha)$$

if $\operatorname{inc}(\mathcal{W} \cup \{(\neg\varphi,1)\}) \geq \alpha$ *holds.*

This definition corresponds to the property of classical logic that a clause φ can be proved from a set of clauses W, if $W \cup \{\neg\varphi\}$ is inconsistent. Note that $\neg\varphi$ need not be a clause in any case. But Definition 3.42 can be applied to arbitrary (sets of) formulae of the kind $(\varphi,\alpha) \in \mathcal{L} \times (0,1]$.

Theorem 3.44 (correctness of possibilistic logic) *Let $\mathcal{W}$ be an uncertain knowledge base and $(\varphi,\alpha) \in \mathcal{K} \times (0,1]$. Then*

$$\mathcal{W} \vdash (\varphi,\alpha) \implies \mathcal{W} \models (\varphi,\alpha).$$

Proof:
Let $\mathcal{W}_{(\varphi,\alpha)} \subseteq \mathcal{W}$ be the set of all uncertain clauses of $\mathcal{W}$ that are used in a possibilistic resolution proof for (φ,α). Obviously it is $\mathcal{W}_{(\varphi,\alpha)} \subseteq \mathcal{K} \times [\alpha,1]$.
We define

$$W_{(\varphi,\alpha)} \stackrel{\text{def}}{=} \{\psi \mid \text{there is a } \beta \in [\alpha,1] \text{ with } (\psi,\beta) \in \mathcal{W}_{(\varphi,\alpha)}\}.$$

Leaving out the numbers from the possibilistic resolution proof for (φ,α), we obtain a proof for φ from $W_{(\varphi,\alpha)}$. This means that the set $W' \stackrel{\text{def}}{=} W_{(\varphi,\alpha)} \cup \{\neg\varphi\}$ is inconsistent (in the sense of classical logic). For every interpretation I it therefore follows that there is a proposition $\psi_I \in W'$, such that $I^*(\psi_I) = false$ holds. Hence, for all interpretations I,

$$\mu_{\mathcal{W} \cup \{(\neg\varphi,1)\}}(I) \leq 1-\alpha$$

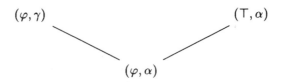

Figure 18   Possibilistic resolution for (φ, α)

and therefore

$$
\begin{aligned}
\mathrm{inc}\left(\mathcal{W} \cup \{(\neg\varphi, 1)\}\right) &= 1 - \sup\left\{\mu_{\mathcal{W}\cup\{(\neg\varphi,1)\}}(I) \mid I \in \mathfrak{P}(\mathcal{P})\right\} \\
&\geq 1 - (1 - \alpha) \\
&= \alpha.
\end{aligned}
$$
□

Theorem 3.45 (completeness of possibilistic logic) *Let* $\mathcal{W}$ *be an un-*
certain knowledge base and $(\varphi, \alpha) \in \mathcal{K} \times (0, 1]$. *Then*

$$
\mathcal{W} \models (\varphi, \alpha) \quad \Longrightarrow \quad \mathcal{W} \vdash (\varphi, \alpha).
$$

Proof:
According to the definition of $\mathcal{W} \models (\varphi, \alpha)$

$$
\alpha \leq \mathrm{inc}\left(\mathcal{W} \cup \{(\neg\varphi, 1)\}\right) = 1 - \sup\left\{\mu_{\mathcal{W}\cup\{(\neg\varphi,1)\}}(I) \mid I \in \mathfrak{P}(\mathcal{P})\right\}.
$$

Hence it follows

$$
\sup\left\{\mu_{\mathcal{W}\cup\{(\neg\varphi,1)\}}(I) \mid I \in \mathfrak{P}(\mathcal{P})\right\} \leq 1 - \alpha,
$$

i.e. for every interpretation I

$$
\mu_{\mathcal{W}\cup\{(\neg\varphi,1)\}}(I) \leq 1 - \alpha
$$

holds. This means that the set

$$
\mathcal{W} = \left\{\psi \mid \text{there is a } \beta \in [\alpha, 1] \text{ with } (\psi,) \in \mathcal{W}\right\} \cup \{\neg\varphi\}
$$

is inconsistent. Therefore the clause φ can be proved from the set $\mathcal{W}\backslash\{\neg\varphi\}$
with the help of ordinary resolution. This resolution proof can be transformed
into a possibilistic resolution proof, in which only clauses $(\psi, \beta) \in \mathcal{W}$ with
$\beta \geq \alpha$ are used, so that we obtain $\mathcal{W} \vdash (\varphi, \gamma)$ with $\gamma \geq \alpha$. With the
possibilistic resolution shown in Figure 18 it results that $\mathcal{W} \vdash (\varphi, \alpha)$.   □

3.5.2 Truth-functional Logic

Having made acquaintance with possibilistic logic as an example of non-truth-functional logic, we now turn to truth-functional logic.

Let us consider the set $\mathcal{P}$ of all atoms again. As an example we restrict ourselves to the logical connectives $\rightarrow$, $\wedge$, and $\vee$. For the set of all expressions that can be built from atoms with the help of $\rightarrow$, $\wedge$, and $\vee$ (and possibly some parentheses), we write $\mathcal{L}$ again.

Instead of the set of truth values $\{true, false\}$ we permit all numbers from the unit interval $[0, 1]$, from 0 for *false* to 1 for *true*. An interpretation I therefore is a mapping $I : \mathcal{P} \rightarrow [0, 1]$. In order to determine the truth value of a compound expression with respect to an interpretation I, evaluation functions for the logical connectives have to be fixed. For the sake of simplicity, we assume that $\rightarrow$ is the Lukasiewicz implication, $\wedge$ a continuous t-norm, and $\vee$ a continuous t-conorm. For the truth value of a compound expression φ with respect to an interpretation I, we write $I^*(\varphi)$ again.

Definition 3.46 *An expression of the form*

$$\varphi \rightarrow p,$$

where $p \in \mathcal{P}$ holds and φ is a logical expression, which is composed of atoms and connectives $\wedge$ and $\vee$, is called an **implication expression**. *$\mathcal{I}$ denotes the set of all implication expressions .*

To be able to prove properties of correctness and completeness, we mainly consider implication expressions and atoms.

Definition 3.47 *A* **fuzzy knowledge base** w *is a mapping*

$$w : \mathcal{I} \cup \mathcal{P} \rightarrow [0, 1].$$

A fuzzy knowledge base w has to be viewed as a lower bound for the mapping $I^* : \mathcal{L} \rightarrow [0, 1]$ induced by an interpretation I, i.e. w expresses that to any formula $\varphi \in \mathcal{I} \cup \mathcal{P}$ a truth value greater than or equal to $w(\varphi)$ should be assigned.

An ordinary knowledge base $W \subseteq \mathcal{I} \cup \mathcal{P}$ can be seen as a fuzzy knowledge base

$$w : \mathcal{I} \cup \mathcal{P} \rightarrow [0, 1], \qquad \varphi \mapsto \begin{cases} 1, & \text{if } \varphi \in W \\ 0, & \text{otherwise.} \end{cases}$$

Definition 3.48 *Let w be a fuzzy knowledge base and $I : \mathcal{P} \rightarrow [0, 1]$ an interpretation. I is* **compatible** *with w, if for all $\varphi \in \mathcal{I} \cup \mathcal{P}$: $w(\varphi) \leq I^*(\varphi)$.*

Definition 3.49 *Let w be a fuzzy knowledge base.*
The mapping $\text{Th}_w : \mathcal{L} \rightarrow [0, 1]$ is defined as

$$\text{Th}_w(\varphi) \stackrel{\text{def}}{=} \inf\{I^*(\varphi) \mid I \text{ is an interpretation compatible with } w\}.$$

Th_w corresponds to the fuzzy set of all expressions that can be deduced semantically from w. Definition 3.49 corresponds to the following definition of classical logic:

$$\text{Th}_W^{(\text{classical})} \overset{\text{def}}{=} \{\varphi \mid \text{For all interpretations } I,$$
$$\text{for which the equation } I^*(\psi) = \text{true}$$
$$\text{is satisfied for all } \psi \in W, \text{ then } I^*(\varphi) = \text{true}\}.$$

$\text{Th}_W^{(\text{classical})}$ denotes the set of all expressions that can be deduced semantically from the (ordinary) knowledge base W.

By this we have defined semantics for our truth-functional logic. To specify a syntax we now generalize the *modus ponens*. Consider a fuzzy knowledge base w, an implication expression $\varphi \to p$, and an interpretation I compatible with w. We define the interpretation I_w, where $I_w(q) = w(q)$ ($q \in \mathcal{P}$). Since I is compatible with w, we have $I_w \leq I$. Since t-norms and t-conorms are monotonic, we obtain $I_w^*(\varphi) \leq I^*(\varphi)$, and since I is compatible with w, it follows

$$w(\varphi \to p) \leq I^*(\varphi \to p)$$
$$= \min\{1 - I^*(\varphi) + I(p), 1\}.$$

From this we conclude

$$w(\varphi \to p) \leq 1 - I^*(\varphi) + I(p),$$

and hence

$$w(\varphi \to p) + I_w^*(\varphi) - 1 \leq w(\varphi \to p) + I^*(\varphi) - 1 \leq I(p),$$

so that from w the new lower bound $w(\varphi \to p) + I_w^*(\varphi) - 1$ can be deduced for the atom p. The next definition formalizes this proof procedure.

Definition 3.50 *Let w and w' be fuzzy knowledge bases.*

*(i) w' can be **deduced directly from** w, if there is an implication $\varphi \to p$, so that*

(a) $w'(p) \leq \max\{w(\varphi \to p) + I_w^*(\varphi) - 1, w(p)\}$,
(b) for all $\psi \in (\mathcal{I} \cup \mathcal{P}) \backslash \{p\}$ it follows $w'(\psi) \leq w(\psi)$.

*(ii) w' can be **deduced from** w, if there is a sequence $w_0, w_1, \ldots, w_n$ of fuzzy knowledge bases with*

(a) $w_0 = w$,
(b) $w_n = w'$,
(c) w_{i+1} can be deduced directly from w_i.

Definition 3.51 *Let w be a fuzzy knowledge base. The mapping* $\text{th}_w : \mathcal{P} \to [0,1]$ *is given as*

$$\text{th}_w(p) \stackrel{\text{def}}{=} \sup\{w'(p) \mid w' \text{ can be deduced from } w\}.$$

th_w is the fuzzy set of all atoms syntactically provable from w.

Theorem 3.52 (correctness) *Let w be a fuzzy knowledge base and $p \in \mathcal{P}$. Then*

$$\text{th}_w(p) \leq \text{Th}_w(p).$$

Proof:
Direct deducibility was defined in such a way that it preserves compatibility, i.e. if the interpretation I is compatible with w, then I is also compatible with the fuzzy knowledge base w' that can be deduced directly from w. By means of complete induction we obtain that an interpretation I, which is compatible with w, is also compatible with any fuzzy knowledge base that can be deduced from w. Hence for all interpretations I involved in the infimum expression in Definition 3.49, and for all fuzzy knowledge bases involved in determining the supremum in Definition 3.51, the inequality $w'(p) \leq I(p)$ holds, from which the assertion of the theorem follows.   □

Theorem 3.53 (completeness) *Let w be a fuzzy knowledge base and $p \in \mathcal{P}$. Then*

$$\text{th}_w(p) \geq \text{Th}_w(p).$$

Proof:
We define $I_0 \stackrel{\text{def}}{=} \text{th}_w$. First we show that the interpretation I_0 is compatible with w. According to the definition of th_w, the inequality $w(q) \leq I_0(q)$ holds for any atom $q \in \mathcal{P}$. Let $\varphi \to q$ be an implication expression. We assume that

$$w(\varphi \to q) > I_0^*(\varphi \to q) = I_0(q) - I_0^*(\varphi) + 1$$

holds, i.e.

$$w(\varphi \to q) + I_0^*(\varphi) - 1 > I_0(q).$$

$p_{i_1}, \ldots, p_{i_n}$ are the atoms appearing in φ. As the t-norm and t-conorm associated with $\wedge$ and $\vee$, respectively, are monotonic and continuous, there are $\varepsilon_1, \ldots, \varepsilon_n > 0$ with the property (3.23). We define the interpretation I_1 by means of

$$I_1(r) \stackrel{\text{def}}{=} \begin{cases} I_0(r) - \varepsilon_j, & \text{if } r = p_{i_j} \\ 0, & \text{otherwise.} \end{cases}$$

Then I_1 satisfies the condition

$$w(\varphi \to q) + I_1^*(\varphi) - 1 > I_0(q). \tag{3.23}$$

We specify the fuzzy knowledge base w' by

$$w'(\psi) = \begin{cases} I_1(\psi), & \text{if } \psi \in \mathcal{P} \\ w(\psi), & \text{if } \psi \in \mathcal{I}. \end{cases}$$

For any of the atoms p_{ij} ($j = 1, \ldots, n$) there is a fuzzy knowledge base w_j that can be deduced from w with $w_j(p_{i_j}) > w'(p_{i_j})$, since $w'(p_{i_j}) < \text{th}_w(p_{i_j})$ holds. Hence w' can also be deduced from w. But from w' we can directly derive the fuzzy knowledge base w'' with

$$w''(\psi) = \begin{cases} w(\varphi \to q) + I_{w'}^*(\varphi) - 1, & \text{if } \psi = q \\ w'(\psi), & \text{otherwise.} \end{cases}$$

This implies the contradiction

$$\begin{aligned} I_0(q) \quad &< \quad w(\varphi \to q) + I_1^*(\varphi) - 1 \\ &= \quad w''(q) \\ &\leq \quad \text{th}_w(q). \end{aligned}$$

Hence I_0 is compatible with w, and it follows for all $p \in \mathcal{P}$ that

$$\begin{aligned} \text{th}_w(p) \quad &= \quad I_0(p) \\ &\geq \quad \inf\{I(p) \mid I \text{ is compatible with } w\} \\ &= \quad \text{Th}_w(p). \qquad \qquad \square \end{aligned}$$

3.6   Supplementary Remarks and References

In the following we add some remarks on general methods of approximate reasoning. We discuss interpretations of possibility distributions and possibilistic inference rules, examine fuzzy measures, and consider extensions of logic-based mechanisms of inference. Finally we briefly present a software tool that can carry out possibilistic reasoning in multi-dimensional spaces of hypotheses. Just as in Chapter 2, these supplementary remarks are accompanied by a large number of important references to literature.

3.6.1   Historical Development: Approximate Reasoning

Since concepts of mathematical logics are of major importance in the field of artificial intelligence, the problem of representing and propagating uncertain expert knowledge was first tackled by assigning numerical weights to symbolic expressions (e.g. logical formulae). The approach which is known best consists in assigning so-called certainty factors [Shortliffe75] to facts and rules. During the process of problem solving these factors are combined according to the reasoning process which refers to an appropriate application of the corresponding rules and facts. As the employed algorithms are heuristic and quite simple, time complexity problems can be avoided. The expert system MYCIN [Buchanan84] showed impressively that such a heuristic procedure can be quite reasonable. Extensions and improvements of this approach were incorporated in the inference systems RUM [Bonissone87], FLOPS [Buckley86], and MILORD [Godo89]. But a problem, which is common to all inference systems similar to MYCIN, is that the combination of uncertain conclusions they use (parallel propagation) can lead to incorrect results [Heckerman88a]. This weakness is related to the non-existence of modularity in knowledge-based systems [Heckerman88b]. A main difficulty lies in stating exact semantics of the numbers assigned to the logical formulae. The necessity of distinguishing between degrees of truth and degrees of uncertainty, as we did in Section 3.5, has been emphasized for some time [Dubois91a].

In contrast to logic-based inference mechanisms, which entered into the AI literature very easily, the different kinds of numerical calculi dealing with uncertain knowledge met with only little attention first (cf. e.g. [Kruse91a]). It is not surprising that probabilistic calculi prevailed [Pearl88, Neapolitan90] at the beginning, since they have a clean mathematical foundation. Other methods were regarded as being superfluous [Cheeseman86]. But in actual applications fuzzy sets and possibilistic reasoning have been applied very successfully, such that in research a lively dialogue takes place between the different scholars of uncertainty management.

The first approaches to generalize a deductive process with the help of a *modus ponens* that was generalized to fuzzy sets are due to L.A. Zadeh [Zadeh73]. In [Zadeh79, Zadeh83b] the mechanism is explained in detail. But underlying concepts can be illustrated much better by using possibility distributions to represent imperfect knowledge. For an extensive coverage of fuzzy logic and related topics see [Zadeh92]. Some applications of fuzzy expert systems are listed in [Kandel91].

The notion of a (possibilistic) F-expert system, which we used in this chapter, refers to a special class of knowledge-based systems for approximate reasoning, that are restricted to focusing a given knowledge base according to available pieces of evidence. There are also other operations that refer to the dynamic behaviour of the knowledge base and thus can possibly modify its structure. These operations are well-known as *updating* and *belief revision*. To gain a deeper understanding of the semantics of these and related concepts we recommend [Gärdenfors88, Gärdenfors92] and [Katsuno91]. The methods considered there are partly also applicable for possibilistic reasoning

[Dubois92b].

Structuring qualitative dependencies within a knowledge base with the help of hypergraphs is widely used in several fields of information science. An exhaustive presentation of general foundations can be found in [Berge73] and [Golumbic80]. Hypergraphs as an important concept for the theory of relational databases are treated, for instance, in [Maier83]. Recent approaches to the management of uncertainty and vagueness in database systems by applying possibility theory are described in [Bosc92, Bosc93]. In artificial intelligence, hypertrees are of major interest for *Bayesian networks*, which are comparable to F-expert systems with the exception that they use probabilistic instead of possibilistic data and an explicit conditioning concept [Pearl86, Lauritzen88, Andreassen87].

Knowledge-based systems that employ a decision-theoretic model as a framework for knowledge representation and inference mechanisms are usually called *normative expert systems*. These models are based on the concept of an *influence diagram* [Shachter88]. In this chapter we have presented only such influence diagrams that do not possess any decision vertices or utility vertices and that are known by the names *belief networks* [Pearl88] and *knowledge maps* [Howard81], respectively.

An important topic of research in this field is qualitative and quantitative learning from databases of samples, which in our case refers to inducing the dependency hypergraph and the (possibilistic) rule base, respectively. Learning causal probabilistic networks is, for example, discussed in [Cooper92].

3.6.2   Possibility Distributions

While fuzzy set theory is strongly related to the consideration of vague concepts, an application of possibility theory refers to the imperfect characterization of an existing element ω_0 of a reference set Ω. In [Zadeh78a] it is suggested that uncertainty about the location of ω_0 might be modelled with the help of the possibility measure $\Pi : \mathfrak{P}(\Omega) \rightarrow [0,1]$, $\Pi(A) = \sup\{\mu(\omega) \mid \omega \in A\}$, if a fuzzy set $\mu : \Omega \rightarrow [0,1]$ is given as a sole characterization of ω_0. This basic idea was refined in the course of time, see e.g. [Higashi83, Delgado87, Zadeh81, Giles82, Nguyen79, Yager83, Yager80b], and finally established in [Dubois87b] as an axiomatic approach, in which possibility distributions and possibility measures are analysed without underlying fuzzy sets. In [Yager92a] the concept of the specificity of a possibility distribution is discussed in detail. Thus possibility theory turns out to be suitable to handle such uncertain knowledge which rests on context-dependent qualitative preferences. Precursors of possibilistic reasoning are the economist G.L.S. Shackle [Shackle61] who was interested in a theory of 'degrees of potential surprise', and the philosophers L.J. Cohen [Cohen73] and N. Rescher [Rescher76] who studied grades of inductive support and plausibility degrees, respectively.

In addition to these approaches the semantic background of possibility theory has been presented in quite different ways. Some applications consider possibility measures as an upper envelope of a set of probabilities: If Π is a possibility measure, then the upper probability $P^* : \mathfrak{P}(\Omega) \to [0,1]$, $P^*(A) = \sup\{P(A) \mid P \in P_\Pi\}$, induced by the set of probabilities $P_\Pi = \{P \mid \forall A \subseteq \Omega : P(A) \leq \Pi(A)\}$, coincides with Π, i.e. we have $P^* = \Pi$ [Dubois88].

For statisticians there are also some obvious relationships to likelihood functions [Loginov66, Stallings77]. A corresponding justification of possibility theory has been given in [Dubois93]. Furthermore, an additional reference to be mentioned is that of Spohn's theory of epistemic states [Spohn88, Spohn90], which has turned out to support one more semantics of possibility theory [Dubois93b].

The concept of a possibility distribution is based on the epistemic interpretation of fuzzy sets and can therefore be related to the interpretation of vague concepts as contour functions of random sets, presented in Section 2.6.1. This approach (see [Nguyen78b, Hestir91, Kampé de Fériet82, Goodman82, Wang83]) has the advantage of clear semantics and is strong enough to motivate some reasoning methods that are essential in practice. We did not examine more refined semantics, based on earlier approaches of [Strassen64, Dempster67, Dempster68, Kampé de Fériet82], and which can justify operations on possibility distributions, the principle of minimum specificity, decision making with the help of possibility measures and other uncertainty measures as well as the special role of the Gödel relation for the interpretation of possibilistic inference rules much more clearly. For more details we refer to [Gebhardt93a, Gebhardt93b, Gebhardt93c].

A central subject of research in possibilistic reasoning is the representation of if-then-rules, since similar formalisms are needed in the domain of fuzzy control as well. From this point of view the main aim is to model a rule of the kind

$$\text{if } \xi \text{ is } \mu_A \text{ then } \eta \text{ is } \mu_B$$

by a possibility distribution

$$\pi(x, y) = I\left(\mu_A(x), \mu_B(y)\right),$$

where I denotes an appropriate implication of a multi-valued logic. There are three classes of well-examined operators analysed in [Trillas85, Dubois84, Weber83] (cf. Section 2.7).

- *S-implications* are implications of the kind

$$I(a, b) = \bot\left(n(a), b\right),$$

where $\perp$ is a t-conorm (for multi-valued disjunction) and n is a strictly decreasing involution $n : [0, 1] \to [0, 1]$ with $n(0) = 1$ (a strict negation). The representatives known best are the *Kleene–Dienes implication* $I_K(a, b) = \max(1 - a, b)$ with $\perp'_K = \max$, the *Reichenbach implication* $I_R(a, b) = 1 - a + ab$ with $\perp_R(a, b) = a + b - ab$, and the *Łukasiewicz implication* $I_L(a, b) = \min(1, 1 - a + b)$ with $\perp_L(a, b) = \min(1, a + b)$. All of these implications are based on reading an implication $p \to q$ as $\neg p \lor q$.

- *R-implications* on the other hand reflect a partial ordering on the set of propositions: $I(a, b) = 1$ is chosen if and only if $a \le b$. R-implications are based on a residual concept (cf. Definition 2.62) in lattices and are defined by

$$I(a, b) = \sup \{c \in [0, 1] \mid \top(a, c) \le b\},$$

where $\top$ is a t-norm.

For $\top = \min$, the *Gödel implication* results, which is $I_G(a, b) = 1$ for $a \le b$, and $I_G(a, b) = b$ for $a > b$. For $\top_L = \max(0, a + b - 1)$, Łukasiewicz implication $I_L(a, b)$ is obtained again.

- So-called *QL-implications* are of the form

$$I(a, b) = \perp (n(a), \top(a, b)),$$

where $\perp$ is a t-conorm, n is a strict negation, and $\top$ is defined by $\top(a, b) = n(\perp(n(a), n(b)))$. *Zadeh's implication* $I_Z(a, b) = \max(1 - a, \min(a, b))$, for instance, emerges from $\perp = \max$ and $n(a) = 1 - a$. In [Mizumoto85, Dubois91a] and [Bandler80] properties of fuzzy implications are analysed. Note that the name QL-implication is somewhat misleading, since strictly speaking it is no implication in the sense of classical logic.

Another approach to interpret a rule of the kind

$$\textbf{if } \xi \textbf{ is } \mu_A \textbf{ then } \eta \textbf{ is } \mu_B$$

consists in defining the possibility distribution π, which is chosen as its interpretation, by restrictions that can be justified semantically. Thus the rules mentioned above can be regarded, for example, as gradual rules 'The more μ_A, the more μ_B' such that π is restricted by the inequalities

$$\min (\mu_A(x), \pi(x, y)) \le \mu_B(y)$$

for all $(x, y) \in X \times Y$. In this case the principle of minimum specificity yields the Gödel implication I_G for π [Dubois92b]. Still another approach arises from deriving possibility measures in an axiomatic way, using the generalized *modus ponens* for fuzzy sets [Fukami80]. Again only the Gödel implication I_G yields satisfactory results [Dubois85]. The fundamental role of I_G is also confirmed in Section 3.3 and, for example, in [Gebhardt92a] and [Gebhardt93c], which go into further detail with respect to semantical foundations.

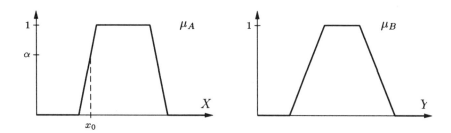

Figure 19   Two possibility distributions

If the possibility distribution employed for an interpretation of the mentioned if-then-rules is described by $\pi_I(x, y) = I\left(\mu_A(x), \mu_B(y)\right)$, where μ_A and μ_B are the possibility distributions sketched in Figure 19, if in addition I stands for the implications I_G, I_K, $I_{\min}$ with $I_{\min}(a, b) = \min(a, b)$, and I_L, and if finally the possibilistic inference mechanism is carried out for the crisp input x_0 with $\mu_A(x_0) = \alpha$ (x_0 is described by the possibility distribution π_{x_0} with $\pi_{x_0}(x_0) = 1$ and $\pi_{x_0}(x) = 0$ for $x \neq x_0$), then we obtain the projections shown in Figure 20 (possibility distributions for the output value η).

In the general case of a propagation in multi-dimensional spaces the ideas put forth in [Shenoy90] for purely probabilistic approaches and Dempster–Shafer theory can be transferred to possibility theory such that local propagation algorithms can be achieved [Fonck92]. An idea is to define a measure of possibility on the product space by means of *conditional* measures of possibility on the subspaces specified by the hyperedges of a dependency hypergraph. For the probabilistic case this is carried out in [Lauritzen88]. If Π is a measure of possibility on X, then using the relation

$$\Pi(A \cap B) = \Pi(B|A) * \Pi(A) \tag{3.24}$$

for all $A, B \subseteq X$ with $\Pi(A) \neq 0$, we obtain a restriction for the definition of a conditional possibility measure $\Pi(\cdot|\cdot)$, where $*$ is an operator that has to be chosen in an appropriate way. The Cox axioms for conditional probabilities [Cox46] are based on an analogous presupposition and result in the choice of the product for $*$. With respect to possibility measures, *min* turns out to be suitable for $*$ [Dubois87a]. The least-specific conditional possibility measure that satisfies the condition for (3.24) $* = \min$, assuming that $\Pi(A) \neq 0$, is

$$\Pi(B|A) = \begin{cases} 1, & \text{if } A \neq \emptyset \wedge \Pi(A) = \Pi(A \cap B) \\ \Pi(A \cap B), & \text{if } \Pi(A \cap B) < \Pi(A) \\ 0, & \text{if } A \cap B = \emptyset. \end{cases}$$

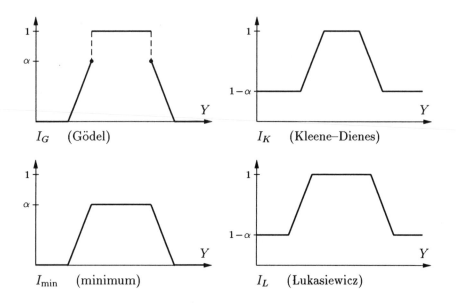

Figure 20   Resulting possibility distributions after applying the possibilistic inference mechanism to the crisp input value x_0

Criticism on these approaches, but also alternative suggestions can be found in [Nguyen78a, Ramer89, Dubois91d, Bouchon87].

The existing controversy between the application domains of possibility and (subjective) probability [Cheeseman86], especially the Bayesian approach, comes from the fact that both calculi aim to model comparable aspects of uncertainty handling. There are, of course, several relationships between the two calculi, but on the other hand one can also find some interpretations of possibility distributions that owe nothing to probability theory. In this connection we mention possibility as *similarity*, where the interpretation is clearly related to metric spaces as already pointed out in section 2.6.2 [Ruspini90], and possibility as *preference*, mathematically justified by comparative possibility relations [Dubois93a].

3.6.3 Fuzzy Measures

An element ω_0 of a reference set Ω can also be described by stating all subsets A of Ω that contain ω_0. Mathematically speaking this can be reflected by the function

$$g_{\omega_0} : \mathfrak{P}(\Omega) \quad \rightarrow \quad \{0,1\},$$
$$g_{\omega_0}(A) = 1 \quad \Longleftrightarrow \quad \omega_0 \in A.$$

In many practical cases, however, ω_0 is not known exactly. Probability theory, for example, deals with those cases where ω_0 is the outcome of a random experiment. But at least the probability that the unknown outcome ω_0 lies in a given set A can be stated and thus the knowledge about the uncertainty of the experiment can be characterized by the probability measure P. P has the properties $P(\emptyset) = 0$, $P(\Omega) = 1$, and σ-additivity, i.e. if $A_1, A_2, \ldots$ is a system of disjunct subsets of Ω, then $P\left(\bigcup_{i=1}^{\infty} A_i \right) = \sum_{i=1}^{\infty} P(A_i)$ holds. The book [Fine73] sheds some light on several mathematical and interpretative aspects of probability theory, in [Halmos50] the basic result in measure theory, which is based on σ-additive measures, is described. Set functions are also used in other uncertainty calculi [Kruse91a, Kruse91b].

Choosing a suitable set function depends much on the considered semantics. Nearly all types of semantics have the following common minimal demands with respect to set functions: a fuzzy measure $g : \mathfrak{P}(\Omega) \rightarrow [0,1]$ is a set function, which satisfies $g(\emptyset) = 0$, $g(\Omega) = 1$, and $A \subseteq B \Longrightarrow g(A) \leq g(B)$ for all $A, B \subseteq \Omega$. For infinite reference sets Ω continuity is demanded additionally: if $\mathcal{A} \subseteq \mathfrak{P}(\Omega)$ is a system of sets with the property that any monotonic sequence $A_1 \subseteq A_2 \subseteq \ldots$ or $A_1 \supseteq A_2 \supseteq \ldots$ of sets $A_i \in \mathcal{A}$ converges to an element of $\mathcal{A}$, then for each of these set sequences

$$\lim_{n \to \infty} g(A_n) = g\left(\lim_{n \to \infty} A_n \right).$$

In the infinite case, $\mathcal{A}$ is usually chosen to be an appropriate σ-algebra.

For applications, the so-called λ-*additive fuzzy measures* are of some importance, for which the property

$$g_\lambda(A \cup B) = g_\lambda(A) + g_\lambda(B) + \lambda g_\lambda(A)g_\lambda(B)$$

holds for all $A, B \in \mathcal{A}$ with $A \cap B = \emptyset$ [Sugeno74, Kruse82a, Kruse82b]. In [Grabisch92, Lamata89] more recent results in this field of research can be found. The monograph [Wang92] is dedicated completely to fuzzy measures. Fuzzy measures on a generalized σ-algebra, a *fuzzy algebra*, are discussed in [Klement80, Klement82, Butnariu93].

In the following we consider some important examples of fuzzy measures, restricting ourselves to finite reference sets Ω.

A *measure of belief* [Shafer76, Smets81] is a set function

$$\text{Bel} : \mathfrak{P}(\Omega) \to [0, 1],$$

that satisfies the properties of fuzzy measures and the additional axiom

$$\text{Bel}\left(\bigcup_{i=1}^{n} A_i\right) \geq \sum_{I : \emptyset \neq I \subseteq \{1,\ldots,n\}} (-1)^{|I|+1} \, \text{Bel}\left(\bigcap_{i \in I} A_i\right)$$

for all $n \in \mathbb{N}$ and any choice $A_1, \ldots, A_n$ of subsets of Ω.

Analogously, a *plausibility measure* is a set function

$$\text{Pl} : \mathfrak{P}(\Omega) \to [0, 1],$$

which is a fuzzy measure and in addition satisfies the axiom

$$\text{Pl}\left(\bigcap_{i=1}^{n} A_i\right) \leq \sum_{I : \emptyset \neq I \subseteq \{1,\ldots,n\}} (-1)^{|I|+1} \, \text{Pl}\left(\bigcup_{i \in I} A_i\right).$$

For an arbitrary $A \subseteq \Omega$, $\text{Bel}(A) = 1 - \text{Pl}(\Omega \backslash A)$.

Measures of belief and measures of plausibility have also been investigated in the theory of random sets [Stoyan87, Matheron75]. If $(X, \mathfrak{S}, P)$ is a probability space, and $\Gamma : X \to \mathfrak{P}(\Omega)$ is a set-valued mapping, then

$$m_\Gamma : \mathfrak{P}(\Omega) \quad \to \quad [0, 1],$$
$$m_\Gamma(A) \qquad \overset{\text{def}}{=} \quad P(\{x \in X \mid \Gamma(x) = A\})$$

is called the *induced mass distribution* on Ω. Given a fixed set $B \subseteq \Omega$, one wants to know how probable it is that the non-empty sets $\Gamma(x)$, induced by Γ, are certainly contained in B (belief), or have a non-empty intersection with B (plausibility). From this it follows that

$$\text{Bel}_\Gamma(B) = P\left(\{x \in X \mid \emptyset \neq \Gamma(x) \subseteq B\}\right) = \sum_{A : A \subseteq B} m_\Gamma(A)$$

and

$$\mathrm{Pl}_\Gamma(B) = P\left(\{x \in X \mid \Gamma(x) \cap B \neq \emptyset\}\right) = \sum_{A:A\cap B\neq\emptyset} m_\Gamma(A).$$

The set functions $\mathrm{Bel}_\Gamma : \mathfrak{P}(\Omega) \to [0,1]$ and $\mathrm{Pl}_\Gamma : \mathfrak{P}(\Omega) \to [0,1]$ are also called *belief functions* and *plausibility functions*, respectively. They can serve as measures of belief and measures of plausibility, respectively [Dempster67]. G. Shafer suggested in [Shafer76] the development of a theory of belief functions that completely dispenses with a consideration of the underlying random sets. Advances in the Dempster–Shafer theory of evidence are considered in [Yager94].

A special type of random sets is formed by those sets for which, for any two sets $\Gamma(x)$ and $\Gamma(x')$, one of the relations $\Gamma(x) \subseteq \Gamma(x')$ or $\Gamma(x) \supseteq \Gamma(x')$ holds. These random sets are called *consonant sets*. The corresponding measure of belief Bel_Γ satisfies the conditions $\mathrm{Bel}_\Gamma(\emptyset) = 0$, $\mathrm{Bel}_\Gamma(\Omega) = 1$, and $\mathrm{Bel}_\Gamma(A \cap B) = \min\left(\mathrm{Bel}_\Gamma(A), \mathrm{Bel}_\Gamma(B)\right)$ for arbitrary $A, B \subseteq \mathfrak{P}(\Omega)$. For the induced measure of plausibility Pl_Γ, then analogously $\mathrm{Pl}_\Gamma(\emptyset) = 0$, $\mathrm{Pl}_\Gamma(\Omega) = 1$, and $\mathrm{Pl}_\Gamma(A \cup B) = \max\left(\mathrm{Pl}_\Gamma(A), \mathrm{Pl}_\Gamma(B)\right)$ for all $A, B \subseteq \mathfrak{P}(\Omega)$. It is obvious that each of these measures of plausibility Pl_Γ is determined uniquely by the possibility distribution $\pi_\Gamma : \Omega \to [0,1]$ and the two formulae $\pi_\Gamma(\omega) = \mathrm{Pl}_\Gamma(\{\omega\})$ and $\mathrm{Pl}_\Gamma(A) = \max_{\omega \in A}\{\pi_\Gamma(\omega)\}$.

The theory of random sets can be examined also for infinite reference sets Ω [Matheron75], though its application is more difficult with respect to possibility distributions, since measures of possibility do not necessarily satisfy the axioms for continuity of fuzzy measures [Puri82].

3.6.4   Extensions of Logic-based Mechanisms of Inference

When considering logic-based mechanisms of inference, in Section 3.5, for the sake of simplicity we restricted ourselves to propositional calculi. The extension of possibilistic logic to a first-order predicate logic can be carried out retaining the results concerning correctness and completeness presented in Section 3.5.1 [Dubois87a, Dubois89a, Dubois91a]. The same is true for the logic discussed in Section 3.5.2, even if the Lukasiewicz implication is replaced by the Gödel implication [Klawonn92c].

Within the framework of possibilistic logic, the unit interval is often regarded only as an ordinal scale [Dubois91c]. But the model introduced in Section 2.6.1 and tailored to set-theoretic approaches can be transferred to logical calculi such that the numbers used in possibilistic logic have an interpretation in terms of measure theory [Klawonn92d].

Besides the approaches mentioned, other logic-based inference mechanisms in connection with truth values or degrees of uncertainty taken from the unit interval were presented in [Lee71, Lee72, Mukaidono82, Yager85, Orci85,

Orci89]. The common basis of these approaches is the resolution principle, so that generalizations of logic programming [Lloyd87] and especially of Prolog systems can be designed [Clocksin91].

The first programming language for the implementation of fuzzy inference, called FUZZY, is due to Le Faivre [Le Faivre74]. A version that incorporates fuzzy truth values was described in [Freksa81]. Later on numerous fuzzy Prolog systems were presented. Among these are Fuzzy-Prolog-ELF [Ishizuka85], FPROLOG [Martin87], and Fuzzy Prolog [Mukaidono89] Some versions deal with linguistic truth values like 'it is more or less true', that are described by fuzzy sets on the interval [0, 1] within a logical programming environment [Umano87]. All of these approaches, however, lack a clean distinction between truth values and degrees of uncertainty. A more detailed overview of logic-based fuzzy systems is given in [Dubois91a].

The methodology of integrating truth values into the formal language of a logical calculus and of modifying the mechanisms of deduction accordingly by adding suitable axioms, as suggested in the fuzzy logic of Pavelka [Pavelka79] and its extensions [Novák90a, Novák90b, Novák92], is also known in the field of probabilistic logic [Nilsson86, Bacchus90]. However, in these probabilistic calculi the numbers from the unit interval are strictly interpreted in the sense of probability theory as subjective probabilities or as relative frequencies.

3.6.5 POSSINFER – An Implementation

In Sections 3.2, 3.3, and 3.4 we discussed in detail the theoretical foundations of a special class of knowledge-based systems, which we called possibilistic F-expert systems. The methods presented there have been applied in a cooperation between the Computer Science Department of the University of Braunschweig and German Aerospace. This cooperation also includes the conception and implementation of a prototype version of the software tool POSSINFER (POSSibilistic INFERence), that was implemented in the programming language C on SUN workstations under SUN-UNIX with the help of X-Windows and OSF/Motif.

Since POSSINFER is based on the concepts introduced in this chapter, we need not enter into theoretical background here, but can restrict ourselves to some remarks concerning the implementation. A POSSINFER session consists of the four phases mentioned in Section 3.2:

(1) Fixing a structure of representation for the expert knowledge by laying down n attributes and a related n-dimensional universe $\mathcal{U}$.

(2) Constructing a knowledge base by first providing a hypergraph and, by this, a modularization $\mathcal{M}$ to declare the existing qualitative dependencies between the involved attributes, and then quantifying these general dependencies by means of a system of possibilistic relations, i.e. the

definition of a rule base $\mathcal{R}(\mathcal{U}, \mathcal{M})$.

(3) Application-dependent choice of a total evidence by specifying an evidence system $\mathcal{E}(\mathcal{U}, \mathcal{N})$, which is based on a special partition $\mathcal{N} = \{\{1\}, \ldots, \{n\}\}$ compatible with $\mathcal{M}$.

(4) Execution of the inference mechanism by starting the propagation algorithm, which determines the i-th restrictions $\kappa^{(i)}$ of the state $\sigma(\mathcal{X}, \mathcal{E}(\mathcal{U}, \mathcal{N}))$ for $\mathcal{X} = (\mathcal{U}, \mathcal{M}, \mathcal{N}, \mathcal{R}(\mathcal{U}, \mathcal{M}))$.

Considering phase (1), it should be noted that the domains $\Omega^{(i)}$, which can be associated with the attributes A_i, correspond to self-defined enumeration types or the standard datatypes *real* and *int*, which are known from imperative programming languages. In addition POSSINFER supports the declaration of a finite set of names and corresponding (parameterized) possibility distributions with respect to particular attributes. They are useful for making the specification of rule bases and evidence systems in phases (2) and (3) more expressive.

Due to practical considerations the possibility distributions must not be chosen arbitrarily. The underlying domains $\Omega^{(i)}$ are normally finite (if they are not *per definitionem*, then certainly because of the approximated internal representation of their elements). However, if the cardinality of $\Omega^{(i)}$ is high (especially if the data type is *real* or *int*), it is recommended to refer operations to the horizontal representation of possibility distributions based on their α-cuts, which means using possibility distributions of the kind $F_{D_k}(\mathbf{R})$ as is done in the software tool SOLD, which was described in Section 2.8.5. But even for a small $k \in \mathbf{N}$, the operations to be carried out on the α-cuts raise problems of efficiency, since unions of multi-dimensional intervals have to be managed with respect to relations $R \in \mathcal{R}(\mathcal{U}, \mathcal{M})$.

Instead of partitioning the set $[0, 1]$ of possibility degrees induced by $F_{D_k}(\mathbf{R})$, POSSINFER therefore also accepts a partition of the domains $\Omega^{(i)}$ into a finite number of intervals, where $\Omega^{(i)}$ contains only the representative values. This allows a simple execution of the mentioned operations (cylindrical extension, intersection, projection) by means of the vertical representation of possibility distributions as mappings from domains into the unit interval. Only the operators *min* and *max* are needed. Furthermore arrays may be employed instead of linked dynamic data structures, by which a costly heap management is rendered dispensable.

Building the knowledge base, which belongs to phase (2), can be started under X-Windows and OSF/Motif by constructing the hypergraph $H_{\mathcal{M}}$ on the screen (Figure 21). The characterization of the vertices of $H_{\mathcal{M}}$ as particular attributes belongs to phase (1). Specifying the rule base $\mathcal{R}(\mathcal{U}, \mathcal{M})$ is connected to characterizing the hyperedges. $\varrho^M \in \mathcal{R}(\mathcal{U}, \mathcal{M})$ can be defined for each

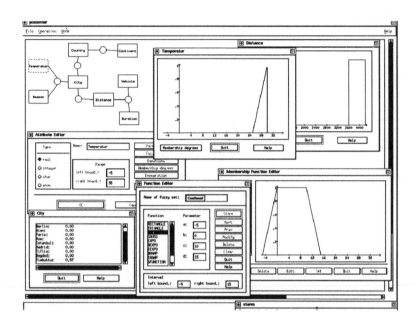

Figure 21   The system POSSINFER

$M \in \mathcal{M}$, either by assigning the possibility degrees for each element, or by fixing a conjunctive system of possibilistic inference rules, based on the possibility distributions declared in phase (1). These possibilistic inference rules are then interpreted with the help of the corresponding Gödel relation.

Whereas phases (1) and (2) define an application environment to be a possibilistic F-expert system $\mathcal{X} = (\mathcal{U}, \mathcal{M}, \mathcal{N}, \mathcal{R}(\mathcal{U}, \mathcal{M}))$ with $\mathcal{M} = \{\{1\}, \ldots, \{n\}\}$, which can be used for several POSSINFER sessions, phases (3) and (4) refer to possibilistic reasoning w.r.t. $\mathcal{X}$. In phase (3), an evidence system $\mathcal{E}(\mathcal{U}, \mathcal{N})$ is specified by stating for all attributes A_i a possibility distribution on $\Omega^{(i)}$, reflecting an imperfect characterization of the corresponding observed attribute value. Finally starting the propagation algorithm in phase (4), yields the i-th restrictions $\kappa^{(i)}$ of the state $\sigma(\mathcal{X}, \mathcal{E}(\mathcal{U}, \mathcal{N}))$ and their graphical presentation.

The evidence system and the possibility distributions calculated by the propagation can, of course, be made available via existing interfaces for a further processing in other programs. In reverse it is possible to read rule bases and evidence systems that have been created with other programs (using database systems if necessary).

3.6.6   Exercises

Exercise 3.1 A meteorologist observes the degree of clouding of the sky at the same place at two different points in time t_1, t_2. At time t_1 he calls it *fair*, at time t_2 *cloudy*. With the help of the possibility distributions $\pi_1, \pi_2 : [0, 100] \rightarrow [0, 1]$ shown in Figure 22, he represents the vague data that serve as an interpretation of these linguistic descriptions.

π_i, $i = 1, 2$, are imperfect characterizations of the actual degrees of clouding ω_i (in percent) at time t_i.

The meteorologist is interested in their average, that is $\Phi(\omega_1, \omega_2)$, defined as

$$\Phi : \mathbf{R} \times \mathbf{R} \rightarrow \mathbf{R}, \quad \Phi(x, y) = \frac{1}{2}(x + y).$$

a) Determine the resulting possibility distribution $\hat{\Phi}(\pi_1, \pi_2)$ by applying the extension principle.

b) Let $A_1 = [20, 30]$, $A_2 = [50, 60]$, and $A_3 = [30, 70]$ be imprecise characterizations of $\Phi(\omega_1, \omega_2)$. Calculate $\mathrm{Nec}_{\hat{\Phi}(\pi_1, \pi_2)}(A_i)$ and $\mathrm{Poss}_{\hat{\Phi}(\pi_1, \pi_2)}(A_i)$, $i = 1, 2, 3$. Give an interpretation of these values.

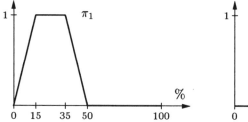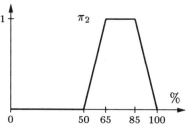

Figure 22   Possibility distributions of Exercise 3.1

Exercise 3.2 Let general knowledge about an existing imperfect relation between age and body-height of persons be given in the form of the two linguistic rules

$$L_1: \quad \textbf{if } age \textbf{ is } baby \textbf{ then } height \textbf{ is } very \; little,$$
$$L_2: \quad \textbf{if } age \textbf{ is } child \textbf{ then } height \textbf{ is } little,$$

which are interpreted by the following system of conjunctively combined possibilistic inference rules:

$$R_1: \quad \textbf{if } \xi^{\{1\}} \textbf{ is } \mu_1 \textbf{ then } \xi^{\{2\}} \textbf{ is } \nu_1,$$
$$R_2: \quad \textbf{if } \xi^{\{1\}} \textbf{ is } \mu_2 \textbf{ then } \xi^{\{2\}} \textbf{ is } \nu_2,$$

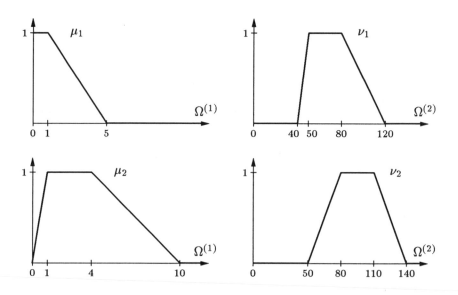

Figure 23   The possibility distributions μ_1, μ_2, ν_1, ν_2 of exercise 3.2

where $\mu_1, \mu_2 \in \text{Poss}(\Omega^{(1)})$, $\nu_1, \nu_2 \in \text{Poss}(\Omega^{(2)})$ with
$\Omega^{(1)} = \text{Dom}(age) \quad = [0, 150] \quad \text{(age in years)}$,
$\Omega^{(2)} = \text{Dom}(height) = [40, 250] \quad \text{(height in cm)}$,
$\Omega = \Omega^{(1)} \times \Omega^{(2)}$.

a) Determine the Gödel relations $\varrho_1 \equiv \varrho^{\text{Gödel}}[\mu_1, \nu_1]$ and $\varrho_2 \equiv \varrho^{\text{Gödel}}[\mu_2, \nu_2]$ induced by R_1 and R_2, respectively.

b) Patrick is a boy and between two and three years old. What can be inferred from the knowledge base $\varrho \equiv \min\{\varrho^{\text{Gödel}}[\mu_1, \nu_1], \varrho^{\text{Gödel}}[\mu_2, \nu_2]\}$ about his body-height?

To answer this question, determine the possibility distributions
$\pi_j \equiv \widehat{\text{infer}}_j(\mathbb{I}_{[2,3]}, \varrho_j)$, $j = 1, 2$, according to Section 3.3, and then calculate the resulting possibility distribution $\pi \equiv \min(\pi_1, \pi_2)$.

Exercise 3.3 Let $\mathcal{X} = (\mathcal{U}, \mathcal{M}, \mathcal{N}, \mathcal{R})$ be a possibilistic F-expert system with
$\mathcal{U} = (\Omega^{(i)})_{i=1}^{4}$,
$\Omega^{(1)} = \{a_1, b_1, c_1\}$, $\Omega^{(2)} = \{a_2, b_2\}$, $\Omega^{(3)} = \{a_3, b_3\}$, $\Omega^{(4)} = \{a_4, b_5\}$,
$\mathcal{M} = \{\{1, 2\}, \{2, 3\}, \{3, 4\}, \{1, 4\}\}$, and $\mathcal{N} = \{\{1\}, \{2\}, \{3\}, \{4\}\}$.
Let $\mathcal{R} = \{\varrho^{\{1,2\}}, \varrho^{\{2,3\}}, \varrho^{\{3,4\}}, \varrho^{\{1,4\}}\}$ be given as shown in Table 8.

x \ y	a_2	b_2
a_1	0.2	0.3
b_1	1	0.2
c_1	0.7	1

$\varrho^{\{1,2\}}(x,y)$

x \ y	a_3	b_3
a_2	0.3	1
b_2	1	0.5

$\varrho^{\{2,3\}}(x,y)$

x \ y	a_4	b_4
a_3	1	0.3
b_3	0.2	1

$\varrho^{\{3,4\}}(x,y)$

x \ y	a_4	b_4
a_1	0.8	1
b_1	0.1	0.6
c_1	1	0.7

$\varrho^{\{1,4\}}(x,y)$

Table 8 Representation of the rule base $\mathcal{R}$

a) $\mathcal{X}$ is called *non-redundant*, if

$$\forall M \in \mathcal{M} : \varrho^M = \text{proj}_M^N \left(\min_{M^* \in \mathcal{M}} \text{ext}_{M^*}^{N_*} \left(\varrho^{M^*} \right) \right)$$

is satisfied. Determine a non-redundant possibilistic F-expert system $\mathcal{X}' = (\mathcal{U}, \mathcal{M}, \mathcal{N}, \mathcal{R}')$ that is equivalent to $\mathcal{X}$.

b) Transform $\mathcal{X}$ into an equivalent possibilistic F-expert system $\mathcal{X}'' = (\mathcal{U}, \mathcal{M}''\mathcal{N}, \mathcal{R}'')$ with dependency hypertree $H_{\mathcal{M}''}$.

c) For $\mathcal{E}(\mathcal{U}, \mathcal{N}) = \{\varepsilon^{\{1\}}, \varepsilon^{\{2\}}, \varepsilon^{\{3\}}\}$ with $\varepsilon^{\{1\}} \equiv \mathbb{I}_{\Omega^{(1)}}$, $\varepsilon^{\{2\}} \equiv \mathbb{I}_{\Omega^{(2)}}$, $\varepsilon^{\{3\}} \equiv \mathbb{I}_{\Omega^{(3)}}$, and $\varepsilon^{\{4\}} : \Omega^{(4)} \to [0, 1]$, $\varepsilon^{\{4\}}(a_4) = 1$, $\varepsilon^{\{4\}}(b_4) = 0.3$, calculate the i-th restrictions $\kappa^{(i)}$ of $\sigma\left(\mathcal{X}, \mathcal{E}(\mathcal{U}, \mathcal{N})\right)$, $i = 1, 2, 3, 4$.

Exercise 3.4 Let the following uncertain propositions be given:

- If Tweety is a bird and if Tweety does not have an injured wing, then it can fly. (0.9)
- If Tweety is a penguin, then it is a bird. (1.0)
- If Tweety is a penguin, then it cannot fly. (1.0)
- Tweety is a bird. (0.7)
- Tweety's wings are not injured. (0.8)

The number in parentheses expresses how far the corresponding proposition is at least regarded to be necessarily true.

Translate these propositions into an appropriate uncertain knowledge base and determine the degree of inconsistency of this knowledge base with the help of possibilistic resolution.

4

Fuzzy Control

The biggest success of fuzzy systems in industrial and commercial applications has been achieved with *fuzzy controllers*. This chapter is devoted to issues in fuzzy control. However, special knowledge in control engineering is not required for the understanding of this chapter. Therefore it is suited for readers who are generally interested in the subject of fuzzy control as well as for readers who are working in the field of control engineering and are looking for a presentation of basic methods including a semantic foundation of fuzzy control. Fuzzy control is seen as a special kind of table-based control method.

After a brief description of the basic differences between classical control approaches, where a physical model of the process is developed, and fuzzy control, that tries to model a human expert, we introduce standard methods of fuzzy control. Here, we do not question the semantics of these concepts. Section 4.3 describes the single steps to be performed, and some of the problems that might be encountered, while developing and optimizing a fuzzy controller.

The topic of Section 4.4 is the development of a fuzzy controller with a proper semantic background, based on the idea of interpolation in the framework of equality relations. Here we will discover that this background may lead us to an intuitively appealing fuzzy controller with a formal foundation.

Section 4.5 deals with relational equations. They give us another fundamental approach to fuzzy control and are closely related to approximate reasoning.

We have chosen the 'cartpole' problem as an ongoing example — not for the special difficulties it presents to classical control engineering but because it is a well-suited, simple, and clear example of a non-linear process.

4.1   Knowledge-based versus Classical Models

We will not present fuzzy controllers from the viewpoint of control engineering, as this would unnecessarily exclude several readers who are not familiar with this field. The most popular methods of fuzzy control are introduced using a quite simple example of a control problem. We put emphasis on a proper foundation and motivation of the described concepts, as far as there is a corresponding semantic background. This does not mean that this chapter is uninteresting or even futile for a control engineer. The contrary is true, a profound understanding of the basic principles of fuzzy control should enable him to appreciate the use of fuzzy controllers in complex processes, to make profitable use of methods, and to question currently existing applications.

Fuzzy control is here seen as a way of defining non-linear table-based control systems, where the definition of the non-linear transition function can be made without the need to specify each entry of the table individually. The development can be viewed as a kind of knowledge-based interpolation technique. The analysis of the table-based controller can be performed by standard methods of control engineering.

Our simplified model can be described as follows. A technical system — for example, an electrical engine moving an elevator, or a heating installation — is considered. We want this system to behave in a certain way — perhaps the engine should maintain a certain number of revolutions per minute or the heating should guarantee a certain room temperature. All these systems have in common a time-dependent variable that should reach a desired value. We call this variable the *output variable*. In the first case the output variable is the number of revolutions, in the second the room temperature. The output is controlled by a *control variable* which can be adjusted. According to these notions, we have to control the output variable of the engine through the adjustment of the current, the output variable of the heating through the thermostat. In addition to the control variables there are *disturbance variables* that have an influence on the output variable. A disturbance variable for the engine driving an elevator can be its load, for the heating element it may be the outside temperature or the sunshine through a window.

In order to determine the actual value, control variable measurements of the current output variable ξ and the change of the output variable $\Delta\xi = \frac{d\xi}{dt}$ are taken. If the output variable is given in finite time intervals, we may set $\Delta\xi(t_{n+1}) = \xi(t_{n+1}) - \xi(t_n)$, so that it is not necessary to measure $\Delta\xi$.

Example 4.1 [cartpole problem] The task in the cartpole problem is to balance an upright standing pole through the appropriate movement of its foot point. The lower end of the pole can be moved unrestrained (to a certain extent) along a horizontal axis (see Figure 1). We have a mass m at the foot and a mass M at the head. The influence of the mass of the shaft itself is

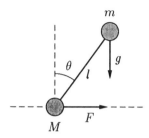

Figure 1 The cartpole problem

negligible. The task is to determine the force that is necessary to balance the pole standing upright. To determine this force we measure the angle θ of the pole in relation to the vertical axis and the change of the angle, i.e. the angular velocity $\dot{\theta} = \frac{d\theta}{dt}$; these are our output variables. The aim is that both take the values zero. Finally we have the force F as the control variable.

We will use the cartpole problem as an ongoing example to illustrate the notions and methods of fuzzy control that will be introduced. Thus, we will return repeatedly to this problem. At this point, we want to stress again that we have chosen this problem because it is a clear and vivid example and not because it is particularly difficult to solve with the methods of classical control engineering.   □

Control engineering, in general, is not restricted to the measurement of just the output value ξ and the change of the output value $\Delta\xi = \frac{d\xi}{dt}$. Sometimes it is useful to look at higher derivatives or further values. Regarding heating elements, for instance, nearly always includes measuring the outside temperature. Instead of computing the control value it is also possible to determine the change of the control value. This leads to the present control value as further input.

In the following we will refer only to the terms *input variable* $\xi_1, \ldots, \xi_n$ and *control variable* η. The measurements are used to determine an actual value of η. We also allow η to specify the change of the control variable. In addition we assume that the measured input ξ_i, $i = 1, \ldots, n$, is a value of the set X_i, and that the control variable is in the set Y. The solution of the control problem should be a *control function* $\varphi : X_1 \times \ldots \times X_n \to Y$ that assigns to each input tuple $(x_1, \ldots, x_n) \in X_1 \times \ldots \times X_n$ an adequate control value $y = \varphi(x_1, \ldots, x_n)$.

Example 4.1 (continuation) We define $X_1 \overset{\text{def}}{=} [-90, 90]$ for the cartpole problem, i.e. the angle θ can vary between $-90°$ and $+90°$. Theoretically, every angular velocity is possible. But extreme angular velocities can only be achieved artificially. Furthermore, the input devices are only able to measure a certain domain. We assume that $-45 \leq \dot{\theta} \leq 45$ holds (unit: degrees per second), i.e. $X_2 \stackrel{\text{def}}{=} [-45, 45]$.

Analogously, we assume that the force F remains in the boundaries of -10 and $+10$ (unit: Newton) and therefore define $Y \stackrel{\text{def}}{=} [-10, 10]$.   □

Classical control engineering is based on a formal description of a technical system, often given in the form of differential equations. The specification of a control function φ simply means the calculation of an appropriate solution of this differential equation. Often this is done with approximation or linearization methods.

Example 4.1 (continuation) The cartpole problem can be described by the differential equation

$$(M + m)\sin^2 \theta \cdot l \cdot \ddot{\theta} + m \cdot l \cdot \sin \theta \cdot \cos \theta \cdot \dot{\theta}^2 - (m + M) \cdot g \cdot \sin \theta = -F \cdot \cos \theta,$$

where we have

$$
\begin{array}{ll}
g : & \text{gravitational constant,} \\
l : & \text{length of the pole,} \\
m : & \text{mass at the head of the pole,} \\
M : & \text{mass at the foot of the pole.}
\end{array}
$$

We have to determine $F(t)$ by this differential equation in such a way that $\theta(t)$ and $\dot{\theta}(t)$ converge within an acceptable time towards zero.

This pure physical analysis of the cartpole problem presumes, of course, that the developed differential equation represents a good model of reality. To obtain such a formal description, the corresponding knowledge about the physical process is necessary.   □

However, it is often impossible to specify an accurate mathematical model of the process, or the description with differential equations is extremely complex. In addition this method requires a profound physical knowledge from the person who develops the controller. And even if a description of the process with differential equations is available, the exact solution of these equations can be extremely difficult. Solving such a differential equation also requires a certain amount of mathematical knowledge by the person who develops the controller.

Thus, at the centre of classical control engineering we have the physical–mathematical model of a system that has to be controlled. Certainly, this is a reasonable approach. But despite this, it should be possible to control a process without this approach. A simple proof of this fact is that a person is obviously able to ride a bike without even knowing of the existence of differential equations.

Alternatively to classical control engineering, it seems to be obvious to simulate the behaviour of a person who is able to control the given process. We call this development of a model of a human 'control expert' *knowledge-based analysis*. To make such an analysis the expert may be questioned directly. The expert then specifies his knowledge in form of *linguistic rules*. Instead of directly interviewing the expert it is also possible to observe his behaviour and extract from the observation protocol the necessary information. The results of this procedure can be used to provide appropriate (linguistic) rules that control the process. In the case of the cartpole problem

'If θ is approximately zero and $\dot{\theta}$ is also approximately zero, then F has to be approximately zero, too.'

could obviously be a reasonable linguistic rule. To obtain the solution of the cartpole problem we need several additional rules (we will return to this point later).

Linguistically generated rules mainly consist of a premise and a conclusion. The premise describes a certain situation in the form of a fuzzy specification of measured values. The conclusion specifies an appropriate fuzzy output value.

Linguistic rules might be suitable to describe the expert's behaviour. On the other hand, given a crisp input value, a fully automatic control is in need of the calculation of an appropriate crisp output value. In general it is not possible to use only linguistic rules. In order to use these linguistic rules for automatic control, we have to develop an adequate mathematical model for linguistic expressions like *approximately zero, positive small*, etc. Furthermore we have to determine to which degree a crisp input value corresponds to a linguistic term, and in which form this term should be represented by a fuzzy set in our mathematical model. Then we have to determine how the rules should be used to process the input values. We would like to obtain a fuzzy set as the primary output, that specifies linguistically a good choice of the output. Of course, it would be desirable to obtain a crisp output directly. But we cannot expect to derive crisp conclusions from fuzzy rules. So finally, we have to compute a crisp value from the fuzzy output.

An adequate architecture for the knowledge-based model of a controller is illustrated in Figure 2.

- The *fuzzification interface* receives the current input value. If necessary, it maps the input value to a suitable domain (e.g. one could choose a normalization that transforms the domains into the unit interval). Furthermore, we can use the fuzzification interface to convert an input value into a linguistic term or into a fuzzy set. In general, the crisp measured input

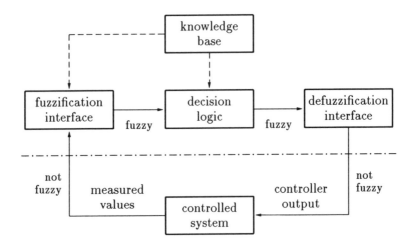

Figure 2   Architecture of a fuzzy controller

x_0 is turned into the fuzzy set $\mathbb{1}_{\{x_0\}}$ where

$$\mathbb{1}_{\{x_0\}}(x) = \begin{cases} 1, & \text{if } x = x_0 \\ 0, & \text{otherwise.} \end{cases}$$

In case we have detailed information about the accuracy of the measurement or if the measurement itself is not crisp, other fuzzy sets than $\mathbb{1}_{\{x_0\}}$ (the characteristic function of the set $\{x_0\}$) can be used.

- The *knowledge base* contains information about the boundaries, possible transformations of the domains, and the fuzzy sets with their corresponding linguistic terms. This information represents the *data base*. In addition the knowledge base contains a *rule base* consisting of linguistic control rules.
- The *decision logic* represents the processing unit. It determines the corresponding output value from the measured input according to the knowledge base.
- The *defuzzification interface* has the task of determining a crisp output value — taking the information about the control variable provided by the decision logic into account. Finally, if necessary, it carries out a transformation of the output value into the appropriate domain.

4.2   Two Approaches to Fuzzy Control

In this section we introduce two basically different types of fuzzy controllers, that are motivated on a very intuitive basis. A more rigorous analysis of the semantics of fuzzy controllers is discussed in Section 4.4 where a formal framework for the understanding and interpretation of Mamdani's approach is provided.

In both approaches the aim is the specification of a control function. Only the models and the methods are fuzzy, the obtained control function itself is always crisp. Otherwise it would not be possible to apply these models in practice.

4.2.1   The Approach of Mamdani

In this approach an expert has to specify his knowledge in the form of linguistic rules. First, he has to determine linguistic terms that are suitable for the description of the states of the variables. Hence, he has to determine for each input domain $X_1, \ldots, X_n$ and the output Y appropriate linguistic terms like *approximately zero*, *positive small*, etc. It is of course possible that *approximately zero* related to ξ_1 means something completely different than *approximately zero* related to ξ_2. For instance, we could use the three linguistic terms *negative*, *approximately zero*, and *positive* to describe the states of the variable ξ_1. In the mathematical model each linguistic term has to be represented by a fuzzy set.

As the linguistic terms simply represent names of the fuzzy sets or the represented concepts, we assume that firstly the fuzzy sets are specified and secondly they are assigned corresponding linguistic terms. To determine suitable rules for the knowledge base of the fuzzy controller, we partition each of the sets $X_1, \ldots, X_n$, and Y with the help of specified fuzzy sets. For this reason we define p_1 distinct fuzzy sets $\mu_1^{(1)}, \ldots, \mu_{p_1}^{(1)} \in F(X_1)$ on the set X_1 and associate a linguistic term with each set. Often, if a set X_1 corresponds to an interval $[a, b]$ of the real line, triangular functions given by

$$
\begin{aligned}
\mu_{x_0, \varepsilon} : [a, b] &\longrightarrow [0, 1], \\
x &\longmapsto 1 - \min\{\varepsilon \cdot |x - x_0|, 1\}
\end{aligned}
$$

are used to describe the fuzzy sets $\mu_1^{(1)}, \ldots, \mu_{p_1}^{(1)} \in F(X_1)$. The value $x_0 \in [a, b]$ indicates the peak of the triangle, i.e. we have $\mu_{x_0, \varepsilon}(x_0) = 1$. The parameter $\varepsilon > 0$ determines, how acute-angled ($\varepsilon \geq 1$) or how obtuse-angled ($0 < \varepsilon \leq 1$) the triangle is. If we have $a < x_1 < \ldots < x_{p_1} < b$, it is common to define just the fuzzy sets $\mu_2^{(1)}, \ldots, \mu_{p_1-1}^{(1)}$ as triangular functions. At the boundaries often

$$
\begin{aligned}
\mu_1^{(1)} : [a, b] &\longrightarrow [0, 1], \\
x &\longmapsto \begin{cases} 1, & \text{if } x \leq x_1 \\ 1 - \min\{\varepsilon \cdot (x - x_1), 1\}, & \text{otherwise} \end{cases}
\end{aligned}
$$

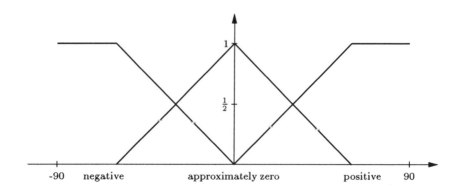

Figure 3   A coarse fuzzy partition

for the left boundary and

$$\mu_{p_1}^{(1)} : [a, b] \longrightarrow [0, 1],$$

$$x \longmapsto \begin{cases} 1, & \text{if } x_{p_1} \leq x \\ 1 - \min\{\varepsilon \cdot (x_{p_1} - x), 1\}, & \text{otherwise} \end{cases}$$

for the right boundary is chosen. Each fuzzy set $\mu_1^{(1)}, \ldots, \mu_{p_1}^{(1)}$ corresponds to a linguistic term like *positive small*. Figure 3 shows a coarse fuzzy partition of the set $[-90, 90]$ with the corresponding linguistic terms *negative*, *approximately zero*, and *positive*. Figure 4 shows a finer fuzzy partition with seven fuzzy sets. Here, the fuzzy sets represent the linguistic terms *negative big*, *negative medium*, *negative small*, *approximately zero*, *positive small*, *positive medium*, and *positive big*.

Above all, triangular functions are used because piecewise linear functions are easy to handle with a computer. Storage and computations are simple. In Section 4.4 we will see that triangular functions can also play an important role as singletons in the sense of Example 2.56.

In principle, it is possible to use any other types of fuzzy sets instead of triangular functions. In order to maintain the concept of interpreting fuzzy sets as fuzzy values or fuzzy intervals, it is recommended to restrict oneself to unimodal functions, representing fuzzy numbers or fuzzy intervals (elements of $F_I(\mathbf{R})$). Often the fuzzy sets of a fuzzy partition are chosen in such a way that they satisfy the condition

$$i \neq j \Rightarrow \sup_{x \in X_1} \{\min\{\mu_i^{(1)}(x), \mu_j^{(1)}(x)\}\} \leq 0.5.$$

Later on we will see that this is equivalent to a disjointness requirement.

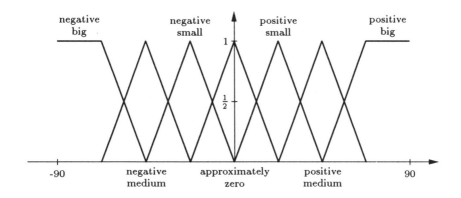

Figure 4   A finer fuzzy partition

In the same way that domain X_1 is divided into p_1 fuzzy sets $\mu_1^{(1)}, \ldots, \mu_{p_1}^{(1)}$, the remaining sets $X_2, \ldots, X_n$, and Y are partitioned by p_i, $i = 2, \ldots, n$, and p fuzzy sets $\mu_1^{(i)}, \ldots, \mu_{p_i}^{(i)} \in F(X_i)$, and $\mu_1, \ldots, \mu_p \in F(Y)$, respectively.

These fuzzy partitions and the linguistic terms associated with fuzzy sets represent the data base in our knowledge base.

Example 4.1 (continuation) For the cartpole problem we choose for the set X_1 a fuzzy partition as shown in Figure 4. The supports of the fuzzy sets (the sets of points with a membership value greater than zero) are intervals with a length of a quarter of the whole range X_1. According to this, the triangles have a width of 45. Similar fuzzy partitions are chosen for the sets X_2 and Y. The corresponding triangles have a width of 22.5 and 5, respectively. □

After determining the fuzzy partitions of the sets $X_1 \times \ldots \times X_n, Y$, we specify the rule base. A single rule has the form

if ξ_1 is $A^{(1)}$ and ... and ξ_n is $A^{(n)}$ then η is B,

where $A^{(1)}, \ldots, A^{(n)}$, and B represent the linguistic terms that correspond to the fuzzy sets $\mu^{(1)}, \ldots, \mu^{(n)}$, and μ, respectively, according to the fuzzy partitions of the set $X_1 \times \ldots \times X_n$, and Y. Therefore, the rule base consists of k control rules

$$R_r: \text{if } \xi_1 \text{ is } A_{i_{1,r}}^{(1)} \text{ and } \ldots \text{ and } \xi_n \text{ is } A_{i_{n,r}}^{(n)} \text{ then } \eta \text{ is } B_{i_r}, \quad r = 1, \ldots, k.$$

We do not consider these control rules as implications, but more in the sense of a piecewise defined function. This means that the k rules correspond to the

$$\theta$$

	nb	nm	ns	az	ps	pm	pb
nb			ps	pb			
nm				pm			
ns	nm		ns	ps			
az	nb	nm	ns	az	ps	pm	pb
ps				ns	ps		pm
pm				nm			
pb				nb	ns		

$\dot{\theta}$ (row labels)

Table 1   Rule base for the cartpole problem

'definition of the function' $\eta = \varphi(\xi_1, \ldots, \xi_n)$, where

$$\eta \,\hat{=}\, \begin{cases} B_{i_1} & \text{if } \xi_1 \hat{=} A^{(1)}_{i_{1,1}} \text{ and } \ldots \text{ and } \xi_n \hat{=} A^{(n)}_{i_{n,1}} \\ \vdots & \vdots \qquad\qquad\qquad\qquad\qquad \vdots \\ B_{i_k} & \text{if } \xi_1 \hat{=} A^{(1)}_{i_{1,k}} \text{ and } \ldots \text{ and } \xi_n \hat{=} A^{(n)}_{i_{n,k}}. \end{cases}$$

Example 4.1 (continuation) We will use the rule base given in Table 1, consisting of 19 rules, for the cartpole problem. The rule table is interpreted in the following way. For instance, the entry in the second row and the fourth column of the table determines the rule

> **if** θ **is** *approximately zero* **and** $\dot{\theta}$ **is** *negative medium*
> **then** F **is** *positive medium.*

The table does not have an entry in every intersection, i.e. a linguistic term for the output value is not specified for every possible pair of linguistic terms (angle × angular velocity). Later on we will see that it is not necessary to fill the table completely with terms, since the entries do not correspond to pairwise excluding cases. In addition, we do not have to determine an output for extreme situations, where it is impossible to prevent the pole from falling according to the limited force we can apply to the pole. □

In the following we explain how the table-based control function can be defined. A tuple $(x_1, \ldots, x_n) \in X_1 \times \ldots \times X_n$ of the current measured input, picked up by the fuzzification interface is forwarded to the decision logic. If necessary, the tuple is transformed, but it is not yet turned into a fuzzy set. Thus a real 'fuzzification' is not carried out.

The decision logic applies each rule R_r separately. Consider the rule if ξ_1 is $A^{(1)}$ and ... and ξ_n is $A^{(n)}$ then η is B.

Firstly the decision logic determines the degree to which a measured input fulfills the premise of the rule (called degree of applicability). For $\nu = 1, \ldots, n$ the value $\mu^{(\nu)}(x_\nu)$ is calculated. It determines how far the measured input x_ν corresponds to the fuzzy set $\mu^{(\nu)}$ and its associated linguistic term. As the premise of the rule demands that $x_1, \ldots, x_n$ fulfil the linguistic terms $A^{(1)}, \ldots, A^{(n)}$, respectively, the values $\mu^{(\nu)}(x_\nu)$, $\nu = 1, \ldots, n$, have to be combined in a conjunctive manner. This can be done by determining the value of $\alpha = \min\{\mu^{(1)}, \ldots, \mu^{(n)}\}$. Thus we obtain for each rule R_r of the k control rules the value

$$\alpha_r \stackrel{\text{def}}{=} \min\{\mu_{i_{1,r}}^{(1)}(x_1), \ldots, \mu_{i_{n,r}}^{(n)}(x_n)\}. \tag{4.1}$$

The value (4.1) gives the degree of applicability of the premise of rule R_r. The output of rule R_r is a fuzzy set of output values obtained by 'cutting off' the fuzzy set μ_{i_r} associated with the conclusion of rule R_r at the level of applicability given by (4.1). Technically speaking, rule R_r implies for the measured input $(x_1, \ldots, x_n)$ the fuzzy set

$$\mu_{x_1, \ldots, x_n}^{\text{output}(R_r)} : Y \quad \longrightarrow \quad [0, 1],$$

$$y \quad \longmapsto \quad \min\{\mu_{i_{1,r}}^{(1)}(x_1), \ldots, \mu_{i_{n,r}}^{(n)}(x_n), \mu_{i_r}(y)\}.$$

In the case of $\mu_{i_{1,r}}^{(1)}(x_1) = \cdots = \mu_{i_{n,r}}^{(n)}(x_n) = 1$, we deduce $\mu_{x_1, \ldots, x_n}^{\text{output}(R_r)} = \mu_{i_r}$. In other words: if the measured input completely fulfills the premise of rule R_r, rule R_r passes its 'conclusion'-fuzzy set as output. In case of $\nu \in \{1, \ldots, n\}$ $\mu_{i_{\nu,r}}^{(\nu)}(x_\nu) = 0$, indicating that the rule is not applicable, we obtain for rule R_r the fuzzy set identical to 0.

Example 4.1 (continuation) The current measured input shall be $\theta = 36°$ for the angle and $\dot{\theta} = -2.25° \text{s}^{-1}$ for the angular velocity. The value (4.1) is non-zero only for the two rules

R_1:   **if** θ is *positive small* **and** $\dot{\theta}$ is *approximately zero* **then** F is *positive small*

and

R_2:   **if** θ is *positive medium* **and** $\dot{\theta}$ is *approximately zero* **then** F is *positive medium*.

Hence, according to formula (4.1) we have fulfilled the premise of rule R_1 to the extent of $0.4 = \min\{0.4, 0.8\}$. This leads us to the output fuzzy set of rule R_1

$$\mu_{36, -2.25}^{\text{output}(R_1)}(y) = \begin{cases} \frac{2}{5} \cdot y, & \text{if } 0 \le y \le 1 \\ 0.4, & \text{if } 1 \le y \le 4 \\ 2 - \frac{2}{5} \cdot y, & \text{if } 4 \le y \le 5 \\ 0, & \text{otherwise.} \end{cases}$$

For rule R_2 the premise is fulfilled to the extent of $0.6 = \min\{0.6, 0.8\}$, and therefore we obtain

$$
\mu_{36,-2.25}^{\text{output}(R_2)}(y) = \begin{cases}
\frac{2}{5} \cdot y - 1, & \text{if} \quad 2.5 \leq y \leq 4 \\
0.6, & \text{if} \quad 4 \quad \leq y \leq 6 \\
3 - \frac{2}{5} \cdot y, & \text{if} \quad 6 \quad \leq y \leq 7.5 \\
0, & \text{otherwise.}
\end{cases}
$$

Figures 5 and 6 illustrate the application of rules R_1 and R_2. The horizontal line indicates up to which degree each premise is fulfilled. The corresponding fuzzy set in the conclusion of the rule is cut at exactly this height. As results we obtain the hatched fuzzy sets.

Since all other rules have a degree of applicability of zero, each of them provides as its result the fuzzy set that is constantly zero.   □

After the decision logic has evaluated each rule, it has to combine all the fuzzy sets obtained from the rules into one output fuzzy set by determining the maximum (union). This corresponds to the union of these fuzzy sets. Finally the decision logic yields the fuzzy set

$$\mu_{x_1,\dots,x_n}^{\text{output}} : Y \quad \rightarrow \quad [0,1],$$

$$y \quad \mapsto \quad \max_{r \in \{1,\dots,k\}} \left\{ \min\{\mu_{i_1,r}^{(1)}(x_1), \dots, \mu_{i_n,r}^{(n)}(x_n), \mu_{i_r}(y)\} \right\}, \quad (4.2)$$

which is passed on to the defuzzification interface to determine a crisp output value.

Example 4.1 (continuation) Given the measured input $\theta = 36°$ and $\dot{\theta} = -2.25°\text{s}^{-1}$, we obtain the fuzzy sets shown on the right side of Figures 5 and 6. The union of these two fuzzy sets is illustrated in Figure 7.

$$
\mu_{36,-2.25}^{\text{output}}(y) = \begin{cases}
\frac{2}{5} \cdot y, & \text{if} \quad 0 \quad \leq y \leq 1 \\
0.4, & \text{if} \quad 1 \quad \leq y \leq 3.5 \\
\frac{2}{5} \cdot y - 1, & \text{if} \quad 3.5 \leq y \leq 4 \\
0.6, & \text{if} \quad 4 \quad \leq y \leq 6 \\
3 - \frac{2}{5} \cdot y, & \text{if} \quad 6 \quad \leq y \leq 7.5 \\
0, & \text{otherwise.}
\end{cases}
$$

The fuzzy sets derived from the other rules are zero and are therefore negligible for the union.   □

The whole method, explained above, determines a mapping which assigns to each tuple $(x_1, \dots, x_n) \in X_1 \times \dots \times X_n$ of measured input a fuzzy set $\mu_{x_1,\dots,x_n}^{\text{output}}$ of Y. The output of the decision logic is not a crisp value, but a description

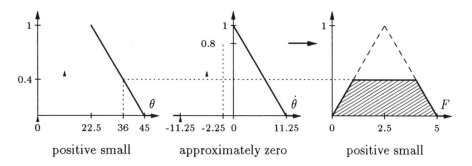

Figure 5   Evaluation of rule R_1

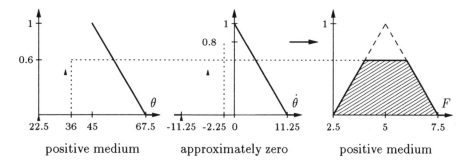

Figure 6   Evaluation of rule R_2

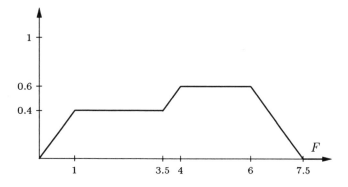

Figure 7   The fuzzy set $\mu_{36,-2.25}^{\text{output}}$

of the output value in the form of a fuzzy set. The defuzzification interface has to derive a crisp value from this fuzzy set $\mu^{\text{output}}_{x_1,\ldots,x_n}$. This procedure is called defuzzification. The following four defuzzification strategies are the most common.

The Max Criterion Method

Using this method, we choose an arbitrary value $y \in Y$ for which the fuzzy set $\mu^{\text{output}}_{x_1,\ldots,x_n}$ reaches a maximum membership. In Figure 7, according to our Example 4.1, we are free to choose any value $y \in [4, 6]$ as output value.

The advantage of this strategy is that it is applicable for arbitrary fuzzy sets and for an arbitrary domain Y of control actions that is not necessarily a subset of the real line. On the other hand we have the disadvantage that the max criterion method is more a class of defuzzification strategies rather than a single method. It is unspecified which value of maximum membership has to be chosen. If we pick a random value after each measurement, we get a non-deterministic controller. In addition, this may lead to discontinuous control actions.

The Mean of Maxima Method (MOM)

A presupposition for the mean of maxima method is that Y is an interval and the set $\text{Max}(\mu^{\text{output}}_{x_1,\ldots,x_n}) \overset{\text{def}}{=} \{y \in Y \mid \forall y' \in Y : \mu^{\text{output}}_{x_1,\ldots,x_n}(y') \le \mu^{\text{output}}_{x_1,\ldots,x_n}(y)\}$ is non-empty and (Borel-)measurable. As crisp output value, we determine the mean value of the set $\text{Max}(\mu^{\text{output}}_{x_1,\ldots,x_n})$. If the set $\text{Max}(\mu^{\text{output}}_{x_1,\ldots,x_n})$ is finite, we obtain the output value by the formula

$$\eta = \frac{1}{\left|\text{Max}(\mu^{\text{output}}_{x_1,\ldots,x_n})\right|} \sum_{y \in \text{Max}(\mu^{\text{output}}_{x_1,\ldots,x_n})} y,$$

which may be written as

$$\eta = \frac{1}{\displaystyle\int_{y \in \text{Max}(\mu^{\text{output}}_{x_1,\ldots,x_n})} dy} \cdot \int_{y \in \text{Max}(\mu^{\text{output}}_{x_1,\ldots,x_n})} y \, dy$$

in the case of an infinite set $\text{Max}(\mu^{\text{output}}_{x_1,\ldots,x_n})$. It is possible that $\eta \notin \text{Max}(\mu^{\text{output}}_{x_1,\ldots,x_n})$. The consequences of this case are described in Example 4.2.

The mean of maxima method results, at least in the case of fuzzy partitions with symmetric triangular functions, in discontinuous control actions. The set $\text{Max}(\mu^{\text{output}}_{x_1,\ldots,x_n})$ is solely determined by the conclusion fuzzy set μ_{i_r} of the rule R_r with the highest degree of applicability of its premise. So we have a symmetric interval $\text{Max}(\mu^{\text{output}}_{x_1,\ldots,x_n})$ around the point y_{i_r}, where μ_{i_r} reaches a membership value of 1.

As long as fuzzy set μ_{i_r} is 'dominating', the output value y_{i_r} remains unchanged. Only when a premise of another rule R_s reaches a higher degree of applicability, does the output value suddenly jump from the former value to the one where the fuzzy set associated with the conclusion of rule R_s reaches a membership degree of one. This unsteady control behaviour may lead to undesirable discontinuous changes in the process being controlled and might stress the controller and the plant.

Centre of Gravity Method (COG)

Using the centre of gravity demands the same presuppositions as the mean of maxima method. We define the value located under the centre of gravity of the area that is given by the function $\mu^{\text{output}}_{x_1,\ldots,x_n}$ as output value η. Technically this value is computed by the formula

$$\eta = \frac{1}{\int\limits_{y \in Y} \mu^{\text{output}}_{x_1,\ldots,x_n}(y)dy} \cdot \int\limits_{y \in Y} y \cdot \mu^{\text{output}}_{x_1,\ldots,x_n}(y)dy. \tag{4.3}$$

One important advantage of this method is that the controller shows nearly always a smooth control behaviour. The main idea of the centre of gravity strategy is to take the rules into consideration according to their degree of applicability. Assuming a continuous behaviour of the process, the centre of gravity strategy guarantees the following feature: if a certain control rule was dominating in one step, because it had reached the highest degree of applicability of its premise, it is not necessarily dominating again the next time. It will, however, maintain a certain influence on the calculation of the centre of gravity. Because of the continuity of the process its influence will not suddenly — from one step to another — decrease to zero.

A disadvantage of the centre of gravity strategy is that it can hardly be justified semantically. Furthermore, its computation may need a longer time than simpler methods. Sometimes its results can also be counterintuitive as shown in Example 4.2.

The centre of gravity method is also called the centre of area method (COA). Instead of choosing the value below the centre of gravity as output, the same result is obtained by taking the value where the area described by the output fuzzy set $\mu^{\text{output}}_{x_1,\ldots,x_n}$ is split into two parts equal in size.

Example 4.1 (continuation) The defuzzification of the fuzzy set in Figure 7 using the mean of maximum yields the value $F = 5$ as output. The centre of gravity method results in $F = \frac{245}{62} \approx 3.95$ (cf. Figure 8).   □

Although both methods appear quite reasonable at first glance, they can lead to some undesirable results. This is illustrated in the following example.

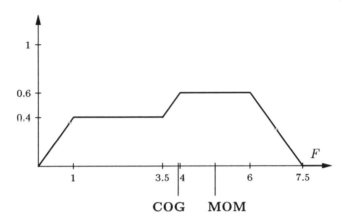

Figure 8 Defuzzification with centre of gravity and mean of maxima

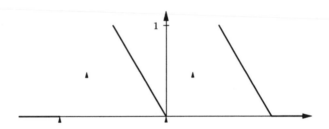

Figure 9 An output fuzzy set that leads to an undesirable result in the case of defuzzifizication by one of the strategies MOM and COG

Example 4.2 Suppose we consider a fuzzy controller steering a model car. It should move automatically around obstacles lying in its way. If an obstacle appears right in front of the car, the fuzzy set shown in Figure 9 could represent the output of the decision logic. The fuzzy set can be interpreted as 'evade the obstacle either by turning to the left or to the right' — a reasonable control instruction.

The centre of gravity as well as the mean of maxima method result in the value 0 for control action to be carried out. In other words, the car bumps directly into the obstacle.   □

The anomaly illustrated in Example 4.2 does not occur while using convex fuzzy sets.

A convex fuzzy set can be interpreted as a representation of a single (fuzzy)

control value or interval. In Example 4.2 we have two contrary absolutely reasonable control actions. The control rules describe a non-deterministic behaviour of the control expert who randomly chooses to go either to the left or to the right.

Defuzzification can be divided into two different tasks:

(i) Conversion of a fuzzy set into crisp values.

(ii) Selection of one of several possible control actions.

If the final fuzzy set represents only one single control value, the second task is superfluous. It is sufficient to determine an appropriate control value from the given fuzzy set. The defuzzification strategies MOM and COG work under this presupposition. If, in contrast, the fuzzy set represents a set of several elements, both tasks (i) and (ii) have to be solved. In this case the order in which they are carried out is of importance.

If we first complete task (ii), we have to choose from a fuzzy set representing a set of several elements a fuzzy subset, representing only a single fuzzy element. This fuzzy set can then be defuzzified into a crisp value by applying a strategy like MOM or COG. This would mean in our Example 4.2 that the fuzzy set in Figure 9 is first reduced to one of the two triangles — say the left one — and then this triangle is defuzzified. This would lead to the instruction 'evade to the left'.

The other possibility would be to complete first task (i). We obtain a set of several elements and then choose one of them. Regarding our Example 4.2, this method would yield at first the two elements which are assigned the membership degree one. After the fuzzy set is turned into a crisp set — in this case a set with only two elements — we have to choose one of these elements.

To avoid the problems of distinguishing these two different steps in defuzzification, the control rules should be defined in such a way that they model a deterministic behaviour of the control expert. In this case we should be able to interpret the output of the decision logic in the form of the fuzzy set $\mu^{\text{output}}_{x_1,\ldots,x_n}$ as a single fuzzy value making task (ii) superfluous. This restriction is not very serious. Firstly, for reasons of reliability, a deterministic behaviour of the controller is of course highly desirable. Secondly, only in situations where several control actions are possible, does the control expert — while specifying the rules — have to decide on one of the possible control actions. Technically speaking, this presupposition demands not a description of a general relation $R \subseteq (X_1 \times \ldots \times X_n) \times Y$ of input/output tuples from the expert, but forces him to describe a function $\varphi : X_1 \times \ldots \times X_n \to Y$, i.e. a relation $R \subseteq (X_1 \times \ldots \times X_n) \times Y$, where for each $(x_1, \ldots, x_n) \in X_1 \times \ldots \times X_n$ there is exactly one element $y \in Y$, such that $((x_1, \ldots, x_n), y) \in R$.

4.2.2   The Approach of Takagi and Sugeno

The approach to fuzzy control first proposed by Takagi and Sugeno can be seen as a modification of Mamdani's model.

For a Sugeno fuzzy controller fuzzy partitions for the input domains have to be specified in the same way as for the Mamdani fuzzy controller. A fuzzy partition of the output domain is not needed, since the control rules are given in the form of

$$R_r : \quad \textbf{if } \xi_1 \text{ is } A_{i_{1,r}}^{(1)} \textbf{ and } \ldots \textbf{ and } \xi_n \text{ is } A_{i_{n,r}}^{(n)}$$

$$\textbf{then } \eta = f_r(\xi_1, \ldots, \xi_n),$$

$$r = 1, \ldots, k,$$

where f_r is a mapping from $X_1 \times \ldots \times X_n$ to Y, $r = 1, \ldots, k$. Generally it is assumed that f_r is linear, i.e. $f_r(x_1, \ldots, x_n) = a_1^{(r)} \cdot x_1 + \ldots + a_n^{(r)} \cdot x_n + a^{(r)}$.

The task of the decision logic is to determine the degree of applicability α_r of each premise and to compute the value $f_r(x_1, \ldots, x_n)$ for the input tuple $(x_1, \ldots, x_n) \in X_1 \times \ldots \times X_n$ for each rule R_r. This is done analogously as in the case of the Mamdani fuzzy controller by the formula (4.1). In addition the decision logic delivers the crisp control value η according to the formula

$$\eta = \frac{\sum\limits_{r=1}^{k} \alpha_r \cdot f_r(x_1, \ldots, x_n)}{\sum\limits_{r=1}^{k} \alpha_r}.$$

Thus, we finally get as the control value the weighted sum of the outputs of the rules, with each rule contributing according to the degree of applicability of its premise. A defuzzification is therefore superfluous for the Sugeno fuzzy controller.

Example 4.3 The task is to pass a bend with a model car [Sugeno85a] at a constant speed, as illustrated in Figure 10.

We can use the following measured input (cf. Figure 11):

ξ_1:   distance of the car to the beginning of the bend,

ξ_2:   distance of the car to the inner barrier,

ξ_3:   direction (angle) of the car,

ξ_4:   distance of the car to the outer barrier.

The control variable η represents the rotation speed of the steering wheel. The domains of the input variables are $X_1 = [0, 150]\,(\text{cm})$, $X_2 = [0, 150]\,(\text{cm})$, $X_3 = [-90, 90]\,(°)$, and $X_4 = [0, 150]\,(\text{cm})$. The corresponding fuzzy partitions are illustrated in Figure 12.

The rules controlling the car are given in the following form:

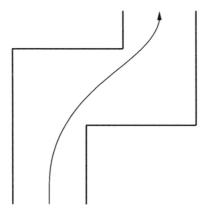

Figure 10   Passing a bend

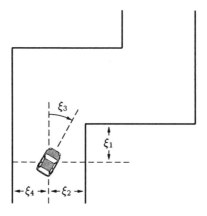

Figure 11   Input variables for the control of the car

R_r: **if ξ_1 is A and ξ_2 is B and ξ_3 is C and ξ_4 is D**

then $\eta = p_0^{(A,B,C,D)} + p_1^{(A,B,C,D)} \cdot \xi_1 + p_2^{(A,B,C,D)} \cdot \xi_2$
$$+ p_3^{(A,B,C,D)} \cdot \xi_3 + p_4^{(A,B,C,D)} \cdot \xi_4.$$

Here we have

$$A \in \{small, medium, big\},$$
$$B \in \{small, big\},$$
$$C \in \{outwards, forward, inwards\},$$
$$D \in \{small\},$$
$$p_0^{(A,B,C,D)}, \ldots, p_4^{(A,B,C,D)} \in \mathbf{R}.$$

Rules in which all four input variables $\xi_1, \ldots, \xi_4$ do not appear simultaneously are also allowed.

The functions $\eta = f_r(\xi_1, \xi_2, \xi_3, \xi_4)$ in the conclusions of the rules (describing the control variable in the rule R_r) are assumed to be linear in the control variables $\xi_1, \ldots, \xi_4$. As a matter of fact, this is common in most applications. The rule base controlling the car is listed in Table 2.

Assume that the car is 10 cm away from the beginning of the bend ($\xi_1 = 10$). The distance of the car to the inner barrier is 30 cm ($\xi_2 = 30$), to the outer barrier 50 cm ($\xi_4 = 50$), and the car's direction is 'forward' ($\xi_3 = 0$). According to formula (4.1) and with respect to rules $R_1, \ldots, R_{20}$, only for the premises of the rules R_4 and R_7 do we then obtain a value different from zero. The degrees of applicability are shown in Table 3.

For the premises of rule R_4 and R_7 we obtain a degree of applicability of $\alpha_4 = \frac{1}{4}$ and $\alpha_7 = \frac{1}{6}$, respectively. This results in a weight of $W_4 = \frac{\frac{1}{4}}{\frac{1}{4}+\frac{1}{6}} = \frac{3}{5}$ for rule R_4, and $W_7 = \frac{\frac{1}{6}}{\frac{1}{4}+\frac{1}{6}} = \frac{2}{5}$ for R_7.

Rules R_4 and R_7 yield

$$
\begin{aligned}
\eta_4 &= 0.303 - 0.026 \cdot 10 + 0.061 \cdot 30 - 0.050 \cdot 0 + 0.000 \cdot 50 \\
&= 1.873
\end{aligned}
$$

and

$$
\begin{aligned}
\eta_7 &= 2.990 - 0.017 \cdot 10 + 0.000 \cdot 30 - 0.021 \cdot 0 + 0.000 \cdot 50 \\
&= 2.820,
\end{aligned}
$$

respectively. The value for the control variable η is obtained computing the sum of the outputs of the rules R_4 and R_7, weighted by the corresponding factors. Finally, the output is

$$\eta = \frac{3}{5} \cdot 1.873 + \frac{2}{5} \cdot 2.820 = 2.2518. \qquad \square$$

rule	ξ_1	ξ_2	ξ_3	ξ_4	p_0	p_1	p_2	p_3	p_4
R_1	–	–	outwards	small	3.000	0.000	0.000	-0.045	-0.004
R_2	–	–	forward	small	3.000	0.000	0.000	-0.030	-0.090
R_3	small	small	outwards	–	3.000	-0.041	0.004	0.000	0.000
R_4	small	small	forward	–	0.303	-0.026	0.061	-0.050	0.000
R_5	small	small	inwards	–	0.000	-0.025	0.070	-0.075	0.000
R_6	small	big	outwards	–	3.000	-0.066	0.000	-0.034	0.000
R_7	small	big	forward	–	2.990	-0.017	0.000	-0.021	0.000
R_8	small	big	inwards	–	1.500	0.025	0.000	-0.050	0.000
R_9	medium	small	outwards	–	3.000	-0.017	0.005	-0.036	0.000
R_{10}	medium	small	forward	–	0.053	-0.038	0.080	-0.034	0.000
R_{11}	medium	small	inwards	–	-1.220	-0.016	0.047	-0.018	0.000
R_{12}	medium	big	outwards	–	3.000	-0.027	0.000	-0.044	0.000
R_{13}	medium	big	forward	–	7.000	-0.049	0.000	-0.041	0.000
R_{14}	medium	big	inwards	–	4.000	-0.025	0.000	-0.100	0.000
R_{15}	big	small	outwards	–	0.370	0.000	0.000	-0.007	0.000
R_{16}	big	small	forward	–	-0.900	0.000	0.034	-0.030	0.000
R_{17}	big	small	inwards	–	-1.500	0.000	0.005	-0.100	0.000
R_{18}	big	big	outwards	–	1.000	0.000	0.000	-0.013	0.000
R_{19}	big	big	forward	–	0.000	0.000	0.000	-0.006	0.000
R_{20}	big	big	inwards	–	0.000	0.000	0.000	-0.010	0.000

Table 2 The control rules for the car

	small	medium	big
$\xi_1 = 10$	0.8	0	0

Membership degrees of the value $\xi_1 = 10$ to the fuzzy sets of domain X_1.

	small	big
$\xi_2 = 30$	0.25	0.167

Membership degrees of the value $\xi_2 = 30$ to the fuzzy sets of domain X_2.

	outwards	forward	inwards
$\xi_3 = 0$	0	1	0

Membership degrees of the value $\xi_3 = 0$ to the fuzzy sets of domain X_3.

	small
$\xi_4 = 50$	0

Membership degree of the value $\xi_4 = 50$ to the fuzzy set of domain X_4.

Table 3 The membership degrees for controlling the car

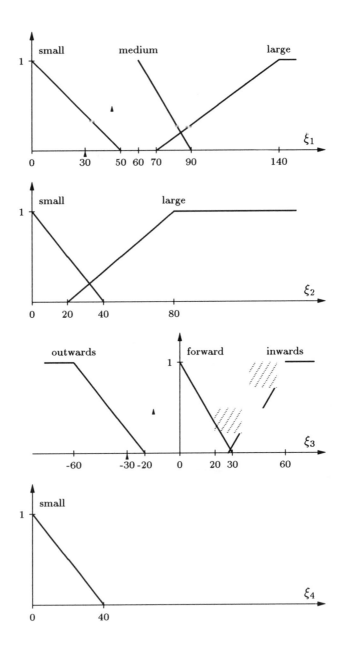

Figure 12   The fuzzy partitions of the sets X_1, X_2, X_3, and X_4

4.3 On the Design Parameters of Fuzzy Controllers

In this section we will briefly explain the individual steps to be carried out when designing a fuzzy controller and some problems that might occur. It should be emphasized that fuzzy control should be seen as a technique for the design of a non-linear transition function which determines the control actions. In this sense, fuzzy control provides a suitable framework for a knowledge-based specification or interpolation of this transition function.

4.3.1 Identification of Input and Control Variables

The first design step for a fuzzy controller is devoted to the data base in the knowledge base. For simple processes like the cartpole problem the input and control variables are determined directly by the process itself. In other cases suitable input and control variables have to be identified. In more complex processes it can be quite difficult to find out which input variables are necessary for a good control behaviour. A fuzzy controller monitoring an air conditioning system obviously needs the room temperature and perhaps the moisture as measured inputs. A measurement of the temperature outside the room might be helpful, too, but we already might question the necessity of it. Similarly, the number of persons in the room can influence the room temperature. But an automatic determination of the number of people that are present is a problem which is technically difficult to realize and would be much too expensive for an air conditioning system.

We restricted our considerations to fuzzy controllers with a single control variable. In the case of more than one control variable a separate fuzzy controller has to be designed for each of these variables. It should be taken into account that the effects of two or more control variables may neutralize or amplify each other.

Often we have a single optimal state for the process to be controlled. Regarding the cartpole problem, this state is described by the values $\theta = 0$ and $\dot{\theta} = 0$, expressing that the pole stands upright and is not about to fall in either direction. For this type of problem it is recommended to consider the error and the change of error with respect to the desired state. For the cartpole problem we have the angle as error and the angular velocity for the change of error. In general the error and the change of error have to be extracted from the measured data. For an air conditioning system the error is simply the difference of the measured temperature and the desired temperature.

4.3.2 The Domains of the Input and Control Variables

After the input and the control variables have been specified, the domains for these variables, i.e. $X_1, \ldots, X_n$, and Y, have to be defined. Some domain boundaries are determined directly by the process itself. For instance, in the cartpole problem, the angle of the pole cannot exceed the bounds of the interval $-90°$ and $+90°$. Other values can only be estimated by experience or heuristic methods. When first looking at the cartpole problem, the possible values for the angular velocity are not known. Furthermore we always have technical restrictions. The force F in the cartpole problem, for example, cannot be completely freely chosen; there are technical and physical limitations.

Often an analogue/digital converter transforms analogue measurements into digital values. In this case a discretization of the domains is convenient. We then can assume the sets $X_1, \ldots, X_n$, and Y to be finite. A discretization corresponds to partitioning an interval into a finite number of subintervals. We can choose any point from a certain subinterval as the representative for all other values in this subinterval. Besides the discretization a normalization for technical reasons, i.e. a transformation to the unit interval or to $[-1, 1]$, is sometimes required. This transformation is in most cases linear, but non-linear transformations are also possible.

4.3.3 Fuzzy Partitions of the Domains

After the identification of appropriate domains for the input and control variables, fuzzy partitions including corresponding linguistic terms have to be specified for these domains. In the case of a Sugeno controller this only has to be done for the input domains.

In addition to a possible horizontal discretization of the domains that we have discussed above, a vertical discretization, replacing the unit interval as the set of values for the membership degrees by a finite subset, can be useful as well.

For each domain to be partitioned, the number and shapes of the corresponding fuzzy sets have to be defined. The shape of fuzzy sets is most conveniently determined by a parametric representation. Symmetrical triangular fuzzy sets, for instance, are defined by the two parameters that specify their width and the position of the peak.

4.3.4 Linguistic Rules

When the data base is determined by the specification of the input and the control variables including the corresponding domains, the control rules have to be defined using the linguistic terms associated with the fuzzy sets that appear in the fuzzy partitions of the domains. There are four different ways to constitute a rule base:

(i) An expert who is able to control the process by 'manual operation' specifies linguistic rules that reflect his behaviour.

(ii) An expert who is able to control the process is observed for a period of time. Data of the measured input and control values, including the corresponding behaviour of the process, are recorded. Then linguistic control rules that model the expert's behaviour have to be extracted from these protocol data. This can be done directly by analysing the data or by clustering — or fuzzy clustering — methods [Bezdek81].

(iii) The process is described by a fuzzy model from which the control rules can be directly derived. Before this method is applicable further research work is necessary.

(iv) The fuzzy controller learns rules by itself, using meta-knowledge that enables the controller to evaluate its behaviour and to decide whether a control action forces the process into a better or worse state. Suggestions for this technique can be found in [Procyk79, Scharf85, Shao88, Tanaka88, Sugeno85b].

If — as described above — it is possible to use the error and the change of error as input variables, we can rely on some standard rules like 'If the error and the change of error are zero, then no control action is needed' or 'If the the error is not zero, but the change of error implies a decrease of error, then maintain the activated control action.'

It should be guaranteed that for all input tuples at least one rule is firing. This always happens if the degree of applicability of at least one rule is bigger than zero and so some control action is activated. It is not always necessary to take the entire space $X_1 \times \ldots \times X_n$ of possible input tuples into account. Looking again at the cartpole problem in Example 4.1, no rule is specified that fires in the case of $\theta = 88°$ and $\dot{\theta} = \pm 44° s^{-1}$. It is likely that for $\dot{\theta} = -44° s^{-1}$ at an angle of $\theta = 88°$ the pole cannot be prevented from falling anymore in this extreme situation. Furthermore, the combination $\theta = 88°$ and $\dot{\theta} = +44° s^{-1}$, which means that the pole has nearly fallen but nevertheless rises again very rapidly, is very unlikely and will never appear in practice.

The completeness of a rule base should guarantee that for all possible situations the controller can determine a control action. On the other hand — not only for reasons of computation speed — the number of rules should not be increased arbitrarily. The disjunctive combination of the output fuzzy sets of all rules leads to the output fuzzy set getting more and more vague with the number of rules that are applicable. This anomaly will be the subject of discussions in the following two sections.

To simplify the defuzzification and in order to get a predictable, reliable fuzzy controller, it should be kept in mind that one should model the deterministic behaviour of a control expert. A rule base that considers for one situation two or more contradicting control actions as equally likely, as in Example 4.2, should be avoided.

4.3.5 Evaluation of Linguistic Rules

A part of the rule base of a fuzzy controller contains the information about how the degree of applicability of a rule has to be computed. The most often used method is the min-operation, as in equation (4.1). In principle, the minimum can be replaced by any other t-norm. In applications that require a smooth control behaviour, only continuous t-norms should be considered.

Usually the degree of applicability α_r of the premise of rule R_r is determined as in equation (4.1). But the 'output' of a single rule R_r in some applications differs from that given by the formula for the fuzzy set $\mu_{x_1,\ldots,x_n}^{\text{output}}$. Instead of taking the minimum of the degree of applicability α_r of rule R_r and the corresponding fuzzy set μ_{i_r} in the conclusion, the output fuzzy set is obtained by multiplying it by the value α_r. The application of the formula $\alpha_r \cdot \mu_{i_r}$ is called *dot method* or *max dot method*, if the output fuzzy sets $\alpha_r \cdot \mu_{i_r}$ of the single rules are combined in a disjunctive manner by the maximum-operator.

For a Mamdani controller the way of combining the fuzzy sets, resulting from the activated rules, has to be specified. Nearly always the maximum (4.2) is chosen for this operation, although it is in principle possible to replace the maximum by any other (continuous) t-conorm as a union operator.

In the case of a Sugeno controller, for each rule R_r an appropriate output function f_r has to be determined. This is the last step in the design of a Sugeno controller. The functions f_r — in most cases linear functions in the control variables — have to be guessed, derived from observations, or obtained by analytical methods. The latter technique is applied when appropriate control functions are known for certain regions of the input domain, e.g. a linearization of the process in each region. In this case a Sugeno controller can be used to switch smoothly between the control functions for the different regions, when the process crosses the border of two regions.

For the Mamdani as well as for the Sugeno controller we have used crisp input values and have avoided a real fuzzification of the measured input. But it is possible that inaccurate measurements with distortions have to be handled. To cope with this problem it is advisable to consider fuzzy sets rather than crisp values as inputs. For, instance a triangular membership function, having the maximum at the measured value, could represent an imprecise measurement. Trapezoidal membership functions are also suitable. Since for such fuzzification techniques there is still no standard technique established, we will not discuss this problem in further detail.

4.3.6   The Choice of the Defuzzification Strategy

For the Mamdani fuzzy controller a suitable defuzzification method has to be chosen. Despite its missing semantic background the centre of gravity method is commonly used, since it guarantees continuous changes of the control actions. In the case of a finite output Y the centre of gravity has to be modified into

$$\eta = \frac{1}{\displaystyle\sum_{y \in Y} \mu_{x_1,\ldots,x_n}^{\text{output}}(y)} \cdot \sum_{y \in Y} y \cdot \mu_{x_1,\ldots,x_n}^{\text{output}}(y).$$

Once again, we emphasize that a consistent defuzzification can only be guaranteed, when the rule base models a deterministic control behaviour, so that the problems in Example 4.2 will not occur.

A short discussion of other defuzzification strategies can be found in Section 4.6.3.

4.3.7   Optimizing a Fuzzy Controller

In general, the first design of a fuzzy controller will not result in an optimal control behaviour. To improve the control behaviour, tuning is necessary. Depending on how well or badly the first design of a controller is able to handle the process, some minor or major changes have to be considered. Starting from small adjustments of single parameters (e.g. shape and position of the fuzzy sets, changing the fuzzy partitions of the domains, adding or deleting fuzzy sets, leading to alterations in the rule base) the tuning process might lead in the worst case to a total redesign of the controller. A redesign may become necessary, when it is realized that the measured inputs are insufficient according to low precision instruments or to missing information that requires the measurement of additional input variables.

Without considering the case of a complete redesign, the strategies for optimizing a controller strongly depend on the nature of the process and the desired accuracy of the control behaviour. In general, due to the principles of a fuzzy controller, it is quite easy to improve a controller, because it is simple to identify the specific rules that lead to an undesirable behaviour. Changing single rules or the corresponding fuzzy sets has mainly local influences, simplifying the tuning procedure.

4.4   Fuzzy Control as Interpolation in the Presence of Imprecision

The methods presented in Section 4.2 are based on the use of heuristic ideas, without a concrete interpretation of the applied techniques. The principle idea behind fuzzy control is the specification of non-linear transition functions in the framework of knowledge-based interpolation. The method described for the Mamdani controller can be interpreted as (fuzzy) interpolation taking imprecision into account. Before we can explain in which sense a Mamdani controller performs fuzzy interpolation, we have to develop an approach to interpolation in the presence of imprecision. It turns out that the operations and results for this interpolation technique are more or less identical with Mamdani's control method

Section 4.4.2 provides the framework for translating a Mamdani controller into our framework of interpolation, in which the soundness of the rule base can be checked.

4.4.1   Equality Relations as a Basis for Fuzzy Control

In this section we will develop a new approach to fuzzy control that is based on a viewpoint differing from those introduced until now. The approach is based on the idea that the following information about the process is available:

- For some inputs a corresponding reasonable control action can be specified, at least approximately.
- An indistinguishability is inherent in the domains for the control and input variables. The indistinguishability characterizes input values or control values which are more or less identifiable in the context of the process, meaning that, for example, a difference of less than 0.00001 in the control action does not really influence the behaviour of the process. The problem specific knowledge about the indistinguishability can be described. We will see that equality relations are capable of modelling such an indistinguishability.

Let us assume that for some input values, the appropriate control values are known approximately. We also expect that small changes in the inputs lead only to small changes in the corresponding control actions, without giving a precise definition of the notion of small changes. Our aim is to derive a control function from some cases for which an appropriate control action is known, together with the formal specification of what the small changes mean in the context of the considered process.

Specification of a Partial Control Function

The control expert is asked to determine the control action for a finite number of input values, i.e. he will provide for the input tuples $(x_1^{(r)}, \ldots, x_n^{(r)}) \in X_1 \times \ldots \times X_n$, $r = 1, \ldots, k$, the corresponding control values $y^{(r)}$. Note that

it is not required that values are really exact, since the indistinguishability
inherent in the process admits small imprecision.

In order to guarantee a deterministic control behaviour, we require that no
two such rules have the same premise, i.e. we have $x_1^{(r)} = x_1^{(s)} \wedge \ldots \wedge x_n^{(r)} = x_n^{(s)} \implies r = s$.

Technically speaking, we obtain a mapping $\varphi_0 : X_0 \to Y$ with $X_0 = \{(x_1^{(r)}, \ldots, x_n^{(r)}) \mid r \in \{1, \ldots, k\}\}$ being the set of input tuples for which a
control value is specified by the expert. φ_0 alone as a partial mapping from
$X_1 \times \ldots \times X_n$ to Y does not enable us to derive a totally defined control
function $\varphi : X_1 \times \ldots \times X_n \to Y$ that assigns to each input tuple an appropriate
output value $y \in Y$.

Specification of the Indistinguishability Structure

However, we should not consider each input tuple $(x_1, \ldots, x_n) \in X_1 \times \ldots \times X_n$
as an isolated point without any relations to the other input tuples. Assuming
that the sets $X_1, \ldots, X_n, Y$ are intervals of real numbers, this isolated point
of view is unrealistic. First of all, the control expert himself is often unable
or even unwilling to distinguish between two input tuples that are very close
together. Secondly, a negligible deviation is not of much importance for a
decision on a certain control action. Of course, the meaning of negligible
depends on the process. While controlling a process, an expert neither solves
complex differential equations nor uses values or measurements with an
arbitrary precision.

We model the indistinguishability that is caused, on the one hand, by the
limited precision of measurements and, on the other hand, by the intended
inexactness of the expert with equality relations $E_1, \ldots, E_n$, and F on the
sets $X_1, \ldots, X_n$, and Y. These relations can be interpreted in the sense of
Example 2.51. We also recommend reference to Examples 2.54 and 2.60 for a
deeper understanding.

We do not require that the equality relations are induced by the standard
metric on the intervals $X_1, \ldots, X_n, Y$. In general, there are ranges, where an
expert distinguishes values very precisely and there are other ranges where
only the magnitude of a value is of interest. If we return to the cartpole
problem, given an angular velocity of $\dot{\theta} = 0°\text{s}^{-1}$, the control action taken for
an angle of $\theta = 65°$ will not be much different from the one taken for an angle
of $\theta = 55°$. In both cases the pole is very oblique, and a strong action has to
be carried out to bring it back to an upright position. If we look, in contrast,
at the angles $\theta = -5°$ and $\theta = +5°$, we face a completely different situation:
Now it is very important to distinguish whether the pole is inclined to the left
or to the right.

It is reasonable to specify an equality relation in terms of scaling functions.

Example 4.4 Let $c : [-90, +90] \to \mathbb{R}_0^+$ be an integrable function

$$f : [-90, +90] \quad \to \quad \mathbb{R},$$
$$\alpha \quad \mapsto \quad \int_{-90}^{\alpha} c(x)dx. \tag{4.4}$$

Using the equality relation $E_{f(\delta)}$ simply means that we magnify the neighbourhood of the angle θ with a scaling factor of $c(\theta)$. A scaling factor smaller than 1 represents a reduction. If we distinguish precisely only in the range between $-10°$ and $+10°$, and otherwise only roughly, we can define

$$c : [-90, +90] \quad \to \quad \mathbb{R}^+,$$
$$\alpha \quad \mapsto \quad \begin{cases} 2, & \text{if } -10 \le \alpha \le 10 \\ \frac{1}{2}, & \text{otherwise} \end{cases}$$

and use the equality relation $E_{f(\delta)}$ where f is defined as in (4.4). In general, we could use a continuous function c that reaches its maximum at zero. $\square$

Scaling factors are one possibility for specifing an equality relation. Other methods will be discussed in Section 4.4.2.

Determination of the Fuzzy-valued Control Function

The information provided by the control expert should consist of k 'rule'-tuples $((x_1^{(r)}, \ldots, x_n^{(r)}), y^{(r)})$, $r = 1, \ldots, k$ including the description of the equality relations that reflect the indistinguishability of the measured input and output values.

Our aim is to derive a control function $\varphi : X_1 \times \ldots \times X_n \to Y$ from this information. Taking the indistinguishability induced by the equality relations into account, we can no longer expect to obtain a crisp function φ. Moreover, we should adapt all our concepts to the equality relations. Therefore we are not trying to determine the control function φ or its graph directly, but are interested in the extensional hull of the graph, i.e. the view of the control function in the presence of the specified indistinguishability.

Hence, what we need is an equality relation on the product space $X_1 \times \ldots \times X_n \times Y$. From Theorem 2.61, we have the equality relation

$$E((x_1, \ldots, x_n, y), (x_1', \ldots, x_n', y'))$$
$$\overset{\text{def}}{=} \quad \min\{E_1(x_1, x_1'), \ldots, E_n(x_n, x_n'), F(y, y')\}. \tag{4.5}$$

The assumption inherent in the use of this equality relation is the independence of the indistinguishabilities on the sets $X_1, \ldots, X_n$, and Y (induced by the single equality relations $E_1, \ldots, E_n, F$). Independence of indistinguishabilities means here: The indistinguishability of tuple $(x_1, x_2, \ldots, y)$ and tuple $(x'_1, x_2, \ldots, y)$ does not depend on the choice of the values $x_2, \ldots, x_n, y$. If this assumption is not satisfied, single spaces X_i and X_j have to be combined to the product space $X_i \times X_j$, and an equality relation that reflects the dependence of the indistinguishabilities has to be defined on the product space. In the worst case, the equality relation E can only be specified on the whole product space $X_1 \times \ldots \times X_n \times Y$. It should be emphasized that a dependency of the measured values ξ_i and ξ_j does not automatically induce a dependency of the indistinguishabilities of the values of X_i and X_j.

In the following, we assume that the equality relation (4.5) correctly represents the indistinguishability on the set $X_1 \times \ldots \times X_n \times Y$.

We define for the sets X_i and Y the sets

$$X_i^{(0)} \stackrel{\text{def}}{=} \left\{ x \in X_i \mid \exists r \in \{1, \ldots, k\} : x = x_i^{(r)} \right\}, \tag{4.6}$$

and

$$Y^{(0)} \stackrel{\text{def}}{=} \varphi_0(X_0) = \left\{ y \in Y \mid \exists r \in \{1, \ldots, k\} : y = y^{(r)} \right\}. \tag{4.7}$$

The sets $X_i^{(0)}$ and $Y^{(0)}$ contain all values that occur in one of the input–output tuples $(x_1^{(r)}, \ldots, x_n^{(r)}, y^{(r)})$, $r = 1, \ldots, k$, given by the control expert.

For each $x_0 \in X_1^{(0)}$ we obtain a singleton μ_{x_0} induced by the equality relation E_1:

$$\mu_{x_0}(x) = E_1(x, x_0). \tag{4.8}$$

If the equality relation E_1 is given as in Example 2.52, the singletons are represented by triangular functions.

The sets X_i, $i = 2, \ldots, n$, and Y and their corresponding equality relations E_i and F, respectively, are treated in a similar way.

We now face the problem of how to determine the accurate control value y for a measured input tuple $(x_1, \ldots, x_n)$. But first we will look at the hypothetical case that the control function $\varphi : X_1 \times \ldots \times X_n \to Y$ is given completely. We describe a way to determine the value $y = \varphi(x_1, \ldots, x_n)$ that appears a little bit complicated at first glance, but will turn out to be very useful when dealing with a partially defined control function φ_0 in connection with equality relations.

Let

$$G_\varphi \stackrel{\text{def}}{=} \{(x_1, \ldots, x_n, y) \in X_1 \times \ldots \times X_n \times Y \mid \varphi(x_1, \ldots, x_n) = y\}$$

denote the graph of φ. We obtain the value $y = \varphi(x_1, \ldots, x_n)$ from the graph G_φ by determining the set $\{y \in Y \mid (x_1, \ldots, x_n, y) \in G_\varphi\}$, consisting of a single element. This method is depicted in Figure 13.

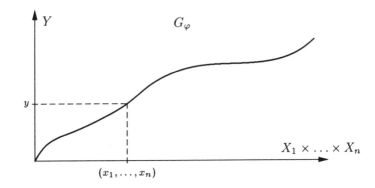

Figure 13   Deriving the value $y = \varphi(x_1, \ldots, x_n)$ from the graph G_φ

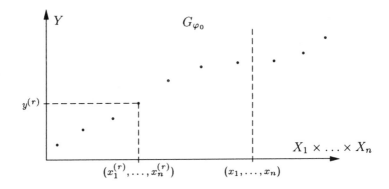

Figure 14   The value $y = \varphi_0(x_1, \ldots, x_n)$ is not defined

Of course, the control function is not completely known, and we consider only a partial mapping φ_0 specified by the control expert. Then the above described method to determine the control value is not applicable, since the set $\{y \in Y \mid (x_1, \ldots, x_n, y) \in G_{\varphi_0}\}$ is empty, unless by chance the input tuple $(x_1, \ldots, x_n)$ is equal to one of the tuples of $(x_1^{(r)}, \ldots, x_n^{(r)})$, $r \in \{1, \ldots, k\}$ (see Figure 14).

However, we can make use of the equality relation (4.5) to fill the gaps in the definition of φ_0: we simply apply the method described for the completely defined function to determine the values $\varphi(x_1, \ldots, x_n)$ and $\varphi_0(x_1, \ldots, x_n)$, respectively, to the extensional hull of the respective graph. The extensional hull of φ_0 is the fuzzy set

$$\mu_{\varphi_0}(x_1', \ldots, x_n', y')$$
$$= \max_{r \in \{1, \ldots, k\}} \left\{ \min\{E_1(x_1^{(r)}, x_1'), \ldots, E_n(x_n^{(r)}, x_n'), F(y^{(r)}, y')\} \right\}. \quad (4.9)$$

μ_{φ_0} is the smallest fuzzy set that contains the graph of φ_0 and respects the equality relation according to Definition 2.55. Obviously we have $\mu_{\varphi_0} \leq \mu_\varphi$, with μ_φ representing the extensional hull of the graph of φ. Since the control function φ is not completely known, we have to determine it with the assistance of μ_{φ_0}.

Applying the method of deriving the value of a function from the graph of the function to the extensional hull of the graph, we obtain from μ_{φ_0} for a measured input tuple $(x_1, \ldots, x_n)$ the fuzzy set

$$\mu_{\varphi_0}^{(x_1, \ldots, x_n)} : Y \longrightarrow [0, 1],$$
$$y \longmapsto \mu_{\varphi_0}(x_1, \ldots, x_n, y) \quad (4.10)$$

that corresponds to the set $\{y \in Y \mid (x_1, \ldots, x_n, y) \in G_{\varphi_0}\}$ (taking the equality relations into account).

Mamdani's Model Revisited

The method of using the extensional hull of the graph of the partial control mapping leads to an output in the form of a fuzzy set. This fuzzy set is identical with the output fuzzy set $\mu_{x_1, \ldots, x_n}^{\text{output}}$ of a corresponding Mamdani fuzzy controller. In order to specify this Mamdani fuzzy controller, we identify the input–output tuples of the partial control function with the linguistic control rules

$$R_r: \quad \textbf{if} \quad \xi_1 \text{ is approximately } x_1^{(r)}$$
$$\textbf{and} \ldots$$
$$\textbf{and } \xi_n \text{ is approximately } x_n^{(r)} \quad (4.11)$$
$$\textbf{then } \eta \text{ is } y^{(r)}.$$

The fuzzy partitions are determined by the singletons that are induced by the values appearing in the rules mentioned above.

Theorem 4.5 *Let $\mu_{x_1, \ldots, x_n}^{\text{output}}$ be the fuzzy set according to (4.2) that is determined by the decision logic of the Mamdani controller using the following knowledge base.*

- *The fuzzy partition of the set X_1 is given by the fuzzy sets μ_{x_0} defined in (4.8), where $x_0 \in X_1^{(0)}$. The fuzzy partitions of the sets $X_2, \ldots, X_n, Y$ are*

defined analogously.
- *The fuzzy set of the form μ_{x_0} is associated with the linguistic term approximately x_0.*
- *The rule base consists of the rules R_r, $r = 1, \ldots, k$, described as in (4.11).*

For the fuzzy set in equation (4.10) we have:

$$\mu_{\varphi_0}^{(x_1, \ldots, x_n)} = \mu_{x_1, \ldots, x_n}^{\text{output}}.$$

Proof:

$$
\begin{aligned}
\mu_{\varphi_0}^{(x_1, \ldots, x_n)}(y) &= \mu_{\varphi_0}(x_1, \ldots, x_n, y) \\[2mm]
&= \max_{r \in \{1, \ldots, k\}} \left\{ \min\{ E_1(x_1^{(r)}, x_1), \ldots, E_n(x_n^{(r)}, x_n), F(y^{(r)}, y) \} \right\} \\[2mm]
&= \max_{r \in \{1, \ldots, k\}} \left\{ \min\{ \mu_{x_1^{(r)}}(x_1), \ldots, \mu_{x_n^{(r)}}(x_n), \mu_{y^{(r)}}(y) \} \right\} \\[2mm]
&= \mu_{x_1, \ldots, x_n}^{\text{output}}(y). \qquad \square
\end{aligned}
$$

The possibility of translating a controller based on equality relations to a Mamdani fuzzy controller does not only provide a background for the interpretation of Mamdani's model. From a practical point of view, this translation opens the wide spectrum of software and hardware, developed for Mamdani fuzzy controllers, to any controller designed on the basis of equality relations.

Determination of a Crisp Control Function

It also is the case that, in the approach based on equality relations, the problem of defuzzification remains open. If we choose a defuzzification method, we finally obtain a mapping $\varphi : X_1 \times \ldots \times X_n \to Y$. In the context of equality relations, it seems reasonable to require the following two properties for the mapping φ:

(i) φ is an extension of the partial mapping φ_0, i.e. we require $\varphi\big|_{X_0} = \varphi_0$, which means $\varphi(x_1, \ldots, x_n) = \varphi_0(x_1, \ldots, x_n)$ for all $(x_1, \ldots, x_n) \in X_0$.

(ii) $\varphi : X_1 \times \ldots \times X_n \to Y$ is extensional with respect to the equality relations

$$E_0 : (X_1 \times \ldots \times X_n)^2 \to [0, 1],$$

$$((x_1, \ldots, x_n), (x_1', \ldots, x_n')) \mapsto \min_{i \in \{1, \ldots, n\}} \{ E_i(x_i, x_i') \}$$

on $X_1 \times \ldots \times X_n$, and F on Y.

The first condition is plausible. To explain the second condition we interpret the equality relations according to Example 2.51. From Example 2.60 we learn that the extensionality of φ is equivalent to φ being error-preserving, which is a reasonable assumption. It means that the equality relations have to be chosen in accordance with each other.

As a consequence of conditions (i) and (ii) we obtain a strong restriction for the defuzzification strategy. Let us examine a measured input tuple $(x_1, \ldots, x_n) \in X_1 \times \ldots \times X_n$ and a rule R_r. We have

$$\alpha_r \overset{\text{def}}{=} \min_{i \in \{1, \ldots, n\}} \{E_i(x_i^{(r)}, x_i)\} = \min_{i \in \{1, \ldots, n\}} \{\mu_{x_i^{(r)}}(x_i)\}. \tag{4.12}$$

Then we can conclude from the extensionality of φ that the inequality

$$\alpha_r \leq F(y^{(r)}, \varphi(x_1, \ldots, x_n)) = \mu_{y^{(r)}}(\varphi(x_1, \ldots, x_n)) \tag{4.13}$$

has to be satisfied. Thus, the control value y has to belong to the α_r-cut of the fuzzy set $\mu_{y^{(r)}}$. Since this is a condition for all rules R_r, the control value y has to be chosen from the set

$$\bigcap_{r=1}^{k} [\mu_{y^{(r)}}]_{\alpha_r} = \bigcap_{r=1}^{k} \left\{ y \in Y \mid \alpha_r \leq \mu_{y^{(r)}}(y) \right\}. \tag{4.14}$$

In the interpretation of extensional mappings as error-preserving mappings this corresponds to the following method. For one measured input tuple $(x_1, \ldots, x_n) \in X_1 \times \ldots \times X_n$ we look at each rule R_r separately. The value $1 - \alpha_r$ represents the smallest error bound for which we are no longer able to distinguish between $(x_1, \ldots, x_n)$ and $(x_1^{(r)}, \ldots, x_n^{(r)})$. As we are looking for a control function φ that is error-preserving, the values $\varphi(x_1, \ldots, x_n)$ and $y^{(r)} = \varphi(x_1^{(r)}, \ldots, x_n^{(r)})$ have to be indistinguishable with respect to the error bound α_r, so that condition (4.13) is fulfilled and therefore the control value has to belong to the set (4.14)).

Let us look again at the fuzzy controller described in Theorem 4.5. We can illustrate the method described above in the following example.

Example 4.6 We consider the 'rules' described by the tuples $(x_1^{(r)}, \ldots, x_n^{(r)}, y^{(r)})$, $r \in \{1, 2, 3\}$. We have the equality relations E_i and F on the sets X_i, $i = 1, \ldots, n$, and Y, respectively. According to Theorem 4.5, we obtain a Mamdani fuzzy controller. Let us furthermore assume that the measured input tuple $(x_1, \ldots, x_n)$ is given. We have

$$\alpha_r = \min\{\mu_{x_1^{(r)}}(x_1), \ldots, \mu_{x_n^{(r)}}(x_n)\}$$

$$= \min\{E_1(x_1^{(r)}, x_1), \ldots, E_n(x_n^{(r)}, x_n)\}$$

as the degree of applicability of the corresponding rule's premise. If the singletons $\mu_{y(r)}$ are represented by triangular functions, we obtain the set (4.14), which is depicted in Figure 15. The control value has to be chosen from this set in order to maintain the extensionality of the control mapping.

□

It is neither reasonable nor necessary for determining the set (4.14) to compute the fuzzy set $\mu^{\text{output}}_{x_1,\ldots,x_n}$, because $\mu^{\text{output}(R_r)}_{x_1,\ldots,x_n}$, $r = 1,\ldots,k$, is required.

Example 4.7 If we are given an angle $\theta = 36°$ and an angular velocity of $\dot\theta = -2.25°s^{-1}$, then the value α_r is different from zero only for the two rules depicted in Figure 5 and Figure 6. The set (4.14) of possible control values consists of a single value: $F = 4$ N.

$F = 4$ N is the only value that is acceptable as output of the defuzzification interface. But neither the centre of gravity nor the mean of maximum strategy yield this value. However, the result of the centre of gravity method is nearly this value, so we obtain at least a good approximation by the centre of gravity method.

□

As illustrated in Example 4.7, it might be useful to determine the set (4.14), instead of using a defuzzification strategy. But this set can only be interpreted reasonably if the fuzzy controller has been designed on the basis of equality relations. The specification of appropriate equality relations might turn out to be problem, because the definition of an equality relation requires a global point of view on the reference set, whereas fuzzy sets have to be specified just for a certain area, i.e. locally. In the following we will show under which conditions a Mamdani controller can be reinterpreted on the basis of equality relations.

4.4.2 Interpretation of a Mamdani Controller Based on Equality Relations

The above considerations have shown that a fuzzy controller based on equality relations with its clear semantic background can be translated to a Mamdani controller which is only motivated on a heuristic basis. The obvious question that arises is whether every Mamdani controller can be interpreted in terms of equality relations. Putting a Mamdani controller in the formal framework of equality relations elucidates its functionality and enables us to check whether the rules are consistent with respect to equality relations, which have a solid underlying semantics.

Let us consider a Mamdani fuzzy controller with the fuzzy partitions $\{\mu_j^{(i)} \mid j \in \{1,\ldots,p_i\}\}$, $i = 1,\ldots,n$, and $\{\mu_j \mid j \in \{1,\ldots,p\}\}$ of the sets X_i and Y and the rule base $\{R_r \mid r \in \{1,\ldots,k\}\}$. To interpret this Mamdani

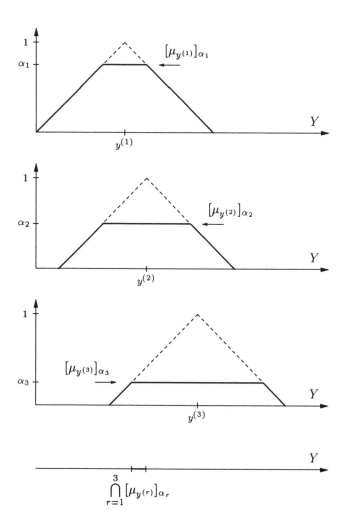

Figure 15 For an extensional control function, the defuzzification has to result in a value from the set $\bigcap_{r=1}^{3} [\mu_{y^{(r)}}]_{\alpha_r}$.

controller as a controller based on equality relations, the following has to be done:

(1) With each fuzzy set $\mu_j^{(i)}$ and μ_j we associate points $x_j^{(i)} \in X_i$ and $y_j \in Y$, respectively.

(2) We define equality relations E_i, $i = 1, \ldots, n$, and F on the sets X_i and Y, respectively, such that the fuzzy sets $\mu_j^{(i)}$ and μ_j can be interpreted as singletons that are induced by the points $x_j^{(i)}$ and y_j, i.e. we have

$$\mu_j^{(i)}(x) \quad = \quad \mu_{x_j^{(i)}}(x) \quad = \quad E_i(x_j^{(i)}, x) \quad \text{and}$$
$$\mu_j(y) \quad = \quad \mu_{y_j}(y) \quad = \quad F(y_j, y), \quad \text{respectively.}$$

(3) The linguistic terms $A_j^{(i)}$ and B_j in the rules have to be replaced by the terms *approximately* $x_j^{(i)}$ and *approximately* y_j, respectively.

The third step should not cause any problems, as far as steps (1) and (2) are feasible. The only condition for step (1) is that all fuzzy sets in the partitions are normal, i.e. for all i, j

$$\left[\mu_j^{(i)}\right]_1 \neq \emptyset \quad \text{and} \quad [\mu_j]_1 \neq \emptyset$$

holds. Under this presupposition we choose for each fuzzy set a value $x_j^{(i)} \in X_i$ and $y_j \in Y$, that fulfills $\mu_j^{(i)}(x_j^{(i)}) = 1$ and $\mu_j(y_j) = 1$, respectively. We associate this value with the corresponding fuzzy set.

The essential step is the second one. It is only feasible under certain conditions. In step (2) each domain $X_1, \ldots, X_n, Y$ is treated separately. Hence, the general problem to be solved for each domain is the following:

We are *given* a family $(\mu_i)_{i \in I}$ of fuzzy sets on X (a fuzzy partition) and a family $(x_i)_{i \in I}$ of elements of X with

$$\forall i \in I : \mu_i(x_i) = 1.$$

We are *searching for* an equality relation E on the set X such that the fuzzy sets μ_i, $i \in I$, correspond to the singletons that are induced by the elements x_i, i.e. we require E to be chosen in the way that

$$\forall i \in I : \mu_i = \mu_{x_i}.$$

holds.

This problem can be solved by making use of the concept of a biimplication.

Theorem 4.8 *Let* $\top$ *be a continuous t-norm. Furthermore, let* $(\mu_i)_{i \in I}$ *be a non-empty family of fuzzy sets of* X *and* $(x_i)_{i \in I}$ *a family of elements of* X *with*

$$\forall i \in I : \mu_i(x_i) = 1.$$

Then the following two statements are equivalent:

(i) There is an equality relation E with respect to $\top$ on X, such that the fuzzy sets μ_i correspond to the singletons that are induced by the points x_i, i.e. we have

$$\mu_i = \mu_{x_i} \qquad (4.15)$$

for all $i \in I$.

(ii) For all $i, j \in I$ the inequality

$$\sup_{x \in X}\{\top(\mu_i(x), \mu_j(x))\} \leq \inf_{y \in X}\{\overleftrightarrow{\top}(\mu_i(y), \mu_j(y))\} \qquad (4.16)$$

holds.

Before proving the theorem, let us examine condition (4.16), which appears to be quite technical. In the sense of Section 2.7 this condition is equivalent to the 'translation' of the statement:

For all $i, j \in I$ we have:

$$(\exists x \in X : (x \in \mu_i \wedge x \in \mu_j)) \to \forall y \in X : (y \in \mu_i \leftrightarrow y \in \mu_j).$$

In other words: the degree of non-disjointness of the fuzzy sets μ_i and μ_j must not exceed their degree of equality. For ordinary sets, this is the usual requirement for two sets of a partition to be either disjoint or equal. This again justifies the use of the term 'fuzzy partition' for the family of fuzzy sets and their associated linguistic terms.

Proof:

(i) $\Rightarrow$ (ii)

Let E be an equality relation that satisfies condition (4.15). We prove (4.16) by showing that the inequalities

$$\sup_{x \in X}\{\top(\mu_i(x), \mu_j(x))\} \leq E(x_i, x_j) \qquad (4.17)$$

and

$$E(x_i, x_j) \leq \inf_{y \in X}\{\overleftrightarrow{\top}(\mu_i(y), \mu_j(y))\} \qquad (4.18)$$

hold.

From the presupposition (4.15), we derive

$$
\begin{aligned}
\sup_{x \in X} \{\mathsf{T}(\mu_i(x), \mu_j(x))\} \quad &= \quad \sup_{x \in X} \{\mathsf{T}(E(x_i, x), E(x_j, x))\} \\
&\underset{E \text{ transitive}}{\leq} \quad \sup_{x \in X} E(x_i, x_j) \\
&= \quad E(x_i, x_j),
\end{aligned}
$$

resulting in (4.17). We prove (4.18) by showing that for all $y \in X$ the inequality

$$
E(x_i, x_j) \leq \overset{\leftrightarrow}{\mathsf{T}}(\mu_i(y), \mu_j(y))
$$

holds. Without loss of generality, let $E(x_i, y) \geq E(x_j, y)$. In combination with (4.15), we obtain

$$
\begin{aligned}
\overset{\leftrightarrow}{\mathsf{T}}(\mu_i(y), \mu_j(y)) \quad &= \quad \sup \big\{ \alpha \in [0, 1] \, \big| \, \mathsf{T}(\max\{E(x_i, y), E(x_j, y)\}, \alpha) \\
&\qquad\qquad\qquad\qquad \leq \min\{E(x_i, y), E(x_j, y)\} \big\} \\
&= \quad \sup \big\{ \alpha \in [0, 1] \, \big| \, \mathsf{T}(E(x_i, y), \alpha) \leq E(x_j, y) \big\} \\
&\geq \quad E(x_i, x_j),
\end{aligned}
$$

since the inequality $\mathsf{T}(E(x_i, y), \alpha) \leq E(x_j, y)$ is satisfied due to the transitivity of E for $\alpha = E(x_j, x_i)$.

(ii) $\Rightarrow$ (i)

We define the equality relation $E : X \times X \rightarrow [0, 1]$ by

$$
E(x, y) \overset{\text{def}}{=} \inf_{i \in I} \left\{ \overset{\leftrightarrow}{\mathsf{T}}(\mu_i(x), \mu_i(y)) \right\}. \tag{4.19}
$$

First we prove that E is an equality relation, without using condition (ii).

(a)
$$
\begin{aligned}
E(x, x) \quad &= \quad \inf_{i \in I} \big\{ \sup \big\{ \alpha \in [0, 1] \, \big| \, \mathsf{T}(\max\{\mu_i(x), \mu_i(x)\}, \alpha) \\
&\qquad\qquad\qquad\qquad \leq \min\{\mu_i(x), \mu_i(x)\} \big\} \big\} \\
&= \quad \inf_{i \in I} \big\{ \sup\{\alpha \in [0, 1] \, \big| \, \mathsf{T}(\mu_i(x), \alpha) \leq \mu_i(x)\} \big\} \\
&= \quad 1.
\end{aligned}
$$

(b) Obviously, $E(x, y) = E(y, x)$ holds, since — taking Theorem 2.63 into account — we have $\overset{\leftrightarrow}{\mathsf{T}}(\alpha, \beta) = \overset{\leftrightarrow}{\mathsf{T}}(\beta, \alpha)$.

(c) $\top\left(E(x,y),E(y,z)\right)$

$$= \top\Bigg(\ \inf_{i\in I}\Big\{\sup\big\{\alpha\in[0,1]\ \ \big|\ \ \top(\max\{\mu_i(x),\mu_i(y)\},\alpha)$$
$$\leq\min\{\mu_i(x),\mu_i(y)\}\big\}\Big\},$$
$$\inf_{j\in I}\Big\{\sup\big\{\beta\in[0,1]\ \ \big|\ \ \top(\max\{\mu_j(y),\mu_j(z)\},\beta)$$
$$\leq\min\{\mu_j(y),\mu_j(z)\}\big\}\Big\}\Bigg)$$

$$\stackrel{\top\ \text{continuous}}{=}\ \inf_{i,j\in I}\Big\{\top\Big(\ \sup\big\{\alpha\in[0,1]\ \ \big|\ \ \top(\max\{\mu_i(x),\mu_i(y)\},\alpha)$$
$$\leq\min\{\mu_i(x),\mu_i(y)\}\big\},$$
$$\sup\big\{\beta\in[0,1]\ \ \big|\ \ \top(\max\{\mu_j(y),\mu_j(z)\},\beta)$$
$$\leq\min\{\mu_j(y),\mu_j(z)\}\big\}\Big)\Big\}$$

$$\leq\ \inf_{i\in I}\Big\{\top\Big(\ \sup\big\{\alpha\in[0,1]\ \ \big|\ \ \top(\max\{\mu_i(x),\mu_i(y)\},\alpha)$$
$$\leq\min\{\mu_i(x),\mu_i(y)\}\big\},$$
$$\sup\big\{\beta\in[0,1]\ \ \big|\ \ \top(\max\{\mu_i(y),\mu_i(z)\},\beta)$$
$$\leq\min\{\mu_i(y),\mu_i(z)\}\big\}\Big)\Big\}$$

$$=\ \inf_{i\in I}\Big\{\sup_{\alpha,\beta\in[0,1]}\big\{\top(\alpha,\beta)\ \ \big|\ \ \top(\max\{\mu_i(x),\mu_i(y)\},\alpha)$$
$$\leq\min\{\mu_i(x),\mu_i(y)\}\ \ \wedge$$
$$\top(\max\{\mu_i(y),\mu_i(z)\},\beta)$$
$$\leq\min\{\mu_i(y),\mu_i(z)\}\big\}\Big\}.$$

$$(4.20)$$

We prove that (4.20) is less than or equal to

$$E(x,z)=\inf_{i\in I}\Big\{\sup\big\{\gamma\in[0,1]\ \ \big|\ \ \top(\max\{\mu_i(x),\mu_i(z)\},\gamma)$$
$$\leq\min\{\mu_i(x),\mu_i(z)\}\big\}\Big\},\qquad(4.21)$$

by showing that for all $i \in I$ and for all $\varepsilon \in [0, 1)$

$$\left(\varepsilon < \sup_{\alpha, \beta \in [0,1]} \{T(\alpha, \beta) \mid T(\max\{\mu_i(x), \mu_i(y)\}, \alpha)\right.$$
$$\leq \min\{\mu_i(x), \mu_i(y)\} \quad \wedge$$
$$T(\max\{\mu_i(y), \mu_i(z)\}, \beta)$$
$$\left.\leq \min\{\mu_i(y), \mu_i(z)\}\}\right) \qquad (4.22)$$

$$\Longrightarrow$$

$$\left(\varepsilon < \sup\{\gamma \in [0, 1] \mid T(\max\{\mu_i(x), \mu_i(z)\}, \gamma)\right.$$
$$\left.\leq \min\{\mu_i(x), \mu_i(z)\}\}\right) \qquad (4.23)$$

holds. Let us assume (4.22). Thus there are $\alpha, \beta \in [0, 1]$ that satisfy the three conditions

$$T(\alpha, \beta) > \varepsilon,$$
$$T(\max\{\mu_i(x), \mu_i(y)\}, \alpha) \leq \min\{\mu_i(x), \mu_i(y)\}, \qquad (4.24)$$
$$T(\max\{\mu_i(y), \mu_i(z)\}, \beta) \leq \min\{\mu_i(y), \mu_i(z)\}. \qquad (4.25)$$

We define $\gamma \stackrel{\text{def}}{=} T(\alpha, \beta)$ and prove

$$T(\max\{\mu_i(x), \mu_i(z)\}, \gamma) \leq \min\{\mu_i(x), \mu_i(z)\}. \qquad (4.26)$$

Without loss of generality, let $\mu_i(x) \leq \mu_i(z)$ such that (4.26) simplifies to

$$T(\mu_i(z), \gamma) \leq \mu_i(x).$$

Case 1: $\mu_i(y) \leq \mu_i(x) \leq \mu_i(z)$.

As a consequence, we have

$$
\begin{aligned}
T(\mu_i(z), \gamma) &= T(\mu_i(z), T(\alpha, \beta)) \\
&= T(T(\mu_i(z), \beta), \alpha) \\
&\underset{(4.25)}{\leq} T(\mu_i(y), \alpha) \\
&\leq \mu_i(y) \\
&\leq \mu_i(x).
\end{aligned}
$$

Case 2: $\mu_i(x) \leq \mu_i(y) \leq \mu_i(z)$.

Analogously to case 1 we obtain

$$\mathsf{T}(\mu_i(z), \gamma) \leq \mathsf{T}(\mu_i(y), \alpha),$$

and considering (4.24) it follows that

$$\mathsf{T}(\mu_i(y), \alpha) \leq \mu_i(x).$$

Case 3: $\mu_i(x) \leq \mu_i(z) \leq \mu_i(y)$.

Equation (4.24) leads to

$$\mathsf{T}(\mu_i(z), \gamma) \leq \mathsf{T}(\mu_i(y), \gamma) \leq \mathsf{T}(\mu_i(y), \alpha) \leq \mu_i(x).$$

As a result, the inequality (4.26) is valid and therefore (4.23) holds, too. Consequently, E is transitive and thus an equality relation.

What remains to be proved is $\mu_i = \mu_{x_i}$. We first show $\mu_{x_i} \leq \mu_i$:

$$
\begin{aligned}
\mu_{x_i}(x) &= E(x, x_i) \\
&= \inf_{j \in I} \left\{ \sup \left\{ \alpha \in [0, 1] \;\Big|\; \mathsf{T}(\max\{\mu_j(x_i), \mu_j(x)\}, \alpha) \right.\right. \\
&\qquad\qquad\qquad\qquad\qquad\qquad \left.\left. \leq \min\{\mu_j(x_i), \mu_j(x)\} \right\} \right\} \\
&\leq \sup \left\{ \alpha \in [0, 1] \;\Big|\; \mathsf{T}(\max\{\mu_i(x_i), \mu_i(x)\}, \alpha) \right. \\
&\qquad\qquad\qquad\qquad \left. \leq \min\{\mu_i(x_i), \mu_i(x)\} \right\} \\
&= \sup\{\alpha \in [0, 1] \mid \mathsf{T}(1, \alpha) \leq \mu_i(x)\} \\
&= \mu_i(x).
\end{aligned}
$$

Finally, we show $\mu_i \leq \mu_{x_i}$ by proving that

$$
\sup \left\{ \alpha \in [0, 1] \;\Big|\; \mathsf{T}(\max\{\mu_j(x_i), \mu_j(x)\}, \alpha) \right. \\
\left. \leq \min\{\mu_j(x_i), \mu_j(x)\} \right\} \quad \geq \quad \mu_i(x) \tag{4.27}
$$

holds for all $j \in I$.

Case 1: $\mu_j(x_i) \leq \mu_j(x)$.

This simplifies the left side of (4.27) to

$$\sup\{\alpha \in [0, 1] \mid \mathsf{T}(\mu_j(x), \alpha) \leq \mu_j(x_i)\}. \tag{4.28}$$

The expression (4.28) is greater than or equal to $\mu_i(x)$, if $T(\mu_j(x), \mu_i(x)) \leq \mu_j(x_i)$ holds. Taking (ii) into account, we obtain

$$T(\mu_j(x), \mu_i(x)) \leq \inf_{y \in X} \Big\{ \sup \big\{ \beta \in [0,1] \big| T(\max\{\mu_j(y), \mu_i(y)\}, \beta)$$
$$\leq \min\{\mu_j(y), \mu_i(y)\} \big\} \Big\}$$
$$\leq \sup \big\{ \beta \in [0,1] \big| T(\max\{\mu_j(x_i), \mu_i(x_i)\}, \beta)$$
$$\leq \min\{\mu_j(x_i), \mu_i(x_i)\} \big\}$$
$$= \mu_j(x_i).$$

Case 2: $\mu_j(x) \leq \mu_j(x_i)$.

In this case the left side of (4.27) simplifies to

$$\sup\{\alpha \in [0,1] \mid T(\mu_j(x_i), \alpha) \leq \mu_j(x)\}.$$

Now we have to prove

$$T(\mu_j(x_i), \mu_i(x)) \leq \mu_j(x). \tag{4.29}$$

If $\mu_i(x) \leq \mu_j(x)$ holds, (4.29) is obviously satisfied. So let $\mu_j(x) \leq \mu_i(x)$. Then, since $\mu_i(x_i) = 1$,

$$\mu_j(x_i) = T(\mu_j(x_i), \mu_i(x_i))$$
$$\leq \sup_{z \in X} \{T(\mu_j(z), \mu_i(z))\}$$
$$\overset{(ii)}{\leq} \inf_{z \in X} \left\{ \overset{\leftrightarrow}{T}(\mu_j(z), \mu_i(z)) \right\}$$
$$\leq \overset{\leftrightarrow}{T}(\mu_j(x), \mu_i(x))$$
$$= \sup \big\{ \alpha \in [0,1] \mid T(\max\{\mu_j(x), \mu_i(x)\}, \alpha)$$
$$\leq \min\{\mu_j(x), \mu_i(x)\} \big\}$$
$$= \sup \big\{ \alpha \in [0,1] \mid T(\mu_i(x), \alpha) \leq \mu_j(x) \big\}$$

follows, and consequently $T(\mu_i(x), \mu_j(x_i)) \leq \mu_j(x)$, as T is monotonous and continuous. $\square$

We did not apply condition (ii) of Theorem 4.8 for proving $\mu_i \geq \mu_{x_i}$, and showing that (4.19) defines an equality relation. Condition (ii) is only needed for the inequality $\mu_{x_i} \geq \mu_i$. The following theorem reveals that (4.19) is the greatest equality relation such that all μ_i are extensional.

Theorem 4.9 *Let* T *be a continuous t-norm. In addition, let* $(\mu_i)_{i \in I}$ *be a family of fuzzy sets of* X. *Then, the equality relation* E, *defined by* (4.19), *is the greatest equality relation with respect to* T *such that* μ_i *is extensional for all* $i \in I$.

Proof:
We first show that μ_i, $i \in I$, is extensional with respect to E.

$$\mathsf{T}(\mu_i(x), E(x, y))$$

$$= \quad \mathsf{T}(\mu_i(x), \inf_{j \in I} \{ \overleftrightarrow{\mathsf{T}}(\mu_j(x), \mu_j(y)) \})$$

$$\leq \quad \mathsf{T}(\mu_i(x), \overleftrightarrow{\mathsf{T}}(\mu_i(x), \mu_i(y)))$$

$$\overset{\mathsf{T} \text{ continuous}}{=} \quad \sup_{\alpha \in [0,1]} \{ \mathsf{T}(\mu_i(x), \alpha) \mid \quad \mathsf{T}(\max\{\mu_i(x), \mu_i(y)\}, \alpha)$$
$$\leq \min\{\mu_i(x), \mu_i(y)\} \}.$$

If $\mu_i(x) \leq \mu_i(y)$, (4.30) simplifies to

$$\sup_{\alpha \in [0,1]} \{ \mathsf{T}(\mu_i(x), \alpha) \mid \mathsf{T}(\mu_i(y), \alpha) \leq \mu_i(x) \}$$

$$\leq \quad \sup_{\alpha \in [0,1]} \{ \mathsf{T}(\mu_i(x), \alpha) \mid \mathsf{T}(\mu_i(x), \alpha) \leq \mu_i(x) \}$$

$$= \quad \mathsf{T}(\mu_i(x), 1)$$

$$= \quad \mu_i(x)$$

$$\leq \quad \mu_i(y).$$

If $\mu_i(y) \leq \mu_i(x)$, then (4.30) becomes

$$\sup_{\alpha \in [0,1]} \{ \mathsf{T}(\mu_i(x), \alpha) \mid \mathsf{T}(\mu_i(x), \alpha) \leq \mu_i(y) \} = \mu_i(y).$$

Hence, μ_i is extensional with respect to E.

Let E' be an equality relation with respect to T such that μ_i is extensional for all $i \in I$. We show that

$$\overleftrightarrow{\mathsf{T}}(\mu_i(x), \mu_i(y)) \geq E'(x, y)$$

holds for all $i \in I$. From this inequality we can deduce $E \geq E'$. Without loss of generality, let $\mu_i(x) \leq \mu_i(y)$. Then it follows that

$$\overleftrightarrow{\mathsf{T}}(\mu_i(x), \mu_i(y)) = \sup\{\alpha \in [0, 1] \mid \mathsf{T}(\mu_i(y), \alpha) \leq \mu_i(x)\}. \tag{4.30}$$

Since μ_i is extensional with respect to E', we have

$$\mathsf{T}(\mu_i(y), E'(x, y)) \leq \mu_i(x),$$

which implies that (4.30) yields a value greater than or equal to $E'(x, y)$. $\square$

Taking Theorem 4.9 into account, we conclude that the equality relation given in (4.19) is the greatest equality relation that fulfills (4.15).

Theorem 4.10 *Let* T *be a continuous t-norm. Furthermore, let* $(\mu_i)_{i \in I}$ *be a non-empty family of fuzzy sets* X, *and* $(x_i)_{i \in I}$ *a family of elements of* X *with*

$$\forall i \in I : \mu_i(x_i) = 1.$$

Let the inequality (4.16) *of Theorem 4.8 be valid for all* $i, j \in I$.
Then the equality relation E, *given in* (4.19), *is the greatest equality relation — with respect to* T — *which fulfills* (4.15).

Proof:
We demonstrated in the proof of Theorem 4.8 that E satisfies condition (4.15). Suppose E' is an arbitrary equality relation with respect to T also satisfying (4.15). Then the fuzzy sets $\mu_i = \mu_{x_i}$, $i \in I$, are singletons and therefore, because they are extensional with respect to E', extensional hulls of sets of cardinality one. Considering Theorem 4.9, we have E as the greatest equality relation for which the fuzzy sets μ_i, $i \in I$, are extensional. $\square$

Theorem 4.10 shows: if there is an equality relation that enables us to interpret the fuzzy sets as singletons, then we can compute the greatest one with formula (4.19). Although condition (4.16) can be interpreted in a reasonable way and represents a necessary and sufficient condition for the existence of such an equality relation, it can actually be quite difficult to verify the condition. This is why we present a more simple sufficient condition that can be verified easily.

Theorem 4.11 *Let* T *be a continuous t-norm. Furthermore, let* $(\mu_i)_{i \in I}$ *be a non-empty family of fuzzy sets* X, *and* $(x_i)_{i \in I}$ *a family of elements of* X *with*

$$\forall i \in I : \mu_i(x_i) = 1.$$

Let for all $i, j \in I, i \neq j$ *and for all* $x \in X$ *the equation*

$$\mathsf{T}(\mu_i(x), \mu_j(x)) = 0 \tag{4.31}$$

be valid. Then the equality relation E, *given in* (4.19), *fulfills the condition* (4.15).

Proof:
If $i \neq j$, then (4.31) implies (4.16) since the left side of (4.16) becomes zero. For $i = j$, we obtain the value 1 on the right side of (4.16), and in this case (4.16) is fulfilled, too. We derive from Theorem 4.8 and its corresponding proof that E satisfies condition (4.15). □

Equation (4.31) is equivalent to the requirement that the fuzzy sets of the family $(\mu_i)_{i \in I}$ have to be pairwise disjoint with respect to the intersection induced by the t-norm T.

Example 4.12 Let $T = T_{Luka}$. By using the formula for the biimplication given in Example 2.66, we can rewrite condition (4.16) of Theorem 4.8 in the form

$$\sup_{x \in X} \{\mu_i(x) + \mu_j(x) - 1\} \leq \inf_{y \in X} \{1 - |\mu_i(y) - \mu_j(y)|\}. \quad (4.32)$$

The right side of (4.32) is exactly the same formula as (2.41), which defines the degree of equality of two fuzzy sets.

For $T = T_{Luka}$, Condition (4.31) of Theorem 4.11 is equivalent to

$$\mu_i(x) + \mu_j(x) \leq 1. \quad (4.33)$$

Note that this condition is in accordance with the common assumption for fuzzy partitions in fuzzy control that for each element of the domain the sum of its membership degrees to the fuzzy sets of the fuzzy partition equals 1. Since the fuzzy partitions in Example 4.1 are chosen such that they satisfy (4.33), we can apply Theorem 4.11.

As an example let us take a closer look at the equality relation E induced by (4.19) on the set $X_1 = [-90, 90]$ of possible angles. The seven fuzzy sets of the corresponding fuzzy partition are referred to as $\mu_{nb}, \mu_{nm}, \mu_{ns}, \mu_{az}, \mu_{ps}, \mu_{pm}$, and μ_{pb}. We have

$$\mu_{nb}(\theta) = \begin{cases} 1, & \text{if} & \theta \leq -67.5 \\ -\frac{\theta}{22.5} - 2, & \text{if} & -67.5 \leq \theta \leq -45 \\ 0, & \text{otherwise}, \end{cases}$$

$$\mu_{pb}(\theta) = \begin{cases} 1, & \text{if} & 67.5 \leq \theta \\ \frac{\theta}{22.5} - 2, & \text{if} & 45 \leq \theta \leq 67.5 \\ 0, & \text{otherwise}, \end{cases}$$

$$\mu_w(\theta) = 1 - \min\left\{\frac{|\theta - \theta_w|}{22.5}, 1\right\}, \quad (w \in \{nm, ns, az, ps, pm\}),$$

with $\theta_{nm} = -45$, $\theta_{ns} = -22.5$, $\theta_{az} = 0$, $\theta_{ps} = 22.5$, $\theta_{pm} = 45$ (see also Figure 4). We define $W = \{nb, nm, ns, az, ps, pm, pb\}$. Consequently we get

$$E(\theta, \theta') = \min\{1 - |\mu_w(\theta) - \mu_w(\theta')| \mid w \in W\}. \qquad (4.34)$$

In Table 4 the degree of equality $E(\theta, \theta')$ is given for some angles θ and θ', together with a fuzzy set of the fuzzy partition which satisfies

$$E(\theta, \theta') = 1 - |\mu_w(\theta) - \mu_w(\theta')|,$$

i.e. a fuzzy set where the minimum in (4.34) is reached.

The entry in the third row and fourth column, for instance, means that $E(16.875, 11.25) = 0.75$ holds. This value is obtained if $w = eN$ is chosen in (4.34), i.e. $1 - |\mu_{eN}(16.875) - \mu_{eN}(11.25)| = 0.75$. In general, the minimum is not only assumed for one w, but for several fuzzy sets; e.g. instead of $w = eN$, the entry $w = pk$ would have been equally possible. Because of the symmetry of the equality relation E, only the upper part of the table is depicted.

The use of triangular functions for the partition (see Figure 4) could lead to the conjecture that the equality relation E is induced by the metric δ where $\delta(\theta, \theta') = \frac{1}{22.5}|\theta - \theta'|$. As in Example 2.56 for the equality relation E_δ (induced by the metric δ) we obtain the triangular functions as singletons. The first row of Table 4 seems to confirm the conjecture, as we can read from it

$$E(\theta, \theta') = E_\delta(\theta, \theta') = 1 - \min\left\{\frac{1}{22.5}|\theta - \theta'|, \ 1\right\}.$$

A closer inspection of the table reveals that, in general, only $E_\delta \leq E$ is fulfilled. On the one hand, the equality relation E is quite coarse at the edges (i.e. for angles smaller than $-67.5°$, or greater than $67.5°$), since for all angles $\theta, \theta' \in [67.5, 90]$ it follows that $E(\theta, \theta') = 1$ (for example, see the table entry for $\theta = 67.5$ and $\theta' = 73.125$). This peculiar phenomenon originates from the choice of the fuzzy sets μ_{pb} and μ_{nb}. On the other hand, some strange-looking values appear: $E(5.625, 45) = 0$, but $E(5.625, 61.875) = 0.25$, although $5.625 < 45 < 61.875$ holds. The reason for this can be found in the insufficient covering property of the fuzzy partition. For example, we have

$$\sup\{\mu_w(11.25) \mid w \in \{nb, nm, ns, az, ps, pm, pb\}\} = 0.5.$$

According to the definition of E, we get as a lower bound for $E(\theta, \theta')$:

$$\begin{aligned}
E(\theta, \theta') &= \min\{1 - |\mu_w(\theta) - \mu_w(\theta')| \mid w \in W\} \\
&\geq 1 - \max\{\max\{\mu_w(\theta), \mu_w(\theta')\} \mid w \in W\} \\
&\quad + \min\{\ \min\{\mu_w(\theta), \mu_w(\theta')\} \mid w \in W\} \\
&\geq 1 - \max\{\max\{\mu_w(\theta), \mu_w(\theta')\} \mid w \in W\}. \qquad (4.35)
\end{aligned}$$

	0	5.625	11.25	16.875	22.5	28.125	33.75	39.375	45	61.875	67.5	73.125
0	1 *az*	0.75 *az*	0.5 *az*	0.25 *az*	0 *az*	0 *az*	0 *az*	0 *az*	0 *az*	0 *az*	0 *az*	0 *az*
5.625		1 *az*	0.75 *az*	0.5 *az*	0.25 *az*	0 *az*	0 *az*	0 *az*	0 *az*	0 *az*	0 *az*	0 *az*
11.25			1 *az*	0.75 *az*	0.5 *az*	0.25 *ps*	0 *pm*	0 *pm*	0 *pm*	0 *pb*	0 *pb*	0 *pb*
16.875				1 *az*	0.75 *az*	0.5 *ps*	0.25 *ps*	0 *pm*	0 *pm*	0 *pb*	0 *pb*	0 *pb*
22.5					1 *az*	0.75 *ps*	0.5 *ps*	0.25 *ps*	0 *pm*	0 *pb*	0 *pb*	0 *pb*
28.125						1 *az*	0.75 *ps*	0.5 *ps*	0.25 *ps*	0 *pm*	0 *pb*	0 *pb*
33.75							1 *az*	0.75 *ps*	0.5 *ps*	0 *pm*	0 *pm*	0 *pb*
39.375								1 *az*	0.75 *ps*	0 *pm*	0 *pm*	0 *pm*
45									1 *az*	0.25 *ps*	0 *ps*	0 *ps*
61.875										1 *az*	0.75 *pm*	0.5 *pm*
67.5											1 *az*	0.75 *az*
73.125												1 *az*

Table 4 Some values of the equality relation computed according to (4.19)

For $\theta = 5.625$ and $\theta' = 61.875$, we obtain from (4.35) a value of 0.25. (4.35) can be interpreted as follows: the fuzzy set

$$\mu : [-90, 90] \quad \rightarrow \quad [0, 1],$$
$$\theta \quad \mapsto \quad \sup\{\mu_w(\theta) \mid w \in W\},$$

that corresponds to the union of the sets μ_w ($w \in W$), determines to which degree each element $\theta \in [-90, 90]$ is covered by the fuzzy partition. (4.35) states that two elements $\theta, \theta' \in [-90, 90]$ (under E) can be distinguished at most as good as they are covered by the fuzzy partition. (4.35) is equivalent to

$$E(\theta, \theta') \geq 1 - \max\{\mu(\theta), \mu(\theta')\}.$$

The choice of the t-norm T_{Luka} for the computation of the equality relations is not only motivated by the Examples 2.51 and 2.56, but it also has practical reasons. For example there are no equality relations for $\mathsf{T} = \mathsf{T}_{\min}$ or $\mathsf{T} = \mathsf{T}_{\text{prod}}$ on the set $[-90, 90]$ such that the fuzzy sets can be interpreted as singletons. The reason for this can be found in Theorem 4.8. It is

$$\sup_{\theta \in [-90, 90]} \{\mathsf{T}(\mu_{eN}(\theta), \mu_{pk}(\theta))\} = \left\{ \begin{array}{ll} 0.5, & \text{if} \quad \mathsf{T} = \mathsf{T}_{\min} \\ 0.25, & \text{if} \quad \mathsf{T} = \mathsf{T}_{\text{prod}}, \end{array} \right.$$

but

$$\inf_{\theta \in [-90, 90]} \left\{ \overleftrightarrow{\mathsf{T}}(\mu_{eN}(\theta), \mu_{pk}(\theta)) \right\} = 0$$

for $\mathsf{T} = \mathsf{T}_{\min}$ as well as for $\mathsf{T} = \mathsf{T}_{\text{prod}}$, so that the condition (ii) of Theorem 4.8 is not fulfilled for these two t-norms.

It should be emphasized that for fuzzy controllers the application of the minimum operator (see (4.9)) does not depend on the t-norm of the equality relations. The minimum operator originates from equation (4.5), which is based on the assumption of the independence of the participating equality relations. So the equality relation for the product space of the measured input and the output, given in equation (2.61), appears to be reasonable. Thus it is correct to use the 'max-min-inference method' described in (4.2) for the Mamdani fuzzy controller, even if the controller is interpreted on the basis of equality relations with respect to an arbitrary t-norm.

The choice of the t-norm should allow us to interpret the equality relations in a reasonable way. As we have already seen in the Examples 2.51 and 2.52, the t-norms $\mathsf{T}_{\min}$, and especially T_{Luka}, can be motivated quite vividly. Nevertheless, the existence of an appropriate equality relation with respect to a fixed t-norm for a given fuzzy partition cannot be guaranteed. We have shown that the fuzzy sets of a fuzzy partition $\{\mu_w \mid w \in W\}$ can be interpreted as singletons in the sense of an equality relation with respect to the t-norm T_{Luka} – but not with respect to $\mathsf{T}_{\min}$ or T_{prod}.

The choice of an appropriate t-norm for the equality relation allows to use a substantially weaker presupposition than (4.33) in order to guarantee the existence of an equality relation according to which the fuzzy sets can be interpreted as singletons. A possible candidate is a t-norm T_p'' of the Yager-family with a very small parameter $p > 0$. In this case, instead of (4.33), we obtain the weaker presupposition

$$(1 - \mu_i(x))^p + (1 - \mu_j(x))^p \geq 1.$$

It remains difficult, however, to find proper semantics for equality relations with respect to these t-norms. $\qquad\qquad\qquad\qquad\qquad\qquad\qquad\qquad\qquad\qquad$ $\square$

A fuzzy partition satisfying certain reasonable conditions induces an equality relation under which the fuzzy sets can be interpreted as singletons. The direct specification of an equality relation can turn out to be difficult. Often a control expert might be able to specify fuzzy sets that can be interpreted as singletons according to the equality relations to be determined. By the singleton for a certain point x_0 (in form of the fuzzy set μ_{x_0}) an equality relation can be specified locally in a certain neighbourhood of the point x_0. From the viewpoint of equality relations it might be quite reasonable to define fuzzy partitions instead of equality relations directly. It is possible, with assistance of Theorem 4.8 and formula (4.19), to determine the corresponding equality relation from a fuzzy partition. According to this, Theorem 4.8 supplies a possibility to verify whether a given partition is reasonable. Note that the rule of thumb requiring that the membership degrees of adjacent fuzzy sets μ_i and μ_j of a fuzzy partition sum up to 1 guarantees that a corresponding equality relation (with respect to the t-norm T_{Luka}) always exists.

4.5   Fuzzy Control and Relational Equations

In this section we apply the theory of (fuzzy) relational equations to fuzzy controllers. From the viewpoint of relational equations it is more convenient to combine the measurement values $\xi_1, \ldots, \xi_n$ and treat them as a single value or rather as a tuple of measurement values $\xi = (\xi_1, \ldots, \xi_n)$. Consequently the range of values of ξ is the product space $X = X_1 \times \ldots \times X_n$. The control rules we have to consider are of the form

$$\textbf{if } \xi \textbf{ is } A \textbf{ then } \eta \textbf{ is } B, \tag{4.36}$$

where A and B represent linguistic terms, each of which correspond to a fuzzy set μ_A of X or ν_B of Y, respectively. Rules of the kind

$$\textbf{if } \xi_1 \textbf{ is } A_1 \textbf{ and } \ldots \textbf{ and } \xi_n \textbf{ is } A_n \textbf{ then } \eta \textbf{ is } B,$$

with which we have made acquaintance in the framework of the Mamdani fuzzy controller, can be translated to the form (4.36) without difficulty: the 'linguistic term' $A = (A_1, \ldots, A_n)$ is associated with the fuzzy set

$$\mu_A : X \;\to\; [0,1],$$
$$(x_1, \ldots, x_n) \;\mapsto\; \min\{\mu_{A_1}(x_1), \ldots, \mu_{A_n}(x_n)\},$$

where μ_{A_i}, $i = 1, \ldots, n$, is the fuzzy set corresponding to the linguistic term A_i.

To give a motivation of fuzzy control from the viewpoint of relational equations in a more lucid way we consider, for the moment, only crisp sets. For ordinary sets, solving the control problem consists in specifying a control function $\varphi : X \to Y$ that assigns to each measurement value $x \in X$ an appropriate control value. Technically speaking, the mapping φ corresponds to the relation

$$R_\varphi = \{(x, \varphi(x)) \mid x \in X\} \subseteq X \times Y. \tag{4.37}$$

For a measured input value $x \in X$ the control value $\varphi(x)$ can be calculated using the equation

$$\{\varphi(x)\} = \{x\} \circ R_\varphi.$$

Here $\circ$ denotes the composition of a set $M \subseteq X$ with a relation $R \subseteq X \times Y$, which is defined as

$$M \circ R \stackrel{\text{def}}{=} \{y \in Y \mid \exists x \in X : (x \in M \wedge (x,y) \in R)\} \subseteq Y. \tag{4.38}$$

We now examine the following generalized problem. Let the sets $M_1, \ldots, M_r \subseteq X$ and $N_1, \ldots, N_r \subseteq Y$ be given. We are searching for a control relation $R \subseteq X \times Y$ with

$$\forall i \in \{1, \ldots, r\} : \quad N_i = M_i \circ R. \tag{4.39}$$

The underlying semantics consists in the assumption that the control value has to be chosen from the set N_i, $i = 1, \ldots, r$, if the measured value comes from the set M_i. After determining the relation R from the given sets $M_1, \ldots, M_r$ and $N_1, \ldots, N_r$, we can calculate for any set $M \subseteq X$ of measured values the corresponding set $N = M \circ R$ of control values. M may be a vague observation of the input values. It is not necessarily required that the relation R is a mapping in the sense of (4.37).

The problem to be solved consists in calculating a relation R satisfying condition (4.39) from given sets M_i and N_i, $i = 1, \ldots, r$. Having stated the problem for crisp sets, we now turn to the generalization to fuzzy sets.

Let fuzzy sets $\mu_1, \ldots, \mu_r$ of X and $\nu_1, \ldots, \nu_r$ of Y be given. We are searching for a *fuzzy relation* ϱ, i.e. a fuzzy set of $X \times Y$ that satisfies

$$\forall i \in \{1, \ldots, r\} : \quad \nu_i = \mu_i \circ \varrho.$$

ϱ	140	160	180	200	220
s	1.0	0.5	0.1	0.0	0.0
m	0.0	0.5	1.0	0.5	0.0
h	0.0	0.0	0.4	0.8	1.0

Table 5   The fuzzy relation ϱ

First we have to explain, how the composition ∘ of a fuzzy set μ of X with a fuzzy relation ϱ of $X \times Y$ is defined. To do this we recall formula (4.38) and 'fuzzify' it according to the rules of Section 2.7. In this way the composition of the fuzzy set μ with the fuzzy relation ϱ yields the fuzzy set

$$\mu \circ \varrho : Y \to [0,1], \quad y \mapsto \sup_{x \in X} \{\min\{\mu(x), \varrho(x,y)\}\} \tag{4.40}$$

of Y. Here the conjunction in (4.38) is evaluated by the minimum.

Example 4.13 Let us consider the set $X = \{s, m, h\}$ of classes of cars, where s, m, h stand for small car, medium quality car and car of high class, respectively. Let $Y = \{140, 160, 180, 200, 220\}$ be the set of all possible maximum speeds (in km/h). The fuzzy relation ϱ (cf. Table 5) of the set $X \times Y$ states for any pair $(x, y) \in X \times Y$, how far it is regarded as being possible that the maximum speed of a car of class x equals y.

We can calculate the fuzzy set ν of Y of all possible speeds of a small or medium quality car with the help of formula (4.40) by choosing the characteristic function of the set $\{s, m\}$ as fuzzy set μ of X, i.e. $\mu = \mathbb{1}_{\{s,m\}}$. For ν we obtain $\nu(140) = 1.0$, $\nu(160) = 0.5$, $\nu(180) = 1.0$, $\nu(200) = 0.5$, $\nu(220) = 0.0$.

If we want to learn something about the maximum speed of a car which was produced by a certain car manufacturer in the south of Germany, the fuzzy set μ with $\mu(s) = 0.0$, $\mu(m) = 0.6$, $\mu(h) = 1.0$ is an appropriate representative for the possible car classes. The calculation of the fuzzy set ν of the possible maximum speeds according to formula (4.40) can be illustrated by a Falk scheme as shown in Table 6. To carry out the calculation, we write the fuzzy set μ as a row vector at the bottom left. The fuzzy relation ϱ is noted in matrix form at the top right as shown in Table 5. Then the fuzzy set $\nu = \mu \circ \varrho$ emerges as a row vector at the right bottom. To compute the value $\nu(140)$, we transpose the row vector that represents the fuzzy set μ, and determine the minimum with the column vector component by component. The maximum of the three values thus calculated is the degree of membership of the element 140 to the fuzzy set ν. The degrees of membership of the other

			1.0	0.5	0.1	0.0	0.0
			0.0	0.5	1.0	0.5	0.0
			0.0	0.0	0.4	0.8	1.0
0.0	0.6	1.0	0.0	0.5	0.6	0.8	1.0

Table 6   Falk scheme to calculate $\nu = \mu \circ \varrho$

speeds can be determined analogously. We obtain $\nu(140) = 0.0$, $\nu(160) = 0.5$, $\nu(180) = 0.6$, $\nu(200) = 0.8$, $\nu(220) = 1.0$.   □

Before we formulate a theorem concerning the solvability of a single relational equation $\nu = \mu \circ \varrho$, we have to remember the Gödel relation $\varrho_{\mu,\nu}^{\text{Gödel}}$, which is based on the Gödel implication. Using the terminology of Section 2.7, $\varrho_{\mu,\nu}^{\text{Gödel}}$ is defined as

$$(x, y) \in \varrho_{\mu,\nu}^{\text{Gödel}} \iff (x \in \mu \; \to \; y \in \nu), \tag{4.41}$$

where the implication $\to$ is evaluated in the sense of the Gödel implication, i.e.

$$\varrho_{\mu,\nu}^{\text{Gödel}}(x, y) = \begin{cases} 1, & \text{if } \mu(x) \le \nu(y) \\ \nu(y), & \text{otherwise.} \end{cases}$$

Theorem 4.14 *Let $\mu \in F(X)$, $\nu \in F(Y)$. Furthermore let $\varrho \in F(X \times Y)$ be a fuzzy relation satisfying the relational equation $\nu = \mu \circ \varrho$. This implies:*

(i) $\varrho \le \varrho_{\mu,\nu}^{\text{Gödel}}$.

(ii) *The Gödel relation is also a solution of the relational equation, i.e.* $\nu = \mu \circ \varrho_{\mu,\nu}^{\text{Gödel}}$.

Proof:

(i) Let $x \in X$, $y \in Y$.

If $\mu(x) \le \nu(y)$ holds, then $\varrho(x, y) \le 1 = \varrho_{\mu,\nu}^{\text{Gödel}}(x, y)$ follows.
Hence let $\nu(y) < \mu(x)$. Thus we obtain

$$
\begin{aligned}
\varrho_{\mu,\nu}^{\text{Gödel}}(x, y) \;&=\; \nu(y) \\
&=\; \sup_{x' \in X} \{\min\{\mu(x'), \varrho(x', y)\}\} \\
&\ge\; \min\{\mu(x), \varrho(x, y)\}.
\end{aligned}
$$

Since $\mu(x) > \nu(y)$ is assumed, $\min\{\mu(x), \varrho(x, y)\} = \varrho(x, y)$ holds.

(ii) Let $y \in Y$. Obviously

$$(\mu \circ \varrho_{\mu,\nu}^{\text{Gödel}})(y) = \sup_{x \in X} \{\min\{\mu(x), \varrho_{\mu,\nu}^{\text{Gödel}}(x,y)\}\} \leq \nu(y)$$

is valid. (i) implies

$$\begin{aligned}
\nu(y) &= (\mu \circ \varrho)(y) \\
&= \sup_{x \in X} \{\min\{\mu(x), \varrho(x,y)\}\} \\
&\leq \sup_{x \in X} \{\min\{\mu(x), \varrho_{\mu,\nu}^{\text{Gödel}}(x,y)\}\} \\
&= (\mu \circ \varrho_{\mu,\nu}^{\text{Gödel}})(y).
\end{aligned}$$

$\square$

Theorem 4.14 states that the Gödel relation is the greatest solution of the relational equation $\nu = \mu \circ \varrho$, if there is a solution at all. The following theorem generalizes this result to a system of r relational equations $\nu_i = \mu_i \circ \varrho$, $i = 1, \ldots, r$. If this system of relational equations has a solution, then the fuzzy relation $\min\limits_{i \in \{1,\ldots,r\}} \{\varrho_{\mu_i,\nu_i}^{\text{Gödel}}\}$ is the greatest solution of this system.

Theorem 4.15 *Let $\mu_i \in F(X)$, $\nu_i \in F(Y)$, $i = 1, \ldots, r$. Furthermore let $\varrho \in F(X \times Y)$ be a fuzzy relation that solves the system of r relational equations $\nu_i = \mu_i \circ \varrho$. With the abbreviation $\varrho_r \overset{\text{def}}{=} \min\limits_{i \in \{1,\ldots,r\}} \{\varrho_{\mu_i,\nu_i}^{\text{Gödel}}\}$ the following two statements are valid:*

(i) $\varrho \leq \varrho_r$,

(ii) $\forall i \in \{1, \ldots, r\}: \quad \nu_i = \mu_i \circ \varrho_r$.

Proof:
(i) Theorem 4.14(i) implies

$$\forall i \in \{1, \ldots, r\}: \quad \varrho \leq \varrho_{\mu_i,\nu_i}^{\text{Gödel}}$$

and hence $\varrho \leq \varrho_r$.

(ii) Let $i \in \{1, \ldots, r\}$, $y \in Y$. We obtain

$$(\mu_i \circ \varrho_r)(y) \leq (\mu_i \circ \varrho_{\mu_i,\nu_i}^{\text{Gödel}})(y) \leq \nu_i(y).$$

In addition we have

$$\nu_i(y) = (\mu_i \circ \varrho)(y) \leq (\mu_i \circ \varrho_r)(y).$$

$\square$

The basic principle to solve systems of relational equations $\nu_i = \varrho \circ \mu_i$, $i = 1, \ldots, r$, on the basis of Theorems 4.14 and 4.15 consists in determining the greatest solution, i.e. the Gödel relation, separately for each relational equation, and subsequently finding a combined solution for the entire system by calculating the minimum of the solutions for the single equations. In the case when the system of relational equations does have a solution at all, this procedure will succeed. Assuming that a solution of the system of relational equations exists, the step by step calculation of the solution ϱ_r by

$$\varrho_0 \stackrel{\text{def}}{=} \mathbb{1}_{X \times Y}, \quad \varrho_{i+1} \stackrel{\text{def}}{=} \min\{\varrho_i, \varrho_{\mu_i, \nu_i}^{\text{Gödel}}\} \quad (i = 0, \ldots, r-1)$$

corresponds to the principle of solving successively the systems of relational equations $\mathcal{R}_i = \{\nu_1 = \mu_1 \circ \varrho, \ldots, \nu_i = \mu_i \circ \varrho\}$, $i = 1, \ldots, r$, by the greatest possible fuzzy relation ϱ_i, and in the next step diminishing this fuzzy relation, so that it solves the subsequent relational equation, too.

It is tempting to try to reverse this principle to solve the system of relational equations by starting from the smallest solution of one relational equation and enlarging it step by step. But this does not always work, since even if there are solutions of the relational equation, there is in general no smallest solution. In such a case one may use a certain 'small' solution of the relational equation, which is specified in the next definition.

Definition 4.16 *Let $\mu \in F(X)$, $\nu \in F(Y)$. The fuzzy relation*

$$\varrho_{\mu, \nu}^{\min} : X \times Y \to [0, 1], \quad (x, y) \mapsto \min\{\mu(x), \nu(y)\}$$

is called the **Cartesian product of the fuzzy sets** μ *and* ν.

If a solution of a relational equation exists, then the Cartesian product of the appearing fuzzy sets is a solution, too.

Theorem 4.17 *Let $\mu \in F(X)$, $\nu \in F(Y)$. Furthermore let $\varrho \in F(X \times Y)$ be a fuzzy relation which satisfies the relational equation $\nu = \mu \circ \varrho$. Then $\nu = \mu \circ \varrho_{\mu, \nu}^{\min}$ holds.*

Proof:
Let $y \in Y$. Obviously

$$(\mu \circ \varrho_{\mu, \nu}^{\min})(y) = \sup_{x \in X} \left\{ \min\{\mu(x), \varrho_{\mu, \nu}^{\min}(x, y)\} \right\}$$

$$= \min \left\{ \sup_{x \in X} \{\mu(x)\}, \nu(y) \right\}. \tag{4.42}$$

In addition we obtain from

$$(\mu \circ \varrho)(y) = \sup_{x \in X} \{\min\{\mu(x), \varrho(x, y)\}\} = \nu(y)$$

that $\sup\{\mu(x) \mid x \in X\} \geq \nu(y)$ is satisfied. This implies that (4.42) equals $\nu(y)$. $\square$

If a system of relational equations $\nu_i = \mu_i \circ \varrho$, $i = 1, \ldots, r$, has to be solved with the help of Theorem 4.17, then 'assembling' the overall solution ϱ from the 'small' solutions $\varrho_{\mu_i,\nu_i}^{\min}$ of the single relational equations is the obvious way. A reasonable approach to derive such an assembled solution is

$$\varrho = \max\{\varrho_{\mu_i,\nu_i}^{\min} \mid i \in \{1, \ldots, r\}\}. \tag{4.43}$$

If a crisp value $x_0 \in X$ is given, which is represented by its characteristic function $\mathbb{1}_{\{x_0\}}$, and if we use the fuzzy relation ϱ to calculate the fuzzy set $\nu \in F(Y)$ for the control value, we obtain

$$\nu(y) \quad = \quad (\mathbb{1}_{\{x_0\}} \circ \varrho)(y) \tag{4.44}$$

$$= \quad \max_{i \in \{1,\ldots,r\}} \left\{ \sup_{x \in X} \left\{ \min\{\mathbb{1}_{\{x_0\}}(x), \varrho_{\mu_i,\nu_i}^{\min}(x,y)\} \right\} \right\} \tag{4.45}$$

$$= \quad \max_{i \in \{1,\ldots,r\}} \left\{ \min\{\mu_i(x_0), \nu_i(y)\} \right\}. \tag{4.46}$$

This result corresponds exactly to the fuzzy set which we would obtain for the control value by using a Mamdani fuzzy controller (cf. formula (4.2)). In this sense the Mamdani fuzzy controller can be motivated also on the basis of relational equations. A more stringent procedure with regard to the solution of relational equations consists in using the Gödel relation instead of the Cartesian product, since for the Gödel relation the more general Theorem 4.15 holds, whereas for Cartesian products only the weaker Theorem 4.17 is satisfied, which corresponds to Theorem 4.14 for the Gödel relation.

With Theorem 4.15 we have provided a criterion for the solvability of systems of relational equations as well as a particular solution in the case when a solution exists. The stated solution is the relation ϱ_r. The criterion for solvability consists in the condition that ϱ_r solves the system of relational equations. But it would be desirable to have a condition for solvability that is more easy to check than this. Such a condition is, for example, the very restricting assumption that the fuzzy sets $\mu_1, \ldots, \mu_r$ are pairwise disjoint, i.e. that for all $i \neq j$

$$\min\{\mu_i, \mu_j\} = 0$$

is satisfied.

The central role played by the Gödel relation as far as the solvability of systems of relational equations is concerned, is due to the fact that we have chosen in (4.40) the minimum operator for the conjunction in (4.38). The minimum could be replaced by any other t-norm $\top$ leading to a different fuzzy relation instead of the Gödel relation. The corresponding fuzzy relation can be derived by evaluating the implication in (4.41) with the help of the residuum associated with $\top$.

4.6   Supplementary Remarks and References

Besides a brief historical survey of the development of the theory and applications of fuzzy control in the past two decades, in this section we also discuss some applications of the concepts and methods presented in this chapter. As an example we examine the implementation of a fuzzy controller for idle speed control of a car motor.

4.6.1   Historical Development: Fuzzy Control

The first laboratory application of fuzzy control was described in [Mamdani75]. In this paper a fuzzy controller for a steam engine, which controlled the heat input as well as the valve position for steam supply, was analysed. In an implicit form some concepts developed by L.A. Zadeh underlie this seminal work [Zadeh72b, Zadeh73, Zadeh71b]. Following these ideas, fuzzy control technology was worked out in [Kickert78, Braae79, Procyk79, Baldwin80, Tong80, Tong84, Hirota80]. The first large scale realization of a fuzzy controller was a control system for a cement kiln [Holmblad82]. In [Sugeno85b] prototypes for different domains of application of fuzzy controllers are discussed. The breakthrough of fuzzy technology in Japan occured in 1987, when the Sendai Underground was put into operation, which is run automatically by fuzzy controllers [Yasunubo83, Yasunubo85]. Fuzzy control technology was then significantly improved by the papers [Takagi85, Pedrycz93, Chiu88, Dubois91b, Berenji92, Palm92, Sugeno93]. The activities in the domain of 'Fuzzy Logic' gained attention on an international scale when the Japanese research institute LIFE (Laboratory for International Fuzzy Engineering Research) was founded in 1989, supported half by the Japanese Ministry for International Trade and Industry (MITI) and half by 49 companies of several trades. The research work carried out in Japan did not claim to have put forth new concepts, but was described as mere applications of the theories of L.A. Zadeh. The new aspect that attracted attention was the fact that Japanese manufacturers predominantly used hardware solutions like fuzzy chips or special micro processors called fuzzy computers.

In 1990 several successful industrial applications were reported from Japan in the domain of consumer products like washing machines, vacuum cleaners and camcorders as well as in container cranes, lifts and power plants. Furthermore the first patents for fuzzy computers were granted (causing, by the way, a little joke: the official German translation in the patent letter was 'fusseliger Computer', which means something like 'a computer covered with fluff'). Several successful applications are presented in [White92, Hirota93, Jamshidi93, Terano92, Kandel91, Munakata94, Asakawa94, Kruse94].

4.6.2 Fuzzy Control

Fuzzy control systems can be interpreted in two different ways. In one interpretation fuzzy control is considered to be a special type of a real-time expert system, implementing a part of the human operator's or process engineer's experience by using situation-action rules — see [Taunton89] for a survey of experimental applications of expert systems in process control. In our book we use the second interpretation, which is preferred by control engineers: a fuzzy controller is a non-linear, time-invariant, dynamic-free feedback law. From this point of view the system under control is a non-fuzzy system, the class of fuzzy controllers is only a particular class of non-linear controller (often denoted as table-based controller), for which the transfer characteristics in principle is a static multi-dimensional non-linear mapping that looks like a characteristic surface over the space of controller inputs (in former time, the last trial to solve a given problem after all attempts failed to find a theoretically sound solution [Pfeiffer93]).

For most successful applications of fuzzy control the Mamdani controller [Mamdani75] and the Sugeno controller [Sugeno85a, Sugeno85c, Sugeno85d] served as a basis. Often modifications of these controllers were developed: a survey of several concepts and techniques of fuzzy control is contained in the monographs [Pedrycz93, Driankov93].

Some misunderstandings — especially in the description of Mamdani fuzzy 'logic' controllers — are caused by a erroneous view, which interprets the rules in the sense of logical implications and hence regards the rule base as a conjunctive system of rules. Names like 'generalized *modus ponens*' for the minimum operator contribute to confusion. Though it is possible in principle to design a fuzzy controller based on an interpretation of the rules as logical implications, this leads to completely different methods of calculation, that do not coincide with those of the Mamdani or Sugeno fuzzy controllers. Regarding the rule base as a possibilistic system of rules, as was discussed in Section 3.3, leads to the application of the Gödel implication, so that a fuzzy controller emerges which is more in the philosophy of solving relational equations [Gottwald86a, Gottwald86b, Gebhardt93b], which we presented in Section 4.5. A strictly logical interpretation of the rules results in a fuzzy controller whose decision logic yields a crisp set instead of a fuzzy set [Klawonn92b].

In Section 4.4 we demonstrated that fuzzy controllers can be embedded in the field of interpolation and approximation theory [Klawonn93a, Klawonn93b]. In this context theoretical results on fuzzy controllers are important. In this connection [Kosko92a, Buckley93b] should be mentioned, where it is proved that with respect to certain fuzzy controllers any continuous function on a compact set can be approximated with an arbitrary degree of precision.

Equality relations, which formed the basis for this approach, are treated extensively in [Höhle90, Höhle91, Coulon92, Höhle92], where several generalizations are discussed, e.g. replacing the demand of total existence or replacing the unit interval by other lattices (Heyting-algebras, MV-algebras, Boolean algebras [Höhle85, Klawonn92a]).

Often equality relations are also called *similarity relations* [Zadeh71a] or *indistinguishability operators* [Valverde85, Jacas92]. The notion of an operator is used instead of relation, because equality relations are regarded as fuzzy relations just as in Section 4.5. In this sense, for the t-norm T_{min}, the extensional hull ν of a set — or, more general — of a fuzzy set μ can be calculated with respect to the equality relation E by $\nu = \mu \circ E$. One result of the papers [Valverde85, Jacas92] is that for a given family $(\mu_i)_{i \in I}$ of fuzzy sets these fuzzy sets are extensional with respect to the equality relation (4.19). Theorem 4.8 adds to this result by allowing the interpretation of the fuzzy sets μ_i as singletons, provided that the prerequisite stated in the theorem is satisfied.

To look upon equality relations as indistinguishability operators leads, as shown above, to fuzzy relational equations. The notion of a fuzzy relational equation for the max-min-composition described in (4.40) is due to E. Sanchez [Sanchez76]. This kind of equation was discussed in detail in [Czogala82] and analysed in [Gottwald86a, Gottwald86b]. [Nola89, Nola83] comprehensively describe this theory as well as generalizations for other relational equations. In [Gottwald93] the notion of approximate solutions for (unsolvable) relational equations is introduced.

An ample description of the references between fuzzy relations and fuzzy control can be found in [Pedrycz93].

4.6.3   Realization of Fuzzy Controllers

The aim of a textbook mainly consists in conveying the basic concepts of a theory or of a model. Therefore we have devoted Chapter 4 to elucidating the concept of knowledge-based control on the basis of modelling a human operator's behaviour, to the elementary fuzzy controllers of Mamdani and Sugeno, and to the semantics of fuzzy controllers on the basis of equality relations or fuzzy relational equations. Some hints regarding the pragmatic aspects of the design and realizations of a fuzzy controller shall be given now.

If a problem of control engineering is to be solved with the help of a fuzzy controller, the concepts presented here need not be implemented starting from scratch. Today there are development tools available from several companies, which support the realization of a fuzzy controller. These tools allow the user

to define and adjust rules and fuzzy sets in a graphic environment or with
the help of a text editor. Some of them even permit the user to choose an
appropriate mechanism of inference for the decision logic. In some tools a
simulator is integrated, others provide interfaces for testing. The result of a
pure software solution is a fuzzy controller in the form of C-, PASCAL-, ADA-,
or FORTRAN-code. Other development tools enable the user to program
directly special fuzzy hardware, which is suitable for real-time applications.

The advantage of these commercial systems is that a comfortable
environment for a fast implementation of a fuzzy controller is provided, which
saves costly development time. A disadvantage is certainly the frequent lack
of flexibility of the products, which generally admit a very restricted set
of parameterized fuzzy sets (triangular functions, trapezodial functions) or
simplified defuzzification strategies. Whether the purchase of such systems
pays off depends on the particular application.

If developing one's own implementation, compromises have to be made
between costly, but flexible models and simple, but fast realizations. In most
cases it is advisable to choose a suitable discretization of the degrees of
membership on the ordinate as well as for the domains of the input and
control variable on the abscissa.

For the pratical use of a fuzzy controller with real-time requirements three
possibilities are feasible.

- The calculations of the decision logic are executed directly if possible on
  the basis of suitable hardware. This leads to high demands for computation
  speed, but only a small memory is needed.

- The control function induced by the implementation of the fuzzy controller,
  is computed in advance and stored in a look-up table, so that for the price
  of a large memory high computation speed can be dispensed with.

- As a compromise between the above possibilities the relation matrix, which
  is to be calculated as demonstrated at the end of Section 4.5, can be stored.

An efficient implementation of a fuzzy controller requires that the applied
operations are simple and can be executed fast. Arbitrary fuzzy sets instead
of piecewise linear membership functions, or t-norms and t-conorms instead
of minimum or maximum, lead to costly computations, so that certain real-
time requirements cannot be met any longer. Therefore often a modification
of the centre of gravity method is used for defuzzification. Instead of the
formula (4.3) which requires intensive calculations, or its version for discrete
sets presented in Section 4.3.6, the abscissa values of the centre of gravity
y_r of the fuzzy sets μ_r for the conclusion in the rule R_r are determined in
advance. The output for this modified version of the centre of gravity method
is the sum of the values y_r, weighted with the degree α_r of applicability of

rule R_r.

$$\eta \overset{\text{def}}{=} \frac{\sum\limits_{r=1}^{k} \alpha_r y_r}{\sum\limits_{r=1}^{k} \alpha_r}, \quad \text{where}$$

$$y_r \overset{\text{def}}{=} \frac{1}{\int\limits_{y \in Y} \mu_r(y) dy} \int\limits_{y \in Y} y\mu_r(y) dy \quad \text{or}$$

$$y_r \overset{\text{def}}{=} \frac{1}{\sum\limits_{y \in Y} \mu_r(y)} \sum\limits_{y \in Y} y\mu_r(y) \, .$$

Defuzzification is a general problem in fuzzy control, which is in most cases solved on a purely heuristic basis. Recently, a number of defuzzification methods generalizing other methods like COG and MOM and involving additional parameters have been proposed [Yager93a, Yager93b]. Some defuzzification strategies are even protected by license [Frank93]. Nevertheless, defuzzification remains a heuristic subject even though some authors formulate conditions that should be satisfied by a reasonable defuzzification method [Driankov93, Runkler93].

The formal framework presented in this chapter should be understood as a description of reasonable approaches, but not as a prescription. For reasons of simplicity we have nearly always assumed that in the premise of a rule all input variables $\xi_1, \ldots, \xi_n$ appear. In applications it may be possible that, for certain ranges of values of some of the measured quantities, the values of the other measured quantities are superfluous for determining the control value. In this case there may appear rules that contain only some variables relevant for a certain range for this rule. Likewise it is conceivable to allow in the rules not only and-combinations, but also connectives like disjunction or negation with their corresponding evaluation functions. Furthermore the fuzzy controller can be followed by an intergral component, so that its output does not determine the control value directly, but only its change.

Having designed a rough draft of the fuzzy controller needed, fine tuning and optimization tasks are generally to be carried out. If this tuning is not intended to be done manually, one has to provide appropriate techniques. For this purpose neural networks seem to be adequate, because they can support learning an optimal control strategy. In the framework of this book, which is dedicated to fuzzy systems, we can give only a few hints to the large number of suggestions for linking fuzzy controllers and neural networks [Nauck94].

The most simple method to apply neural networks to improve a fuzzy controller is to let a neural network learn certain parameters of the fuzzy controller. For example, if the fuzzy sets are described in a parameterized form, a neural network can optimize these parameters and hence can learn the fuzzy sets corresponding to the linguistic terms [Ichihashi91, Nomura92a]. Other approaches allow arbitrary fuzzy sets and learn the degree of applicability of a rule with a multilayer perceptron [Hayashi92, Takagi91a, Takagi91b] by a suitable clustering algorithm. Techniques of fuzzy cluster analysis in connection with neural networks [Bezdek92b] are also presented in [Kosko92b].

When combining fuzzy controllers and neural networks it is not necessary to separate strictly the controller and the network. Such a separation means that the neural network has to learn certain parameters or parts of the fuzzy controller. H.R. Berenji [Berenji92] suggests a special architecture of a neural network that is interpreted as a fuzzy controller. In reverse there are some approaches interpreting a fuzzy controller as a neural network [Eklund92b, Nauck92a, Nauck92b, Nauck93]. General suggestions for linking neural networks and fuzzy systems can be found in [Eklund92a, Keller92a, Keller92b], and especially for pattern recognition in [Pao89].

In order to automatically optimize fuzzy controllers, besides neural networks, genetic algorithms are also applicable, by which for instance (parameterized) fuzzy sets [Karr89, Karr92a, Karr92b, Karr92c, Nomura92b, Quian92, Lee93], the number of rules [Karr91], or even suitable fuzzy relations in the sense of Section 4.5 [Pham91] can be learned.

Often fuzzy controllers are used to model a human operator's or process engineer's experience by specifying a sufficiently large, complete, and non-contradictory set of rules. This approach has already turned out to be a difficult problem in artificial intelligence (AI) with respect to the development of rule-based systems. The difference between fuzzy controllers and AI expert systems is that AI solutions normally operate in a symbolic framework whereas fuzzy systems are numerical devices that are more suitable for processing dynamic data. The innovation and advantage of fuzzy controllers is the comfortable and easily understandable way of specifying the multi-dimensional non-linear control mapping by using the implicit parameters described in Section 4.3.

In contrast to several classical control engineering techniques only a few methods are available that guarantee or check stability for fuzzy controllers. If no mathematical plant model exists due to the lack of precise sensor information, the complexity and coupling structure of the plant, the time-varying process behaviour, and difficulties in predicting system dynamics, then it is not possible to check stability for classical control methods nor for fuzzy control. In practice, validation is performed with simulations and massive tests. In case of the existence of a linear plant model a good controller can be designed in a systematic way by classical methods (i.e. PID controller). The same holds to some extent for non-linear plant models [Isermann92], but studies like [Boverie91, Boverie92], in which fuzzy, PID, and adaptive controllers were compared using a benchmark test, show that there are often very good reasons to use fuzzy controllers [Pfeiffer93]. For simple instances, there already exist methods for the proof of stability (see e.g. [Bouslama92, Tanaka92, Hwang92, Driankov93, Kiendl93]), but it should be stated that there is still a lack of systematic design methods for fuzzy controllers. From an industrial perspective fuzzy control is suitable

- if no satisfactory model of the process to control exists, or
- if the development of such a model is too expensive,

but an expert is able to describe, instead of a formal process model, his knowledge about the process in the form of linguistic rules, as it was the case for the cement kiln [Holmblad82] or in the case study which we will present in the next section.

Fuzzy control can also be applied to simple, robust technical processes for which no high precision requirements have to be met and for which cheap standard hardware solutions (e.g. fuzzy chips) can be used, e.g. consumer products like vacuum cleaners. In such applications the use of fuzzy controllers can lead to considerable savings in development and maintenance time.

4.6.4   Control of the Engine Idle Speed of a Spark Ignition Engine — An Implementation

The methods and concepts of fuzzy control presented in this chapter were put into practice during a cooperation between the Computer Science Department of the University of Braunschweig and Volkswagen AG Wolfsburg in order to solve the problem of controlling the idling speed of the VW 2000cc 116hp motor (Golf GTI). It became apparent that fuzzy methods enable a more simple design and an improved quality of control compared to the available production controller [Gebhardt93d, Kruse93].

The motivation for this application emerged from the fact that the recently tightened requirements for the reduction of fuel consumption and pollutant emission of passenger car enginess have enforced a decrease of the engine idle speed. It has to be noticed that the increased use of luxury equipment that raises motor load, like air conditioning system or power steering, at decreased RPM can lead to critical speed drops, which require a most flexible control of the idling speed.

Because of difficulties concerning the controlled system (e.g. dead time of the motor with regard to a change of filling ratio as well as a poor quality of the engine speed information), conventional control engineering led to solutions that were complex and difficult to maintain and were even in part unsatisfactory. The Braunschweig/Volkswagen collaboration did not therefore attempt to develop a physical model, as detailed as possible of the process to control, but concentrated on designing and implementing a controller based on a cognitive analysis.

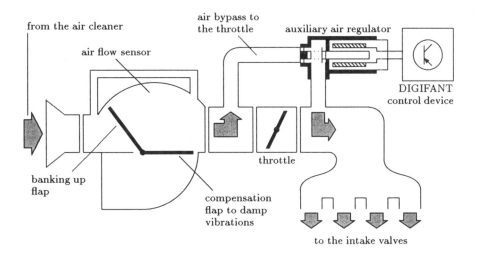

Figure 16   Principle of idling speed control

Elements of Engine Idle Speed Control

In general two methods of engine idle speed control are distinguished, namely sparking advance and volumetric control. For reasons of efficiency the developed fuzzy controller was concerned only with the volumetric control while the sparking advance of the standard production car was retained.

The control principle is shown in Figure 16. In a bypass to the throttle an auxiliary air regulator is located for the control of the fuel inlet during idling, which changes the cross-section of the bypass if necessary. When the car is idling, a sudden drop of engine speed can be caused by one or more of the following loads:

- Switching on electrical accessories, that put an additional load on the motor via the three-phase alterator,
- Switching on the air conditioning system, which puts an additional load on the motor via the air conditioning compressor,
- Activating the power steering, which adds the hydraulic circuit pump to the motor load.

It is the task of volumetric control to compensate the speed drop by increasing the cross-section in such a way that the number of revolutions reaches its set value as fast and as precisely as possible.

One of the main problems is that the torque is small because of the reduced number of revolutions, so that in extreme situations (e.g. combined switch-on of air conditioning system and power steering) sudden severe speed drops can occur. Furthermore at the aggregate at hand the number of revolutions is determined from the Hall-signal, which means, that by manufacturing tolerances, gear clearances, and torsional vibrations the measured number of revolutions can differ up to 30 rpm from the actual value, so that the controller has to work with relatively bad information about the number of revolutions. Still another problem results from numerous stochastic processes causing noise like incomplete combustion or deviation of the ignition or injection system, since a fuel inlet controller must not overreact to such effects — in some cases it must not react at all — in order to avoid oscillations. The most serious problem, however, is the delay time of the controlled system. After changing the fuel inlet, about 10 ignitions pass without change before the motor delivers its changed torque (air transfer time).

Development of the Fuzzy Controller

To cope with the difficulties mentioned with regard to the controlled system, an integration of two controllers was relied on for supervising the filling factor. The kernel of this conception is formed by a Mamdani fuzzy controller (MFC), which was constructed on the mathematical basis of equality relations and which is embedded into a 'meta controller'.

The whole controller is implemented externally on a laptop in the programming language C. Data communication proceeds over the control device of the motor management system. As input variables the number of revolutions REV0_LO, the engine speed set value AARSREV, and a status flag dAIRCON of the air conditioning system (on/off) are processed.

The electrical current AARCURIN for the auxiliary air regulator is the output value. The meta controller consists of three components: data preparation, state detection (with MFC activation), and a range of control delimitation. The data preparation provides for modified information about the number of revolutions by taking the mean of the incoming noisy data. The state recognition activates the MFC if required, causing a new output value to be calculated. To do the calculations the prepared data as well as the original are processed. Connected at the outlet side of the MFC is a range of control delimitation as a security stage. At a switch-on of the air conditioning system then instead of the output value calculated by the MFC a pilot value is set.

The MFC receives as input variables (measured values) the deviation of the number of revolutions dREV and the gradient of the number of revolutions gREV and as output variable (final control value) the change of the current for the auxiliary air regulator dAARCUR. The centre of gravity (COG) strategy serves for defuzzification. Figure 18 shows the performance characteristics induced by the MFC.

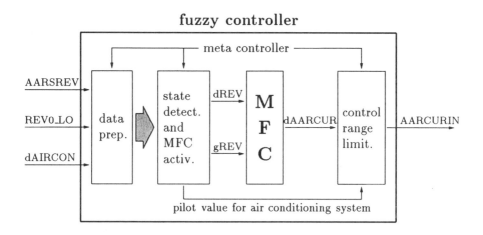

Figure 17 Structure of the fuzzy controller

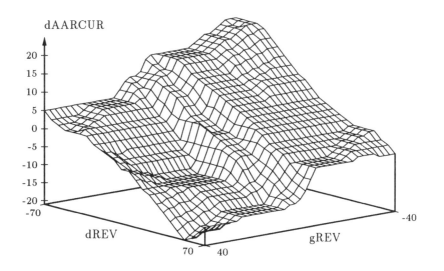

Figure 18 Performance characteristics of the MFC

Implementing the Concept of Equality Relations

For this fuzzy controller, which is based on equality relations, the following measured variables were chosen:

- the deviation dREV from the set value, and

- the change gREV of the number of revolutions between two ignitions in revolutions per minute

From the cognitive analysis it turned out that dREV can assume values from the interval $\Omega^{(dREV)} = [-70, 70]$ and that gREV can assume values from the interval $\Omega^{(gREV)} = [-40, 40]$. The manipulated variable dAARCUR describes the change of current of idling volumetric control. dAARCUR can range over the interval $\Omega^{(dAARCUR)} = [-25, 25]$. For technical reasons, due to the regulating facility, the unit of the quantity dAARCUR is 3.125 mA.

The equality relations E_{dREV}, E_{gREV}, and $E_{dAARCUR}$ on $\Omega^{(dREV)}$, $\Omega^{(gREV)}$ or $\Omega^{(dAARCUR)}$, respectively, were specified by integrable scaling functions c_v : $\Omega^{(v)} \to R_0^+$ (where $v \in \{dREV, gREV, dAARCUR\}$), just as in Example 4.4. To achieve this the sets $\Omega^{(v)}$ were separated in suitable areas, for which a constant scaling or magnification factor was determined. The comparatively low precision of engine speed measurement made it look meaningful to make no distinction in the areas around zero for all quantities. Otherwise even small diversions due to imprecision of measurement would have entailed unnecessary control actions. In particular the functions c_v were chosen as follows:

$$c_{dREV} : \Omega^{(dREV)} \to R_0^+, \qquad x \mapsto \begin{cases} \frac{1}{20}, & \text{if} \quad -70 \le x < -30 \\ \frac{1}{28}, & \text{if} \quad -30 \le x < -2 \\ 0, & \text{if} \quad -2 \le x < 2 \\ \frac{1}{28}, & \text{if} \quad 2 \le x < 30 \\ \frac{1}{20}, & \text{if} \quad 30 \le x \le 70 \end{cases}$$

$$c_{gREV} : \Omega^{(gREV)} \to R_0^+, \qquad x \mapsto \begin{cases} \frac{1}{33}, & \text{if} \quad -40 \le x < -7 \\ 0, & \text{if} \quad -7 \le x < -4 \\ 1, & \text{if} \quad -4 \le x < -2 \\ 0, & \text{if} \quad -2 \le x < 2 \\ 1, & \text{if} \quad 2 \le x < 4 \\ 0, & \text{if} \quad 4 \le x < 7 \\ \frac{1}{33}, & \text{if} \quad 7 \le x \le 40 \end{cases}$$

gREV

	φ_0	−40	−6	−3	0	3	6	40
dREV	−70	20	15	15	10	10	5	5
	−50	20	15	10	10	10	5	0
	−30	15	10	5	5	5	0	0
	0	5	5	0	0	0	−10	−5
	30	0	0	0	−5	−5	−10	−15
	50	0	−5	−5	−10	−15	−15	−20
	70	−5	−5	−10	−15	−15	−15	−15

Table 7 The partial mapping φ_0 to control the idling speed

$$c_{\text{dAARCUR}} : \Omega^{(\text{dAARCUR})} \longrightarrow \mathbf{R}_0^+, \qquad x \mapsto \begin{cases} 0, & \text{if } -25 \le x < -20 \\ \frac{1}{5}, & \text{if } -20 \le x < -5 \\ \frac{1}{4}, & \text{if } -5 \le x < -1 \\ 0, & \text{if } -1 \le x < 1 \\ \frac{1}{4}, & \text{if } 1 \le x < 5 \\ \frac{1}{5}, & \text{if } 5 \le x < 20 \\ 0, & \text{if } 20 \le x \le 25. \end{cases}$$

To make the fuzzy controller complete, besides the equality relations a partial mapping $\varphi_0 : \Omega^{(\text{dREV})} \times \Omega^{(\text{gREV})} \longrightarrow \Omega^{(\text{dAARCUR})}$ is required, which assigns to certain tuples of measured values an appropriate output value. On the basis of measurements at a test engine the partial mapping φ_0 as shown in Table 7 resulted.

With the help of the equality relations and the partial mapping φ_0 a fuzzy controller can be constructed, which, as described in Section 4.4.1, for the tuple of measured values $(x, x') \in \Omega^{(\text{dREV})} \times \Omega^{(\text{gREV})}$ yields as output the fuzzy set $\mu_{\varphi_0}^{(x,x')}$ (cf. (2.35)) of $\Omega^{(\text{dAARCUR})}$. To get a definite value from this fuzzy set the COG defuzzification strategy was employed. The mapping obtained is not extensional. There is no extensional continuation φ of the partial mapping φ_0. To make this clear, we inspect the measured values dREV $= 15$ and gREV $= 40$. If we compare these values with the tuple $(0, 40)$, a degree of equality of

$$\min\{E_{\text{dREV}}(15, 0),\ E_{\text{gREV}}(40, 40)\} = E_{\text{dREV}}(15, 0) = 0.5$$

results, so that in case of extensionality

	nh	nb	nm	ns	ze	ps	pm	pb	ph
$\Omega^{(\text{dREV})}$	–	−70	−50	−30	0	30	50	70	–
$\Omega^{(\text{gREV})}$	–	−40	−5	−3	0	3	5	40	–
$\Omega^{(\text{dAARCUR})}$	−20	−15	−10	−5	0	5	10	15	20

Table 8   The linguistic terms associated with the singletons induced by the values resulting from the mapping

$$
\begin{aligned}
E_{\text{dAARCUR}}\left(\varphi(15,40),\varphi(0,40)\right) &= E_{\text{dAARCUR}}\left(\varphi(15,40),\varphi_0(0,40)\right) \\
&= E_{\text{dAARCUR}}\left(\varphi(15,40),-5\right) \\
&\geq 0.5
\end{aligned}
$$

follows. This means that $\varphi(15,40) \in [-7.5,-2.5]$ must hold. If we replace the tuple $(0,40)$ by the tuple $(30,40)$, we again obtain a degree of equality of 0.5 with tuple $(15,40)$. From this we can infer analogously

$$
E_{\text{dAARCUR}}\left(\varphi(15,40),\varphi(30,40)\right) = E_{\text{dAARCUR}}\left(\varphi(15,40),-15\right) \geq 0.5
$$

and hence $\varphi(15,40) \in [-17.5,-12.5]$, which leads to a contradiction.

The fuzzy controller on the basis of equality relations can be transformed in the sense of Section 4.4.2 into a Mamdani fuzzy controller. For the fuzzy partitions of the sets $\Omega^{(\text{dREV})}$, $\Omega^{(\text{gREV})}$, and $\Omega^{(\text{dAARCUR})}$ we employ the fuzzy sets, which are induced as singletons by the values resulting from the mapping. To each of these fuzzy sets we associate a linguistic term of the kind negative huge (nh), negative big (nb), negative medium (nm), negative small (ns), approximately zero (ze), positive small (ps), positive medium (pm), positive big (pb), or positive huge (ph). Table 8 shows the assignment of values to these linguistic terms for the variables dREV, gREV, and dAARCUR.

Figure 19 illustrates the partitions defined by these fuzzy sets.

For the Mamdani fuzzy controller the partial mapping φ_0 has to be translated to linguistic rules of the form

if dREV is A and gREV is B then dAARCUR is C.

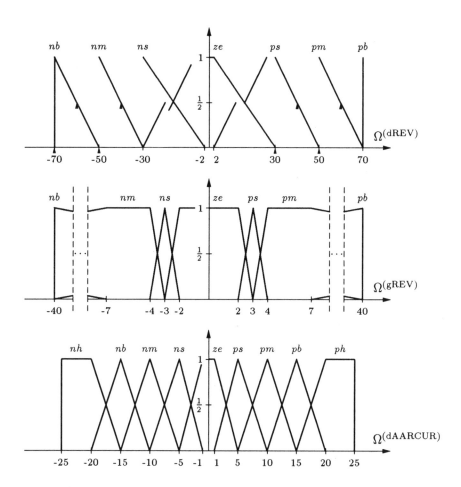

Figure 19 The fuzzy partitions of the sets
$\Omega^{(\mathrm{dREV})}$, $\Omega^{(\mathrm{gREV})}$, and $\Omega^{(\mathrm{dAARCUR})}$

Table 9 shows the rules according to the scheme of Table 7.

gREV

		nb	nm	ns	ze	ps	pm	pb
	nb	ph	pb	pb	pm	pm	ps	ps
	nm	ph	pb	pm	pm	ps	ps	ze
dREV	ns	pb	pm	ps	ps	ze	ze	ze
	ze	ps	ps	ze	ze	ze	nm	ns
	ps	ze	ze	ze	ns	ns	nm	nb
	pm	ze	ns	ns	nm	nb	nb	nh
	pb	ns	ns	nm	nb	nb	nb	nb

Table 9   The rules to control the engine idle speed

Comparing Fuzzy Controller and Standard Controller

The following figures facilitate a direct comparison between the behaviour of both controllers in the case of an increased load caused by power steering or air conditioning. A measuring point corresponds to a stroįke, i.e. the time for 180° of crank angle. AARCUROUT and AARCURIN describe the time history of the output of the standard controller or the cognitive controller, respectively. Figures 20 and 21 show the situation after activating the power steering. Figures 22 and 23 illustrate the behaviour due to switching on the air conditioning system. The over- or undershot at the point of switching on or off, respectively, cannot be regulated to disappear entirely because of the delay time.

Generally speaking the cognitive controller shows a very smooth control behaviour compared to the production line controller. In addition it can be observed that the desired value is reached precisely and quickly and that control is very stable even with slowly increasing load. The number of revolutions does not show any vibrations even after extreme changes of load. Further advantages consist in clean separation of states, simple voting treatment, and short developmental period.

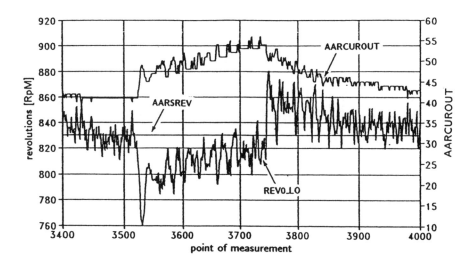

Figure 20   Power steering with standard controller

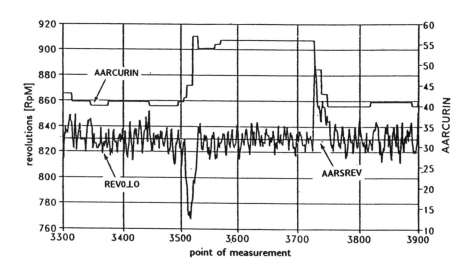

Figure 21   Power steering with fuzzy controller

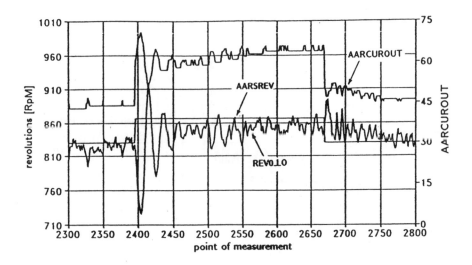

Figure 22   Air conditioning system with standard controller

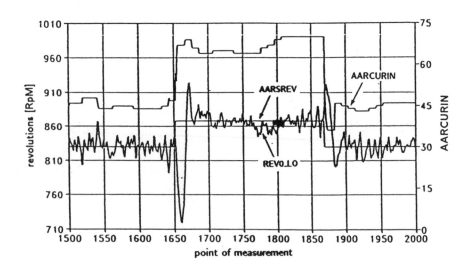

Figure 23   Air conditioning system with fuzzy controller

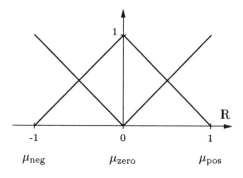

Figure 24   The fuzzy partitions of the sets X_1, X_2 and Y

ξ_1

	neg	eN	pos
neg	neg		eN
eN		eN	
pos	eN		pos

with ξ_2 labeling the rows (neg, eN, pos).

Table 10   The rule base for the Mamdani fuzzy controller

4.6.5   Exercises

Exercise 4.1 Let us consider a Mamdani fuzzy controller with two input variables $\xi_1 \in X_1 = [-1, 1]$ and $\xi_2 \in X_2 = [-1, 1]$ and one control variable $\eta \in Y = [-1, 1]$. The fuzzy partitions of the sets X_1, X_2, and Y are shown in Figure 24, the rule base is given in Table 10.

(a) For the following tuples of measured values (ξ_1, ξ_2) calculate the control value η that results from an application of the COG defuzzification strategy.
$(\xi_1, \xi_2) \in \{(0, 0), (1, 1), (1, 0), (0, 1), (-1, 0),$
$(0.25, 0.5), (-0.25, 0.5), (0.75, 0.75)\}$

(b) Determine the control function $\varphi : X_1 \times X_2 \to Y$, which is induced by the MOM defuzzification method.

(c) Prove that the fuzzy partition shown in Figure 24 satisfies condition (i) of Theorem 4.8 with respect to the t-norm T_{Luka}. Which equality relation E

ξ_1	ξ_2	η
neg	eN	$-0.5\xi_1 + 0.2\xi_2 + 0.1$
eN	neg	$0.3\xi_1 - 0.3\xi_2$
eN	eN	$0.1\xi_1 + 0.1\xi_2$
eN	pos	$0.3\xi_1 + 0.4\xi_2 - 0.2$
pos	eN	$-0.5\xi_1 - 0.5\xi_2$

Table 11 The rule base for the Sugeno fuzzy controller

determined by equation (4.24) is induced by this fuzzy partition. Translate the rule base into a partial function mapping $X_1 \times X_2$ to Y by regarding the fuzzy sets as singletons with respect to the equality relation E.

(d) The fuzzy sets of Figure 24 can be interpreted as singletons with respect to the equality relation $E_\varrho(x, x') = 1 - \min\{|x - x'|, 1\}$. The rule base in Table 10 can be translated into a partial function just as in part (c). If the control function $\varphi : X_1 \times X_2 \to Y$ is to be an extensional extension of this partial function, then the control value $\eta \in Y$ assigned to a measured input tuple $(\xi_1, \xi_2) \in X_1 \times X_2$ must belong to the set defined in (4.16). Calculate this set for the tuples of measured values stated in part (a). Show that the function

$$\varphi : X_1 \times X_2 \to Y, \quad (x_1, x_2) \mapsto \frac{1}{2}(x_1 + x_2)$$

is an extensional extension of the partial function induced by the rule base.

Exercise 4.2 We consider a Sugeno fuzzy controller with the same measured and controlled variables as in Exercise 4.1. Let the fuzzy partitions of the sets X_1 and X_2 be given as shown in Figure 24. The rule base is stated in Table 11. Calculate the control values corresponding to the tuples of measured values of Exercise 4.1(a).

Exercise 4.3 Determine equality relations with respect to the t-norm T_{Luka}, so that the fuzzy sets shown in Figure 12 can be interpreted as singletons. Define the equality relations with the help of scaling functions c as demonstrated in Example 4.4.

Exercise 4.4 Let $X = \{a, b, c\}$, $Y = \{y, z\}$. The fuzzy sets μ_1, μ_2, μ_3 of X and ν_1, ν_2, ν_3 of Y are shown in Tables 12 and 13, respectively.

	a	b	c
μ_1	1.0	0.6	0.2
μ_2	0.0	0.8	1.0
μ_3	0.9	0.1	0.0

Table 12   Three fuzzy sets of X

	y	z
ν_1	1.0	0.4
ν_2	0.6	1.0
ν_3	0.9	0.5

Table 13   Three fuzzy sets of Y

(a) Show that the system of the two relational equations $\nu_i = \mu_i \circ \varrho$, $i = 1, 2$, is solvable and give the greatest fuzzy relation that solves the system. Does the fuzzy relation which is defined by equation (4.43) also yield a solution?

(b) Prove that the system $\nu_i = \mu_i \circ \varrho$ ($i = 1, 2, 3$) of relational equations has no solution.

Symbols

$\subseteq$	set inclusion	7
$\subset$	proper set inclusion	7
$\mathbb{1}_M$	characteristic function of M	8
μ	fuzzy set	9
$F(X)$	set of all fuzzy sets of X	9
$L(X)$	set of all L-sets of X	13
μ_α	α-cut of μ	15
$\mu_{\underline{\alpha}}$	strict α-cut of μ	35
$\top$	t-norm	23
$\bot$	t-conorm	23
$\cap$	intersection (also for fuzzy sets)	24
$\cup$	union (also for fuzzy sets)	24
$^-$ (e.g. $\overline{\mu}$)	complement (also for fuzzy sets)	24
$\mathrm{acc}(a)$	gradual truth value of the proposition a	28
$\mathfrak{P}(X)$	powerset (set of all subsets) of X	29
$\hat{\phi}$	extension of the mapping ϕ	31
$F_N(\mathbf{R})$	class of all normal fuzzy sets of $\mathbf{R}$	31
$F_C(\mathbf{R})$	class of all upper semicontinuous fuzzy sets of $\mathbf{R}$	31
$F_I(\mathbf{R})$	class of all fuzzy intervals of $\mathbf{R}$	31
$\oplus$	addition of fuzzy sets	32
$\odot$	multiplication of fuzzy sets	32
$F_{D_K}(\mathbf{R})$	subclass of $F_N(\mathbf{R})$ with finite range of values	41
(P, Γ)	random set	44
$|A|$	cardinality (number of elements) of the set A	44
E	equality relation	47
μ_{x_0}	extensional hull of the set $\{x_0\}$	52
μ_M	extensional hull of the set M	52

$[\![\varphi]\!]$	truth value of the proposition φ	56
π	possibility distribution	85
Ω	universe (set of all object states possible)	85
POSS(Ω)	set of all possiblity distributions on Ω	85
Poss$_\pi$	measure of possibility of a possibility distribution π	86
Nec$_\pi$	measure of necessity of a possibility distribution π	86
$\equiv$	identical functions	76
$\sqsubseteq$	as least as specific as (for possiblity distributions)	92
$\sqsubset$	more specific than (for possibility distributions)	92
$\mathbf{N}_n$	set $\{1, 2, \ldots, n\}$ of the first n natural numbers	96
Ω^I	subspace of Ω with regard to the index set I	96
red_S^T	pointwise projection (of sets)	96
Π_S^T	projection (of sets)	96
$\hat{\Pi}_S^T$	cylindrical extension (of sets)	96
$\mathcal{U}$	universe	96
$\mathcal{M}$	modularization	97
$\mathcal{N}$	partition	98
$\mathcal{R}(\mathcal{U}, \mathcal{M})$	(possibilistic) rule base w.r.t. $\mathcal{U}$ and $\mathcal{M}$	97,101
ϱ	(possibilistic) knowledge base w.r.t. $\mathcal{U}$ and $\mathcal{M}$	91,101
$\mathcal{E}(\mathcal{U}, \mathcal{N})$	(possiblistic) system of evidence w.r.t. $\mathcal{U}$ and $\mathcal{N}$	98,102
$\mathrm{ext}_M^{\mathbf{N}_n}$	cylindrical extension (of possibility distributions)	101
$\mathrm{proj}_M^{\mathbf{N}_n}$	projection (of possibility distributions)	102
$\mathcal{X}$	(possibilistic) F-expert system	99,103
$\kappa^{(i)}$	i-th restrictions ($i = 1, \ldots, n$) of the state of a (possibilistic) F-expert system	98,99
$\sigma(\mathcal{X}, \mathcal{E}(\mathcal{U}, \mathcal{N}))$	state of $\mathcal{X}$ w.r.t. $\mathcal{E}(\mathcal{U}, \mathcal{N})$	99,102
$H_\mathcal{M}$	hypergraph induced by $\mathcal{M}$	115
$\mathcal{P}$	set of atomic propositions	132
$\mathcal{L}$	set of all proposition-logical expressions	132
$\mathcal{K}_0$	set of all proposition-logical clauses	132
$\mathcal{K}$	$\mathcal{K}_0 \cup \{\top, \bot\}$	132
$\top$	true proposition	132
$\bot$	false proposition	132
N	measure of necessity	132
Π	measure of possibility	133
$\mathcal{W}$	uncertain knowledge base	133

$\vdash$	deducibility	134
I	interpretation	135
$\mu_{(\varphi,\alpha)}$	fuzzy set on the interpretations induced by the uncertain clause (φ, α)	136
$\mu_{\mathcal{W}}$	fuzzy set on the interpretations induced by the uncertain knowledge base $\mathcal{W}$	136
$\mathrm{cons}(\mathcal{W})$	degree of consistency of the uncertain knowledge base $\mathcal{W}$	136
$\mathrm{inc}(\mathcal{W})$	degree of inconsistency of the uncertain knowledge base $\mathcal{W}$	136
$\models$	semantical deducibility	137
$\mathcal{I}$	set of all implications	139
w	fuzzy knowledge base	139
Th_w	fuzzy set of all propositions that are deducible sematically from w	139
th_w	fuzzy set of all propositions that are deducible syntactically from w	141

References

Alsina83   C. Alsina, E. Trillas, and L. Valverde. On some Logical Connectives for Fuzzy Set Theory. *J. of Math. Anal. and Applications*, 93:15–26, 1983.

Andersen89   S.K. Andersen, K.G. Olesen, V.F. Jensen, and F. Jensen. HUGIN — A Shell for Building Bayesian Belief Universes for Expert Systems. *Proc. 11th IJCAI*, Detroit, 1080–1085, Morgan Kaufman, San Mateo, CA, 1989.

Andreassen87 S. Andreassen, M. Woldbye, B. Falckand, and S.K. Andersen. MUNIN — A Causal Probabilistic Network for Interpretation of Electromyographic Findings. *Proc. 10th IJCAI*, Milan, 366–372, Morgan Kaufman, San Mateo, CA, 1987.

Arbib77   M. Arbib. Book Reviews, Vol. 3, *Bul. of the AMS*, 83:946–951, 1977

Artstein75   Z. Artstein and R.A. Vitale. A Strong Law of Large Numbers for Random Compact Sets. *Ann. Probability*, 3:879–882, 1975.

Asakawa94   K. Asakowa and H. Takagi. Neural Networks in Japan. *Communications of the ACM*, Vol. 37, 3:106–112, 1994.

Bacchus90   F. Bacchus. *Representing and Reasoning with Probabilistic Knowledge*. MIT Press, Cambridge, MA, 1990.

Baldwin79   J.F. Baldwin. A New Approach to Approximate Reasoning Using a Fuzzy Logic. *Fuzzy Sets and Systems*, 2:309–325, 1979.

Baldwin80   J.F. Baldwin and N.C.F. Guild. Modelling Controllers using Fuzzy Relations. *Kybernetes*, 9:223–229, 1980.

Bandemer92   H. Bandemer and W. Näther. *Fuzzy Data Analysis*. Series: *Mathematical and Statistical Methods*. Kluwer, Dordrecht, 1992.

Bandemer93   H. Bandemer, ed. *Modelling Uncertain Data*. Mathematical Research Vol. 68, Akademie Verlag, Berlin 1993.

Bandler80   W. Bandler and L.J. Kohout. Fuzzy Power Sets and Fuzzy Implication Operators. *Fuzzy Sets and Systems*, 4:13–30, 1980.

Bellman66   R.E. Bellman, R. Kalaba, and L.A. Zadeh. Abstraction and Pattern Classification. J. Math. Anal. and Appl. 2:581–586, 1966

Bellman70   R.E. Bellman and L.A. Zadeh. Decision-Making in a Fuzzy Environment. *Management Sciences*, 17:141–164, 1970.

Bellman73   R. Bellman and M. Giertz. On the Analytic Formalism of the Theory of Fuzzy Sets. *Information Sciences*, 5:149–156, 1973.

Berenji92   H.R. Berenji. A Reinforcement Learning Based Architecture for Fuzzy Logic Control. *Int. J. of Approximate Reasoning*, 6:267–292, 1992.

Berge73   C. Berge. *Graphs and Hypergraphs*. North-Holland, Amsterdam, 1973.

Bezdek81   J.C. Bezdek. *Pattern Recognition with Fuzzy Objective Function Algorithms*. Plenum Press, New York, 1981.

Bezdek92a   J.C. Bezdek and S.K. Pal, eds. *Fuzzy Models for Pattern Recognition*. IEEE Press, New York, 1992.

Bezdek92b   J.C. Bezdek, E.C. Tsao and N.K. Pal. Fuzzy Kohonen Clustering Networks. *Proc. IEEE Int. Conf. on Fuzzy Systems*, San Diego, 1035–1043, IEEE Press, New York, 1992.

Black37   M. Black. Vagueness: an exercise in logical analysis. *Philosophy of Science*, 4:427–455, 1937.

Bocklisch86   S.F. Bocklisch, S. Orlovski, M. Peschel, and Y. Nishiwaki, eds. *Fuzzy Sets Applications, Methodological Approaches and Results*. Akademie Verlag, Berlin, 1986.

Bolc92   L. Bolc and P. Borowik. *Many-Valued Logics*. Springer, Berlin, 1992.

Bonissone87   P.R. Bonissone, S.S. Gans, and K.S. Decker. RUM: A Layered Architecture for Reasoning with Uncertainty. *Proc. 10th IJCAI*, Milan, 891–898, Morgan Kaufman, San Mateo, NY, 1987.

Bosc88   P. Bosc, M. Galibourg, and G. Hamon. Fuzzy Querying with SQL: Extensions and Implementation Aspects. *Fuzzy Sets and Systems* 28:333-349, 1988.

Bosc92   P. Bosc and O. Pivert. Fuzzy Querying in Conventional Databases. In: L.A. Zadeh and J. Kacprzyk, eds., *Fuzzy Logic for the Management of Uncertainty*, 645–672. John Wiley and Sons, New York, 1992.

Bosc93   P. Bosc and H. Prade. An Introduction to Fuzzy Set and Possibility Theory-Based Approaches to the Treatment of Uncertainty and Imprecision in Data Base Management Systems. *Proc. Workshop on Uncertainty Management in Information Systems: From Needs to Solutions*, Santa Catalina, Avalon, CA, 1993.

Bouchon87   B. Bouchon. Fuzzy Inferences and Conditional Possibility Distributions. *Fuzzy Sets and Systems* 23:33-41, 1987.

Bouchon-Meunier92 B. Bouchon-Meunier. Fuzzy Logic and Knowledge Representation Using Linguistic Modifiers. In: L.A. Zadeh and J. Kacprzyk, eds., *Fuzzy Logic for the Management of Uncertainty*, 399–414. John Wiley and Sons, New York, 1992.

Bouslama92   F. Bouslama and A. Ichikawa. Application of Limit Fuzzy Controllers to Stability Analysis. *Fuzzy Sets and Systems*, 49:103–120, 1992.

Boverie91   S. Boverie, B. Demaya, and A. Titli. Fuzzy Logic Control Compared with Other Automatic Control Approaches. *Proc. 30th IEEE Conf. on Decision and Control*, Brighton, 1991.

Boverie92   S. Boverie, B. Demaya, R. Ketata, and A. Titli. Performance
                 Evaluation of Fuzzy Controller. *Proc. Symposium on Intelli-*
                 gent Components and Instruments for Control Applications,
                 Malaga, 1992.
Braae79   M. Braae and D.A. Rutherford. Selection of Parameters from
                 a Fuzzy Logic Controller. *Fuzzy Sets and Systems,* 2:185–199,
                 1979.
Buchanan84   B.G. Buchanan and E.H. Shortliffe. *Rule Based Expert*
                 Systems: The MYCIN Experiment of the Stanford Heuristic
                 Programming Project. Addison Wesley, Reading, MA, 1984.
Buckles82   B.P. Buckles and F.E. Petry. A Fuzzy Representation of Data
                 for Relational Databases. *Fuzzy Sets and Systems,* 7:213–226,
                 1982.
Buckley86   J.J. Buckley, W. Siler, and D.M. Tucker. A Fuzzy Expert
                 System. *Fuzzy Sets and Systems,* 20:1–16, 1986.
Buckley92   J.J. Buckley. Theory of the Fuzzy Controller: An Introduc-
                 tion. *Fuzzy Sets and Systems,* 51:249–258, 1992.
Buckley93a   J.J. Buckley. Controllable Processes and the Fuzzy Controller.
                 Fuzzy Sets and Systems, 53:27–32, 1993.
Buckley93b   J.J. Buckley. Sugeno Type Controllers are Universal Con-
                 trollers. *Fuzzy Sets and Systems,* 53:299–304, 1993.
Butnariu93   D. Butnariu and E.P. Klement. *Triangular Norm-based Mea-*
                 sures and Games with Fuzzy Coalitions. Kluwer, Dordrecht,
                 1993.
Chang58   C.C. Chang. Algebraic Analysis of Many-Valued Logics.
                 Trans. Amer. Math. Soc., 88:467–490, 1958.
Cheeseman86   P. Cheeseman. Probability versus Fuzzy Reasoning. In: L.N.
                 Kanal and J.F. Lemmer, eds., *Uncertainty in Artificial*
                 Intelligence, 85–102. North-Holland, Amsterdam, 1986.
Chen92   S.J. Chen and C.L. Hwang, eds. *Fuzzy Multiple Attribute*
                 Decision Making. Springer, Berlin, 1992.
Chiu88   S. Chiu and M. Togai. A Fuzzy Logic Programming
                 Environment for Real Time Control. *Int. J. of Approximate*
                 Reasoning 2:163–175, 1988.
Clocksin91   W.S. Clocksin and C.S. Mellish. *Programmieren in Prolog.*
                 Springer, Berlin, 1991.
Cohen73   L.J. Cohen. A Note on Inductive Logic. *The J. of Philosophy*
                 LXX:27-40, 1973.
Cooper92   G. Cooper and E. Herskovits. A Bayesian Method for the
                 Induction of Probabilistic Networks from Data. *Machine*
                 Learning, 9:309–347, 1992.
Coulon92   J. Coulon, J.-L. Coulon, and U. Höhle. Classification of
                 Extremal Subobjects over SM-SET. In: S.E. Rodabaugh,
                 E.P. Klement, and U. Höhle, eds., *Applications of Category*
                 Theory to Fuzzy Subsets, 9–31. Kluwer, Dordrecht, 1992.
Cox46   R. Cox. Probability, Frequency, and Reasonable Expectation.
                 American J. of Physics, 14:1–13, 1946.

Czogala82   E. Czogala, J. Drewniak, and W. Pedrycz. Fuzzy Relation
                   Equations on a Finite Set. *Fuzzy Sets and Systems*, 7:89–101,
                   1982.
Czogala86   E. Czogala and K. Hirota. *Probabilistic Sets: Fuzzy and
                   Stochastic Approach to Decision, Control and Recognition
                   Processes.* Verlag TÜV Rheinland, Köln, 1986.
Delgado87   M. Delgado and S. Moral. On the Concept of Possibility
                   Probability Consistency. *Fuzzy Sets and Systems* 21:311–318,
                   1987.
Dempster67   A.P. Dempster. Upper and Lower Probabilities Induced by a
                   Multivalued Mapping. *Ann. Math. Stat.*, 38:325–339, 1967.
Dempster68   A.P. Dempster. Upper and Lower Probabilities Induced by a
                   Random Closed Interval. *Ann. Math. Stat.*, 39:957–966, 1968.
Desoer63   C.A. Desoer and L.A. Zadeh. *Linear System Theory — The
                   State Space Approach.* McGraw-Hill, New York, 1963.
Dombi82   J. Dombi. A General Class of Fuzzy Operators, the De Morgan
                   Class of Fuzzy Operators and Fuzziness Measures Induced by
                   Fuzzy Operators. *Fuzzy Sets and Systems*, 8:149–163, 1982.
Driankov93   D. Driankov, H. Hellendoorn, and M. Reinfrank. *An
                   Introduction to Fuzzy Control.* Springer, Berlin, 1993.
Dubois80a   D. Dubois and H. Prade. *Fuzzy Sets and Systems — Theory
                   and Applications.* Academic Press, New York, 1980.
Dubois80b   D. Dubois and H. Prade. New Results about Properties and
                   Semantics of Fuzzy Set Theoretic Operators. In: P.P. Wang
                   and S.K. Chang, eds., *Fuzzy Sets: Theory and Application
                   to Policy Analysis and Information Systems*, 59–75. Plenum
                   Press, New York, 1980.
Dubois82   D. Dubois and H. Prade. A Class of Fuzzy Measures Based on
                   Triangular Norms. *Int. J. of General Systems*, 8:43–61, 1982.
Dubois84   D. Dubois and H. Prade. Fuzzy Logics and the Generalized
                   Modus Ponens Revisited. *Int. J. of Cybernetes and Systems*,
                   15:293–331, 1984.
Dubois85   D. Dubois and H. Prade. The Generalized Modus Ponens
                   under sup-min Composition — A Theoretical Study. In:
                   M.M. Gupta, A. Kandel, W. Bandler, and J.B. Kiszka, eds.,
                   Approximate Reasoning in Expert Systems, 217–232. North-
                   Holland, Amsterdam, 1985.
Dubois87a   D. Dubois and H. Prade. Necessity Measures and the
                   Resolution Principle. *IEEE Trans. on Systems, Man, and
                   Cybernetics*, 17:474–478, 1987.
Dubois87b   D. Dubois and H. Prade. *Possibility Theory — An Approach
                   to Computerized Processing of Uncertainty.* Plenum Press,
                   New York, 1987.
Dubois88   D. Dubois and H. Prade. Modelling Uncertainty and Inductive
                   Inference. *Acta Psychologica*, 68:53–78, 1988.
Dubois89a   D. Dubois, J. Lang, and H. Prade. Automated Reasoning Us-
                   ing Possibilistic Logic: Semantics, Belief Revision and Vari-
                   able Certainty Weights. *Proc. 5th Workshop on Uncertainty
                   in Artificial Intelligence*, Windsor, Ontario, 81–87, Morgan
                   Kaufman, San Mateo, CA, 1989.

Dubois89b	D. Dubois and H. Prade. Fuzzy Sets, Probability and Measurement. *European J. of Operational Research*, 40:135–154, 1989.
Dubois90	D. Dubois, H. Prade, and J.M. Toucas. Inference with Imprecise Numerical Quantifiers. In: Z. Ras and M. Zemankova, eds., *Intelligent Systems: State of the Art and Future Directions*, 52–72. Ellis Horwood, Chichester, 1990.
Dubois91a	D. Dubois, J. Lang, and H. Prade. Fuzzy Sets in Approximate Reasoning, Part 2: Logical Approaches. *Fuzzy Sets and Systems*, 40:203–244, 1991.
Dubois91b	D. Dubois and H. Prade. Basic Issues on Fuzzy Rules and their Application to Fuzzy Control. *Proc. IJCAI'91 Workshop on Fuzzy Control*, 5–17, Sydney, 1991.
Dubois91c	D. Dubois and H. Prade. Epistemic Entrenchment and Possibilistic Logic. *Artificial Intelligence*, 50:223–239, 1991.
Dubois91d	D. Dubois and H. Prade. Fuzzy Sets in Approximate Reasoning, Part 1: Inference with Possibility Distributions. *Fuzzy Sets and Systems*, 40:143–202, 1991.
Dubois92a	D. Dubois and H. Prade. Belief Change and Possibility Theory. In: P. Gärdenfors, ed., *Belief Revision*, 142–182. Cambridge University Press, Cambridge, 1992.
Dubois92b	D. Dubois and H. Prade. Possibility Theory as a Basis for Preference Propagation in Automated Reasoning. *Proc. IEEE Int. Conf. on Fuzzy Systems*, San Diego, 821–832, IEEE Press, New York, 1992.
Dubois93	D. Dubois, S. Moral, and H. Prade. A Semantics for Possibility Theory Based on Likelihoods. *CEC-ESPRIT III BRA 6156 DRUMS II*, Annual Report 1993.
Dubois93a	D. Dubois and H. Prade. Fuzzy Sets and Probability: Misunderstandings, Bridges and Gaps. *Proc. 2nd IEEE Int. Conf. on Fuzzy Systems*, San Francisco, 1059–1068, IEEE Press, New York, 1993.
Dubois93b	D. Dubois, H. Prade, and R.R. Yager, eds. *Readings in Fuzzy Sets for Intelligent Systems*. Morgan Kaufman, San Mateo, CA, 1993.
Dummett77	M. Dummett. *Elements of Intuitionism*. Clarendon Press, Oxford, 1977.
Eklund92a	P. Eklund and F. Klawonn. Neural Fuzzy Logic Programming. *IEEE Trans. on Neural Networks*, 3:815–818, 1992.
Eklund92b	P. Eklund, F. Klawonn, and D. Nauck. Distributing Errors in Neural Fuzzy Control. *Proc. Int. Conf. on Fuzzy Logic and Neural Networks*, 2:1139–1142, Iizuka, 1992.
Esogbue83	A.O. Esogbue and R.C. Elder. Measurement and Valuation of a Fuzzy Mathematical Model for Medical Diagnosis. *Fuzzy Sets and Systems*, 10:223–242, 1983.
Evans89	G.W. Evans, W. Karowski, and M.R. Wilhelm. *Applications of Fuzzy Set Methodologies in Industrial Engineering*. North-Holland, Amsterdam, 1989.
Fedrizzi91	M. Fedrizzi, J. Kacprzyk, and M. Roubens. *Interactive Fuzzy Optimization*. Springer, Berlin, 1991.

Feller66	W. Feller. *An Introduction to Probability Theory and its Application* (two volumes). John Wiley and Sons, New York, 1966.
Fine73	T.L. Fine. *Theories of Probability: An Examination of Foundations.* Academic Press, New York, 1973.
de Finetti74	B. de Finetti. *Theory of Probability.* John Wiley and Sons, London, 1974.
Fonck92	P. Fonck. Propagating Uncertainty in a Directed Acyclic Graph. *Proc. IPMU'92*, 17–20, Palma de Mallorca, 1992.
Frank70	M.J. Frank. On the Simultaneous Associativity of $F(x, y)$ and $x + y - F(x, y)$. *Aequationes Math.*, 19:137–152, 1979.
Frank93	H. Frank. FUZZY-BOX: A New Fuzzy Controller Design System. *Proc. EUFIT'93*, 839–844, Aachen, 1993.
Freksa81	C. Freksa. *Linguistic Pattern Characterization and Analysis.* Ph.D. Thesis, University of California, Berkeley, CA, USA, 1981.
French84	S. French. Fuzzy Decision Analysis; Some Criticism. *Management Sciences*, 20:29–44, 1984.
French88	S. French. *Decision Theory: An Introduction to the Mathematics of Rationality.* Ellis Horwood, Chichester, 1988.
Fukami80	S. Fukami, M. Mizumoto, and K. Tanaka. Some Considerations on Fuzzy Conditional Inference. *Fuzzy Sets and Systems*, 4:243–273, 1980.
Gaines76	B.R. Gaines. Foundations of Fuzzy Reasoning. *Int. J. of Man Machine Studies*, 8:623–668, 1976.
Gaines77	B.R. Gaines and L.J. Kohout. The Fuzzy Decade: A Bibliography of Fuzzy Systems and Closely Related Topics. *Int. J. of Man Machine Studies*, 9:1–68, 1977.
Gärdenfors88	P. Gärdenfors. *Knowledge in Flux, Modelling the Dynamics of Epistemic States.* MIT Press, Cambridge, MA, 1988.
Gärdenfors92	P. Gärdenfors, ed. *Belief Revision.* Cambridge University Press, Cambridge, 1992.
Gebhardt90	J. Gebhardt and R. Kruse. Some New Aspects of Testing Hypothesis in Fuzzy Statistics. *Proc. NAFIPS'90 Conf.*, 185–187, Toronto, 1990.
Gebhardt92a	J. Gebhardt. On the Epistemic View of Fuzzy Statistics. In: H. Bandemer, ed., *Modelling Uncertain Data.* Series: *Mathematical Research*, 68:136–141. Akademie–Verlag, Berlin, 1992.
Gebhardt92b	J. Gebhardt and R. Kruse. A Possibilistic Interpretation of Fuzzy Sets by the Context Model. *Proc. IEEE Int. Conf. on Fuzzy Systems*, 1089–1096, San Diego, 1992. IEEE.
Gebhardt93a	J. Gebhardt and R. Kruse. The Context Model — An Integrating View of Vagueness and Uncertainty. *Int. J. of Approximate Reasoning* 9, 283–314, 1993.
Gebhardt93b	J. Gebhardt and R. Kruse. A New Approach to Semantic Aspects of Possibilistic Reasoning. In: M. Clarke, R. Kruse and S. Moral, eds., *Symbolic and Quantitive Approaches to Reasoning and Uncertainty*, 151–159. Springer, Berlin, 1993
Gebhardt93c	J. Gebhardt and R. Kruse. A Comparative Discussion of Combination Rules in Numerical Settings. *CEC-ESPRIT III BRA 6156 DRUMS II*, 37p. Annual Report, 1993.

Gebhardt93d	J. Gebhardt, R. Kruse, M. Otte, and M. Schröder. A Fuzzy Idle Speed Controller. *26th Int. Symp. on Automative Technology and Automation*, 459–463. Aachen, 1993.
Gentilhomme68	Y. Gentilhomme. Les ensembles flous en linguistique. *Cahiers de Ling. Théor. et Appl.*, 5:47–65, 1968.
Gil88	M.A. Gil. On the Loss of Information due to Fuzziness in Experimental Observations. *Ann. Inst. Statist. Math.*, 40:627–639, 1988.
Giles82	R. Giles. Foundations for a Theory of Possibility. In: M.M. Gupta and E. Sanchez, eds., *Fuzzy Information and Decision Processes*, 183–195. North-Holland, Amsterdam, 1982.
Godo89	L. Godo, R. Lopez de Mantaras, C. Sierra, and A. Verdaguer. MILORD: The Architecture and the Management of Linguistically Expressed Uncertainty. *Int. J. of Intelligent Systems*, 4:471–501, 1989.
Goguen67	J.A. Goguen. L-Fuzzy Sets. *J. of Mathematical Analysis and Applications*, 18:145–174, 1967.
Goguen69	J.A. Goguen. The Logic of Inexact Concepts. *Synthese*, 19:325–373, 1969.
Goldblatt79	R. Goldblatt. *Topoi, The Categorial Analysis of Logic*. North-Holland, Amsterdam, 1979.
Golumbic80	M.C. Golumbic. *Algorithmic Graph Theory and Perfect Graphs*. Academic Press, London, 1980.
Goodman82	I.R. Goodman. Fuzzy Sets as Equivalence Classes of Random Sets. In: R.R. Yager, ed., *Recent Developments in Fuzzy Sets and Possibility Theory*, 327–343. Pergamon Press, New York, 1982.
Goodman91a	I.R. Goodman. Algebraic and Probabilistic Bases for Fuzzy Sets and the Development of Fuzzy Conditioning. In: I.R. Goodman, M.M. Gupta, H.T. Nguyen, and G.S. Koges, eds., *Conditional Logic in Expert Systems*, 1–69. North-Holland, Amsterdam, 1991.
Goodman91b	I.R. Goodman, M.M. Gupta, H.T. Nguyen, and G.S. Koges, eds. *Conditional Logic in Expert Systems*. North-Holland, Amsterdam, 1991.
Gottwald86a	S. Gottwald. Characterizations of the Solvability of Fuzzy Equations. *Elektronische Informationsverarbeitung und Kybernetik*, 22:67–91, 1986.
Gottwald86b	S. Gottwald and W. Pedrycz. Solvability of Fuzzy Relational Equations and Manipulation of Fuzzy Data. *Fuzzy Sets and Systems*, 18:45–65, 1986.
Gottwald93	S. Gottwald. *Fuzzy Sets and Fuzzy Logic*. Series: *Artificial Intelligence*. Vieweg, Wiesbaden, 1993.
Grabisch92	M. Grabisch, T. Murofushi, and M. Sugeno. Fuzzy Measure of Fuzzy Events Defined by Fuzzy Integrals. *Fuzzy Sets and Systems*, 50:293–313, 1992.
Gupta77	M.M. Gupta, G.N. Saridis, and B.R. Gaines, eds. *Fuzzy Automata and Decision Processes*. North-Holland, New York, 1977.

Gupta79   M.M. Gupta, R.K. Ragade, and R.R. Yager, eds. *Advances in Fuzzy Set Theory and Applications*. North-Holland, New York, 1979.

Gupta82a   M.M. Gupta and E. Sanchez, eds. *Approximate Reasoning in Decision Analysis*. North-Holland, New York, 1982.

Gupta82b   M.M. Gupta and E. Sanchez, eds. *Fuzzy Information and Decision Processes*. North-Holland, Amsterdam, 1982.

Gupta85   M.M. Gupta, A. Kandel, W. Bandler, and J.B. Kiszka, eds. *Approximate Reasoning in Expert Systems*. North-Holland, New York, 1985.

Halmos50   P.R. Halmos. *Measure Theory*. Van Nostrand, Princeton, NJ, 1950.

Hamacher78   H. Hamacher. Über logische Verknüpfungen unscharfer Aussagen und deren zugehörige Bewertungsfunktionen. In: R. Trappl, G.J. Klir, and L. Ricciardi, eds., *Progress in Cybernetics and Systems Research 3*, 276–288. Hemisphere, Washington, DC, 1978.

Hayashi92   I. Hayashi, H. Namura, H. Yamasaki, and N. Wakami. Construction of Fuzzy Inference Rules by Neural Network Driven Fuzzy Reasoning with Learning Functions. *Int. J. of Approximate Reasoning*, 6:241–266, 1992.

Heckerman88a   D.E. Heckerman. Probabilistic Interpretation for MYCIN's Certainty Factors. In: J. Lemmer and L.N. Kanal, eds., *Uncertainty in Artificial Intelligence (2)*, 167–196. North-Holland, Amsterdam, 1988.

Heckerman88b   D.E. Heckerman and E.J. Horritz. The Myth of Modularity in Rule Based Systems for Reasoning with Uncertainty. In: J.F. Lemmer and L.N. Kanal, eds., *Uncertainty in Artificial Intelligence (2)*, 23–34. North-Holland, 1988.

Heckerman90   D.E. Heckerman. *Probabilistic Similarity Networks*. MIT Press, Cambridge, MA, 1990.

Hestir91   K. Hestir, H.T. Nguyen, and G.S. Rogers. A Random Set Formalism for Evidential Reasoning. In: I.R. Goodman, M.M. Gupta, H.T. Nguyen, and G.S. Rogers, eds., *Conditional Logic in Expert Systems*, 309–344. North-Holland, 1991.

Higashi83   M. Higashi and G.J. Klir. Measures of Uncertainty and Information based on Possibility Distributions. *Int. J. of General Systems* 9:43–58, 1983.

Hirota80   K. Hirota, Y. Arai, and S. Hachisu. Fuzzy Controlled Robot Arm Playing Ping Pong Game. *Fuzzy Sets and Systems*, 3:193–219, 1980.

Hirota81   K. Hirota. Concepts of Probabilistic Sets. *Fuzzy Sets and Systems*, 5:31–46, 1981.

Hirota93   K. Hirota, ed. *Industrial Applications of Fuzzy Technology*. Springer, Tokyo, 1993.

Hisdal78   E. Hisdal. Conditional Possibilities, Independence, and Non-Interaction. *Fuzzy Sets and Systems*, 1:283–297, 1978.

Hisdal88   E. Hisdal. Are Grades of Membership Probabilities? *Fuzzy Sets and Systems*, 35:325–347, 1988.

Höhle85   U. Höhle. The Category of B-Fuzzy Topological Spaces. *Analysis of Fuzzy Information*, 1:185–194, 1985.

Höhle90   U. Höhle. The Poincaré Paradox and the Cluster Problem.
                   In: A. Dress and A. v. Haeseler, eds., *Trees and Hierarchi-*
                   cal Structures, Lecture Notes in Biomathematics 84, 117–124.
                   Springer, 1990.
Höhle91   U. Höhle and L.N. Stout. Foundations of Fuzzy Sets. *Fuzzy*
                   Sets and Systems, 40:257–296, 1991.
Höhle92   U. Höhle. M-Valued Sets and Sheaves over Integral Commu-
                   tative CL-Monoids. In: S.E. Rodabaugh, E.P. Klement, and
                   U. Höhle, eds., *Applications of Category Theory to Fuzzy Sub-*
                   sets, 33–72. Kluwer, Dordrecht, 1992.
Holmblad82   L.P. Holmblad and J.J. Østergaard. Control of a Cement Kiln
                   by Fuzzy Logic. In: M.M. Gupta and E. Sanchez, eds., *Fuzzy*
                   Information and Decision Processes, 389–400. North-Holland,
                   Amsterdam, 1982.
Hopf94   J. Hopf and F. Klawonn. Learning the Rule Base of a Fuzzy
                   Controller by a Genetic Algorithm. In: R. Kruse, J. Gebhardt
                   and R. Palm, eds. *Fuzzy Systems in Computer Science*,
                   Vieweg, Wiesbaden, 1994.
Howard81   R.A. Howard and J.E. Matheson. Influence Diagrams. In:
                   R.A. Howard and J.E. Matheson, eds., *Readings on the*
                   Principles and Applications of Decision Analysis, Vol. 2, 721–
                   762. Strategic Decisions Group, Menlo Park, CA, 1981.
Hwang92   G.-C. Hwang and S. Lin. A Stability Approach to Fuzzy
                   Control. *Fuzzy Sets and Systems*, 48:279–288, 1992.
Ichihashi91   H. Ichihashi. Iterative Fuzzy Modeling and a Hierachical
                   Network. In: R. Lowen and M. Roubens, eds., *Proc. 4th IFSA*
                   Congress, Engineering, 49–52, Brussels, 1991.
Isermann92   R. Isermann, K.H. Lachmann, and D. Matko. *Adaptive*
                   Control Systems. Prentice Hall, Englewood Cliffs, NJ, 1992.
Ishizuka85   M. Ishizuka and N. Kaisai. Prolog-ELF Incorporating Fuzzy
                   Logic. *Proc. 9th IJCAI*, Los Angeles, 701–703, Morgan
                   Kaufman, San Mateo, CA, 1985.
Jacas92   J. Jacas and J. Recasens. Eigenvectors and Generators of
                   Fuzzy Relations. *Proc. IEEE Int. Conf. on Fuzzy Systems*,
                   687–694, San Diego, 1992. IEEE.
Jamshidi93   M. Jamshidi, N. Vadiee, and T.J. Ross, eds. *Fuzzy Logic and*
                   Control. Prentice Hall, Englewood Cliffs, 1993.
Kacprzyk83   J. Kacprzyk and R.R. Yager, eds. *Management Decision*
                   Support Systems Using Fuzzy Sets and Possibility Theory.
                   Verlag TÜV Rheinland, Köln, 1983.
Kacprzyk88   J. Kacprzyk and M. Fedrizzi, eds. *Combining Fuzzy Im-*
                   precision with Probabilistic Uncertainty in Decision Making.
                   Springer, Berlin, 1988.
Kacprzyk90   J. Kacprzyk and M. Fedrizzi. *Multiperson Decision Making*
                   Models Using Fuzzy Sets and Possibility Theory. Kluwer,
                   Dordrecht, 1990.
Kampé de Fériet82 J. Kampé de Fériet. Interpretation of Membership Functions
                   in Terms of Plausibility and Belief. In: M.M. Gupta and
                   E. Sanchez, eds., *Fuzzy Information and Decision Making*
                   1982, 93–98, 1982

Kandel86   A. Kandel. *Fuzzy Mathematical Techniques with Applications.*
                    Addison Wesley, Reading, MA, 1986.

Kandel91   A. Kandel and G. Langholz, eds. *Fuzzy Control Systems.* CRC
                    Press, Boca Raton, 1991.

Karr89   C. Karr, L. Freeman, and D. Meredith. Improved Fuzzy
                    Process Control of Spacecraft Autonomous Rendezvous Using
                    a Genetic Algorithm. *Proc. SPIE Conf. on Intelligent Control
                    and Adaptive Systems*, 274–283, SPIE, 1989.

Karr91   C. Karr. Applying Genetics to Fuzzy Logic. *AI Expert*, 6:26–
                    33, 1991.

Karr92a   C. Karr. Design of an Adaptive Fuzzy Logic Controller Using
                    a Genetic Algorithm. *Proc. Int. Conf. of Genetic Algorithms*,
                    450–457, 1992.

Karr92b   C. Karr and E. Gentry. A Genetic-Based Adaptive pH Fuzzy
                    Logic Controller. *Proc. Int. Fuzzy Systems and Intelligent
                    Control Conf.*, 255–264, 1992.

Karr92c   C. Karr, S. Sharma, W. Hatcher, and T. Harper. Control
                    of an Exothermic Chemical Reaction Using Fuzzy Logic and
                    Genetic Algorithms. *Proc. Int. Fuzzy Systems and Intelligent
                    Control Conf.*, 246–254, 1992.

Katsuno91   H. Katsuno and A.O. Mendelzon. Proportional Knowledge
                    Base Revisions and Minimal Change. *Artificial Intelligence*
                    52:263–294, 1991.

Kaufmann85   A. Kaufmann and M.M. Gupta. *Introduction to Fuzzy
                    Arithmetic.* Van Nostrand Reinhold Company, New York,
                    1985.

Keller85   J.M. Keller, G. Hobson, J. Wooton, A. Nofarich, and
                    K. Luetkemeyer. Fuzzy Confidence Measures in Midlevel
                    Vision. *IEEE Trans. on Systems, Man, and Cybernetics*,
                    15:580–585, 1985.

Keller92a   J.M. Keller and H. Tahani. Implementation of Conjunctive
                    and Disjunctive Fuzzy Logic Rules with Neural Networks. *Int.
                    J. of Approximate Reasoning*, 6:221–240, 1992.

Keller92b   J.M. Keller, R.R. Yager, and H. Tahani. Neural Network
                    Implementation of Fuzzy Logic. *Fuzzy Sets and Systems*, 45:1–
                    12, 1992.

Kendall74   D.G. Kendall. Foundations of a Theory of Random Sets. In:
                    E.F. Harding and D.G. Kendall, eds., *Stochastic Geometry*,
                    322–376. John Wiley and Sons, Chichester, 1974.

Kickert78   W.J.M. Kickert and E.H. Mamdani. Analysis of a Fuzzy
                    Controller. *Fuzzy Sets and Systems* 1:29–44, 1978.

Kiendl93   H. Kiendl and J.J. Rüger. Fast Real-time Fuzzy Controller
                    Realization and Computer-aided Proof of Stability. *First
                    European Congress on Fuzzy and Intelligent Technologies*,
                    124–129. Aachen, 1993.

Kinzel94   J. Kinzel, F. Klawonn and R. Kruse. Modifications of Genetic
                    Algorithms for Designing and Optimizing Fuzzy Controllers.
                    Proc. IEEE Conference on Evolutionary Computation,
                    Orlando, 1994

Klawonn92a   F. Klawonn. Fuzzy Unit Interval and Fuzzy Paths. In:
             S.E. Rodabaugh, E.P. Klement and U. Höhle, eds., *Applica-
             tions of Category Theory to Fuzzy Subsets*, 245–256. Kluwer,
             Dordrecht, 1992.

Klawonn92b   F. Klawonn. On a Lukasiewicz Logic Based Controller. *Proc.
             MEPP'92 Int. Seminar on Fuzzy Control through Neural
             Interpretations of Fuzzy Sets*, 53–56, Turku, 1992. *Reports
             on Computer Science & Mathematics, Series B*, No. 14, Åbo
             Akademie.

Klawonn92c   F. Klawonn. Prolog Extensions to Many-Valued Logics.
             In: U. Höhle and E.P. Klement, eds., *Proc. 14th Linz
             Seminar on Fuzzy Set Theory: Non-Classical Logics and their
             Applications*, 42–45, Linz, 1992. Johannes Kepler Universität.

Klawonn92d   F. Klawonn, J. Gebhardt, and R. Kruse. Logical Approaches
             to Uncertainty and Vagueness in the View of the Context
             Model. *Proc. IEEE Int. Conf. on Fuzzy Systems*, 1375–1382,
             San Diego, 1992. IEEE.

Klawonn93a   F. Klawonn and R. Kruse. Equality Relations as a Basis for
             Fuzzy Control. *Fuzzy Sets and Systems*, 54:147–156, 1993.

Klawonn93b   F. Klawonn and R. Kruse. Fuzzy Control as Interpolation on
             the Basis of Equality Relations. *Proc. IEEE Int. Conf. on
             Fuzzy Systems*, San Francisco, IEEE Press, New York, 1993.

Klement80   E.P. Klement. Fuzzy σ–Algebras and Fuzzy Measurable
             Functions. *Fuzzy Sets and Systems*, 4:83–93, 1980.

Klement82   E.P. Klement. A Theory of Fuzzy Measures: A Survey. In:
             M.M. Gupta and E. Sanchez, eds., *Fuzzy Information and
             Decision Processes*, 59–65. North-Holland, Amsterdam, 1982.

Klir87   G.J. Klir. Where Do We Stand on Measures of Uncertainty,
             Ambiguity, Fuzziness and the Like? *Fuzzy Sets and Systems*,
             24, 1987.

Klir88   G.J. Klir and T.A. Folger. *Fuzzy Sets, Uncertainty and
             Information*. Prentice Hall, Englewood Cliffs, NJ, 1988.

Knopfmacher75   J. Knopfmacher. On Measures of Fuzziness. *J. of Mathemat-
             ical Analysis and Applications*, 49:529–534, 1975.

König92   D. König and V. Schmidt. *Zufällige Punktprozesse*. Series:
             Teubner Skripten zur Mathematischen Stochastik. B.G.
             Teubner, Stuttgart, 1992.

Kosko92a   B. Kosko. Fuzzy Systems as Universal Approximators. *Proc.
             IEEE Int. Conf. on Fuzzy Systems*, 1153–1162, San Diego,
             1992. IEEE.

Kosko92b   B. Kosko. *Neural Networks and Fuzzy Systems*. Prentice Hall,
             Englewood Cliffs, NJ, 1992.

Kraft83   D.H. Kraft and D.A. Buell. Fuzzy Sets and Generalized
             Boolean Retrieval Systems. *Int. J. of Man-Machine Studies*
             19:45–56, 1983.

Krantz71a   D.H. Krantz, R.D. Luce, P. Suppes, and A. Tversky.
             Foundations of Measurement, Vol. 1. Academic Press, New
             York, 1971.

Krantz71b   D.H. Krantz, R.D. Luce, P. Suppes, and A. Tversky.
             Foundations of Measurement, Vol. 2. Academic Press, New
             York, 1971.

Krishnapuram93a   R. Krishnapuram and J.M. Keller, A Possibilistic Approach to Clustering. *IEEE Transactions on Fuzzy Systems*, 1:98–110, 1993.

Krishnapuram93b   R. Krishnapuram, J.M. Keller, and Y. Ma, Quantitative Analysis of Properties and Spatial Relations of Fuzzy Image Regions. *IEEE Transactions on Fuzzy Systems*, 1:222–233, 1993.

Kruse82a   R. Kruse. A Note on λ-additive Fuzzy-Measures. *Fuzzy Sets and Systems*, 8:219–222, 1982.

Kruse82b   R. Kruse. On the Construction of Fuzzy Measures. *Fuzzy Sets and Systems*, 8:323–327, 1982.

Kruse82c   R. Kruse. The Strong Law of Large Numbers for Fuzzy Random Variables. *Information Sciences*, 28:233–241, 1982.

Kruse83   R. Kruse. On the Entropy of Fuzzy Events. *Kybernetes*, 12:53–57, 1983.

Kruse84   R. Kruse. Statistical Estimation with Linguistic Data. *Information Sciences*, 33:197–207, 1984.

Kruse87a   R. Kruse. On a Software Tool for Statistics with Linguistic Data. *Fuzzy Sets and Systems*, 24:377–383, 1987.

Kruse87b   R. Kruse. On the Variance of Random Sets. *J. of Mathematical Analysis and Applications*, 122:469–473, 1987.

Kruse87c   R. Kruse. Parametric Statistics in the Presence of Vague Data. *Proc. 2nd IFSA Congress*, Tokyo, 1987.

Kruse87d   R. Kruse and K.D. Meyer. *Statistics with Vague Data*. Series B: *Mathematical and Statistical Methods*. Reidel, Dordrecht, 1987.

Kruse89   R. Kruse and J. Gebhardt. On a Dialog System for Modelling and Statistical Analysis of Linguistic Data. *Proc. IFSA Congress*, 157–160, Seattle, 1989.

Kruse90   R. Kruse and E. Schwecke. Fuzzy Reasoning in a Multidimensional Space of Hypothesis. *Int. J. of Approximate Reasoning*, 4:47–68, 1990.

Kruse91a   R. Kruse, E. Schwecke, and J. Heinsohn. *Uncertainty and Vagueness in Knowledge Based Systems: Numerical Methods.* Series: *Artificial Intelligence*. Springer, Berlin, 1991.

Kruse91b   R. Kruse, E. Schwecke, and F. Klawonn. On a Tool for Reasoning with Mass Distributions. *Proc. 12th IJCAI*, Sydney, 1190–1195, Morgan Kaufman, San Mateo, CA, 1991.

Kruse91c   R. Kruse and P. Siegel, eds. *Symbolic and Quantitative Approaches to Uncertainty.* Springer, 1991.

Kruse92   R. Kruse. On the Semantic Foundations of Fuzzy Probability Theory and Fuzzy Statistics. In: H. Bandemer, ed., *Modelling Uncertain Data*, 131–135. Akademie Verlag, Berlin, 1992.

Kruse93   R. Kruse and M. Schröder. An Application of Equality Relations to Idle Speed Control. *Proc. 1st European Congress on Fuzzy and Intelligent Technologies*, 370–376, Aachen, 1993.

Kruse94   R. Kruse, J. Gebhardt, and R. Palm, eds. *Fuzzy Systems in Computer Science*. Vieweg, Wiesbaden, 1994.

Kwakernaak78a   H. Kwakernaak. Fuzzy Random Variables Part 1: Definitions and Theorems. *Information Sciences*, 15:1–15, 1978.

Kwakernaak78b   H. Kwakernaak. Fuzzy Random Variables Part 2: Algorithms and Examples for the Discrete Case. *Information Sciences*, 17:253–278, 1978.

Lamata89   M.T. Lamata and S. Moral. Classification of Fuzzy Measures. *Fuzzy Sets and Systems*, 33:243–253, 1989.

Larsen80   P.M. Larsen. Industrial Applications of Fuzzy Logic Control. *Int. J. of Man Machine Studies*, 12:3–10, 1980.

Lauritzen88   S.L. Lauritzen and D.J. Spiegelhalter. Local Computations with Probabilities on Graphical Structures and their Application to Expert Systems. *J. of the Royal Stat. Soc., Series B*, 50:157–224, 1988.

Le Faivre74   R.A. Le Faivre. The Representation of Fuzzy Knowledge. *J. of Cybernetics*, 4:57–66, 1974.

Lee71   R.C.T. Lee and C.L. Chang. Some Properties of Fuzzy Logic. *Information and Control*, 19:417–431, 1971.

Lee72   R.C.T. Lee. Fuzzy Logic and the Resolution Principle. *J. Association of Computation and Machines*, 19:109–119, 1972.

Lee90a   C.C. Lee. Fuzzy Logic in Control Systems: Fuzzy Logic Controller Part I. *IEEE Trans. on Systems, Man, and Cybernetics*, 20:404–418, IEEE Press, New York, 1990.

Lee90b   C.C. Lee. Fuzzy Logic in Control Systems: Fuzzy Logic Controller Part II. *IEEE Trans. on Systems, Man, and Cybernetics*, 20:419–443, 1990.

Lee93   M.A. Lee and H. Takagi. Integrating Design Stages of Fuzzy Systems Using Genetic Algorithms. *Proc. IEEE Int. Conf. on Fuzzy Systems*, San Francisco, 1993.

Lindley87   D.V. Lindley. The Probability Approach to the Treatment of Uncertainty in Artificial Intelligence and Expert Systems. *Stat. Sci.* 2:17–24, 1987.

Lloyd87   J.W. Lloyd. *Foundations of Logic Programming*. Springer, Berlin, 1987.

Loginov66   V.I. Loginov. Probability Treatment of Zadeh's Membership Function and their Use in Pattern Recognition. *Eng. Cybernetics*, 68–69, 1966

Lowen76   R. Lowen. Fuzzy Topological Spaces and Fuzzy Compactness. *J. of Mathematical Analysis and Applications*, 56:621–633, 1976.

de Luca72   A. de Luca and S. Termini. A Definition of a Nonprobabilistic Entropy in the Setting of Fuzzy Set Theory. *Information and Control*, 20:301–312, 1972.

Luce90   R.D. Luce, D.H. Krantz, P. Suppes, and A. Tversky. *Foundations of Measurement*, Vol. 3. Academic Press, New York, 1990.

Ludwig90   G. Ludwig. *Die Grundstrukturen einer physikalischen Theorie*. Springer, Berlin, 1990.

Lukasiewicz30   J. Lukasiewicz. *Philosophical Remarks on Many-valued Systems of Propositional Logic*, 1930. Reprinted in *Selected Works*, L. Borkowski, eds., *Studies in Logic and the Foundations of Mathematics*, North-Holland, Amsterdam, 1970, 153–179.

MacColl78   H. MacColl. The Calculus of Equivalent Statements and
                 Integration. *Proc. Limits. of the London Mathematical Society*
                 9, 1878.

Maier83   D. Maier. *The Theory of Relational Databases*. Pitman,
                 London, 1983.

Mamdani74   E.H. Mamdani. Applications of Fuzzy Algorithms for Simple
                 Dynamic Plant. *Proc. IEEE* 121:1585–1588, 1974.

Mamdani75   E.H. Mamdani and S. Assilian. An Experiment in Linguistic
                 Synthesis with a Fuzzy Logic Controller. *Int. J. of Man*
                 Machine Studies, 7:1–13, 1975.

Mamdani76   E.H. Mamdani. Advances in the Linguistic Synthesis of Fuzzy
                 Controllers. *Int. J. of Man Machine Studies*, 8:669–678, 1976.

Mamdani77   E.H. Mamdani. Application of Fuzzy Logic to Approximate
                 Reasoning Using Linguistic Systems. *IEEE Trans. on*
                 Computers, 26:1182–1191, 1977.

Mamdani81   E.H. Mamdani and B.R. Gaines, eds. *Fuzzy Reasoning and its*
                 Applications. Academic Press, London, 1981.

Martin87   T.P. Martin, J.F. Baldwin, and B.W. Pilsworth. The
                 Implementation of FProlog — A Fuzzy Prolog Interpreter.
                 Fuzzy Sets and Systems, 23:119–129, 1987.

Matheron75   G. Matheron. *Random Sets and Integral Geometry*. John
                 Wiley and Sons, New York, 1975.

Menger42   K. Menger. Statistical Metrics. *Proc. Nat. Acad. Sci. USA*,
                 28:535–537, 1942.

Menger51   K. Menger. Ensembles flous et functions aléatoires. *C. R.*
                 Acad. Sci., 232:2001–2003, 1951.

Miyamoto90   S. Miyamoto. *Fuzzy Sets in Information Retrieval and Cluster*
                 Analysis. Kluwer, Dordrecht, 1990.

Mizumoto76   M. Mizumoto and K. Tanaka. Some Properties of Fuzzy Sets
                 of Type 2. *Information and Control*, 31:312–340, 1976.

Mizumoto85   M. Mizumoto. Extended Fuzzy Reasoning. In: M.M. Gupta,
                 A. Kandel, W. Bandler, and J.B. Kiszka, eds., *Approximate*
                 Reasoning in Expert Systems, 71–85. North-Holland, Amster-
                 dam, 1985.

Moore66   R.E. Moore. *Interval Analysis*. Prentice Hall, Englewood
                 Cliffs, NJ, 1966.

Moore79   R.E. Moore. *Methods and Applications of Interval Analysis*.
                 SIAM, Philadelphia, 1979.

Mukaidono82   M. Mukaidono. Fuzzy Inference in Resolution Style. In: R.R.
                 Yager, ed., *Fuzzy Sets and Possibility Theory — Recent*
                 Developments, 224–231. Pergamon Press, Oxford, 1982.

Mukaidono89   M. Mukaidono, Z.L. Shen, and L. Ding. Fundamentals of
                 Fuzzy Prolog. *Int. J. of Approximate Reasoning*, 3:179–193,
                 1989.

Munakata94   T. Munakata and Y. Jani. Fuzzy Systems: An Overview.
                 Communications of the ACM, Vol. 37, 3:69–76, 1994.

Näther90   W. Näther and M. Albrecht. Linear Regression with Random
                 Fuzzy Observations. *Statistics*, 21:521–531, 1990.

Nauck92a D. Nauck and R. Kruse. Interpreting Changes in the Membership Functions of a Self–Adaptive Neural Fuzzy Controller. *Proc. 2nd Int. Workshop Industrial Applications of Fuzzy Control and Intelligent Systems (IFIS'92)*, 146–152, College Station, 1992.

Nauck92b D. Nauck and R. Kruse. Neural Fuzzy Controller Learning by Fuzzy Error Propagation. *Proc. NAFIPS'92 Conf.*, 388–397, Puerto Vallarta, 1992.

Nauck93 D. Nauck and R. Kruse. A Fuzzy Neural Network Learning Fuzzy Control Rules and Membership Functions by Fuzzy Error Backpropagation. *Proc. IEEE-ICNN'93*, San Francisco, IEEE Press, New York, 1993.

Nauck94 D. Nauck, F. Klawonn, and R. Kruse. *Neuronale Netze und Fuzzy Systeme*. Vieweg, Wiesbaden, 1994.

Neapolitan90 R.E. Neapolitan. *Probabilistic Reasoning in Expert Systems: Theory and Algorithms*. John Wiley and Sons, New York, 1990.

Negoita75 C.V. Negoita and D.A. Ralescu. Representation Theorems for Fuzzy Concepts. *Kybernetes*, 4:169–174, 1975.

Nguyen78a H.T. Nguyen. On Conditional Probability Distributions. *Fuzzy Sets and Systems*, 1:299–309, 1978.

Nguyen78b H.T. Nguyen. On Random Sets and Belief Functions. *J. of Mathematical Analysis and Applications*, 65:531–542, 1978.

Nguyen79 H.T. Nguyen. Towards a Calculus of the Mathematical Notion of Possibility. In: M.M. Gupta, R.K. Ragade, and R.R. Yager, eds. *Advances in Fuzzy Set Theory and Applications*, 235–246. North-Holland, Amsterdam, 1979.

Nguyen84 H.T. Nguyen. Using Random Sets. *Information Science*, 34:265–274, 1984.

Nilsson86 N.J. Nilsson. Probabilistic Logic. *Artificial Intelligence*, 28:71–87, 1986.

Nola83 A.D. Nola and S. Sessa. On the Set of Solutions of Composite Fuzzy Relation Equations. *Fuzzy Sets and Systems*, 9:275–285, 1983.

Nola86 A.D. Nola and A.G.S. Ventre, eds. *The Mathematics of Fuzzy Systems*. Verlag TÜV Rheinland, Köln, 1986.

Nola89 A.D. Nola, S. Sessa, W. Pedrycz, and E. Sanchez. *Fuzzy Relation Equations and their Application to Knowledge Engineering*. Kluwer, Dordrecht, 1989.

Nomura92a H. Nomura, I. Hayashi, and N. Wakami. A Learning Method of Fuzzy Inference Rules by Descent Method. *Proc. IEEE Int. Conf. on Fuzzy Systems*, San Diego, 371–378, IEEE Press, New York, 1992.

Nomura92b H. Nomura, I. Hayashi, and N. Wakami. A Self-Tuning Method of Fuzzy Reasoning by Genetic Algorithm. *Proc. Int. Fuzzy Systems and Intelligent Control Conf.*, 236–245, 1992.

Norwich84 A.M. Norwich and I.B. Turksen. A Model for the Measurement of Membership and the Consequences of its Empirical Implementation. *Fuzzy Sets and Systems*, 12:1–25, 1984.

Novák90a — V. Novák. On the Syntactico–Semantical Completeness of First Order Fuzzy Logic, Part I: Syntax and Semantics. *Kybernetica*, 26:47–66, 1990.

Novák90b — V. Novák. On the Syntactico–Semantical Completeness of First Order Fuzzy Logik, Part II: Main Results. *Kybernetica*, 26:134–154, 1990.

Novák92 — V. Novák. On the Logical Basis of Approximate Reasoning. In: V. Novák, J. Ramík, M. Mareš, M. Černý, and J. Nekola, eds., *Fuzzy Approach to Reasoning and Decision Making*, 17–27. Kluwer, Dordrecht, 1992.

Orci85 — I.P. Orci. Programming in Fuzzy Logic for Expert System Design. *Proc. 5th Int. Workshop on Expert Systems and their Applications*, 1179–1190, Avignon, 1985.

Orci89 — I.P. Orci. Programming in Possibilistic Logic. *Int. J. of Expert Systems*, 2:79–96, 1989.

Orlov78 — A.I. Orlov, Fuzzy and Random Sets (in Russian). *Prikladnoï Mnogomiernii Statistcheskii Analyz*, 262–280. Nauka, Moscow, 1978.

Ovchinnikov84 — S.V. Ovchinnikov. Representation of Transitive Fuzzy Relations. In: H.J. Skala, S. Termini, and E. Trillas, eds., *Aspects of Vagueness*, 105–118. Reidel, Dordrecht, 1984.

Palm92 — R. Palm. Sliding Mode Control. *Proc. IEEE Int. Conf. on Fuzzy Systems*, San Diego, 519–526, IEEE Press, New York, 1992.

Pao89 — Y.-H. Pao. *Adaptive Pattern Recognition and Neural Networks*. Addison Wesley, Reading, MA, 1989.

Pavelka79 — J. Pavelka. On Fuzzy Logic I, II, III. *Zeitschr. Math. Logik Grundl. Math.*, 25:45–52, 119–134, 447–464, 1979.

Pearl86 — J. Pearl. Fusion, Propagation and Structuring in Belief Networks. *Artificial Intelligence*, 29:241–288, 1986.

Pearl88 — J. Pearl. *Probabilistic Reasoning in Intelligent Systems. Networks of Plausible Inference*. Morgan Kaufman, San Mateo, CA, 1988.

Pedrycz81 — W. Pedrycz. An Approach to the Analysis of Fuzzy Systems. *Int. J. of Control*, 34:403–421, 1981.

Pedrycz93 — W. Pedrycz. *Fuzzy Control and Fuzzy Systems*. Research Studies Press, Taunton, 2nd edition, 1993.

Pfeiffer93 — B.M. Pfeiffer and R. Isermann. Criteria for Successful Applications of Fuzzy Control. *First European Congress on Fuzzy and Intelligent Technologies*, 1403–1409. Aachen, 1993.

Pham91 — D.T. Pham and D. Karaboga. Optimum Design of Fuzzy Logic Controllers Using Genetic Algorithms. *J. of Systems Engineering*, 1:114–118, 1991.

Poincaré02 — H. Poincaré. *La Science et L'Hypothèse*. Flammarion, Paris, 1902.

Poincaré04 — H. Poincaré. *La Valeur de la Science*. Flammarion, Paris, 1904.

Prade86 — H. Prade and C.V. Negoita, eds. *Fuzzy Logic in Knowledge Engineering*. Verlag TÜV Rheinland, Köln, 1986.

Procyk79 — T.J. Procyk and E.H. Mamdani. A Linguistic Self-Organizing Process Controller. *Automatica*, 15:15–30, 1979.

Puri82	M.L. Puri and D.A. Ralescu. A Possibility Measure is not a Fuzzy Measure. *Fuzzy Sets and Systems*, 7:311–313, 1982.
Puri86	M.L. Puri and D.A. Ralescu. Fuzzy Random Variables. *J. of Mathematical Analysis and Applications*, 114:409–422, 1986.
Quian92	Y. Quian, P. Tessier, and G. Dumont. Fuzzy Logic Based Modeling and Optimization. *Proc. 2nd Int. Conf. on Fuzzy Logic and Neural Networks*, 349–352, Iizuka, 1992.
Ralescu82	D.A. Ralescu. Fuzzy Logic and Statistical Estimation. *Proc. 2nd World Conf. on Mathematics at the Service of Man*, 1982.
Ramer89	A. Ramer. Conditional Possibility Measures. *Proc. 3rd IFSA Congress*, 412–415, Seattle, 1989.
Rasiowa70	H. Rasiowa and R. Sikorski. *The Mathematics of Metamathematics*. Polish Scientific Publishers, Warsaw, 1970.
Rescher69	N. Rescher. *Many-Valued Logic*. McGraw-Hill, New York, 1969.
Rescher76	N. Rescher. *Plausible Reasoning: An Introduction to the Theory and Practice of Plausible Inference*. Van Gorcom, Assen, Amsterdam, 1976.
Rodabaugh92	S.E. Rodabaugh, E.P. Klement, and U. Höhle, eds. *Applications of Category Theory to Fuzzy Subsets*. Kluwer, Dordrecht, 1992.
Rosser52	J.B. Rosser and A.R. Turquette. *Many-Valued Logics*. North-Holland, Amsterdam, 1952.
Runkler93	T.A. Runkler and M. Glesner. A Set of Axioms for Defuzzification Strategies: Towards A Theory of Rational Defuzzification Operators. *Proc. 2nd IEEE International Conf. on Fuzzy Systems*, San Francisco, 1161–1161, IEEE Press, New York, 1993.
Ruspini69	E.H. Ruspini. A New Approach to Clustering. *Information and Control*, 15:22–32, 1969.
Ruspini70	E.H. Ruspini. Numerical Methods for Fuzzy Clustering. *Information Sciences*, 2:319–350, 1970.
Ruspini90	E.H. Ruspini. Similarity Based Modells for Possibilistic Logics. *Proc. 3rd Int. Conf. on Information Processing and Management of Uncertainty in Knowledge Based Systems*, 56–58, 1990.
Ruspini91	E.H. Ruspini. On the Semantics of Fuzzy Logic. *Int. J. of Approximate Reasoning*, 5:45–88, 1991.
Saaty74	T.L. Saaty. Measuring the Fuzziness of Sets. *J. of Cybernetics*, 4:53–61, 1974.
Saaty78	T.L. Saaty. Exploring the Interface Between Hierarchies, Multiple Objectives, and Fuzzy Sets. *Fuzzy Sets and Systems*, 1:57–68, 1978.
Sanchez76	E. Sanchez. Resolution of Composite Fuzzy Relation Equations. *Information and Control*, 30:38–48, 1976.
Savage72	L.J. Savage. *The Foundations of Statistics*. Dover Publ., New York, 1972.
Scharf85	E.M. Scharf and N.J. Mandve. The Application of a Fuzzy Controller to the Control of a Multi-Degree-Freedom Robot Arm. In: M. Sugeno, ed., *Industrial Applications of Fuzzy Control*, 41–62. North-Holland, Amsterdam, 1985.

Schweizer61   B. Schweizer and A. Sklar. Associative Functions and
               Statistical Triangle Inequalities. *Publicationes Mathematical
               Debrecen*, 8:169–186, 1961.

Schweizer63   B. Schweizer and A. Sklar. Associative Functions and
               Abstract Semigroups. *Publicationes Mathematical Debrecen*,
               10:69–81, 1963.

Schweizer83   B. Schweizer and A. Sklar. *Probabilistic Metric Spaces*. North-
               Holland, Amsterdam, 1983.

Shachter88   R.D. Shachter. Evaluating Influence Diagrams. *Operations
               Research*, 33, 1988.

Shackle61   G.L.S. Shackle. *Decision, Order, and Time in Human Affairs*.
               Cambridge University Press, Cambridge, 1961.

Shafer76   G. Shafer. *A Mathematical Theory of Evidence*. Princeton
               University Press, Princeton, NJ, 1976.

Shafer90   G. Shafer and J. Pearl. *Readings in Uncertain Reasoning*.
               Morgan Kaufman, San Mateo, CA, 1990.

Shao88   S. Shao. Fuzzy Self-Organizing Controller and its Application
               for Dynamic Processes. *Fuzzy Sets and Systems*, 26:151–164,
               1988.

Shenoy90   P.P. Shenoy and G. Shafer. Axioms for Probability and
               Belief Function Propagation. In: R.D. Shachter, T.S. Levitt,
               L.N. Kanal, and J.F. Lemmer, eds., *Uncertainty in Artificial
               Intelligence (4)*, 169–198. North-Holland, Amsterdam, 1990.

Shortliffe75   E. Shortliffe and B.G. Buchanan. A Model of Inexact
               Reasoning in Medicine. *Mathematical Biosciences*, 23:351–
               379, 1975.

Skala78   H.J. Skala. On Many-valued Logics. *Fuzzy Sets and Systems*,
               1:129–149, 1978.

Skala84   H.J. Skala, S. Termini, and E. Trillas, eds. *Aspects of
               Vagueness*. Reidel, Dordrecht, 1984.

Smets81   P. Smets. Medical Diagnosis: Fuzzy Sets and Degrees of Belief.
               Fuzzy Sets and Systems, 5:259–266, 1981.

Smets88   P. Smets, A. Mamdani, D. Dubois, and H. Prade. *Non
               Standard Logics for Automated Reasoning*. Academic Press,
               London, 1988.

Smithson89   M. Smithson. *Ignorance and Uncertainty.* Springer, New York,
               1989.

Spohn88   W. Spohn, Ordinal Conditional Functions: A Dynamic
               Theory of Epistemic States. In: W. Harper and B. Skryms,
               eds., *Causation in Decision, Belief Change, and Statistics*,
               105–134, Kluwer, Dordrecht, 1988.

Spohn90   W. Spohn, A General Non-Probabilistic Theory of Inductive
               Reasoning. In: R.D. Shachter, T.S. Levitt, L.N. Kanal, and
               J.F. Lemmer, eds., *Uncertainty in Artificial Intelligence*, 149–
               158. North-Holland, Amsterdam 1990

Stallings77   W. Stallings. Fuzzy Set Theory versus Bayesian Statistics.
               IEEE Trans. on Systems, Man, and Cybernetics, 7:216–219,
               1977.

Stout91   L.N. Stout. A Survey of Fuzzy Set and Topos Theory. *Fuzzy
               Sets and Systems*, 42:3–14, 1991.

Stoyan87	D. Stoyan, W.S. Kendall, and J. Mecke. *Stochastic Geometry and its Applications.* John Wiley and Sons, Chichester, 1987.
Strassen64	V. Strassen. Meßfehler und Information. *Wahrscheinlichkeitstheorie und verwandte Gebiete* 2:273–305, 1964.
Sugeno74	M. Sugeno. *Theory of Fuzzy Integral and its Application.* Ph. D. Thesis, Tokyo Inst. of Technology, Tokyo, 1974.
Sugeno85a	M. Sugeno, ed. *Industrial Applications of Fuzzy Control.* North-Holland, Amsterdam, 1985.
Sugeno85b	M. Sugeno. An Introductory Survey of Fuzzy Control. *Information Sciences*, 36:59–83, 1985.
Sugeno85c	M. Sugeno and K. Murakami. An Experimental Study on Fuzzy Parking Control Using a Model Car. In: M. Sugeno, ed., *Industrial Applications on Fuzzy Control*, 125–138. North-Holland, Amsterdam, 1985.
Sugeno85d	M. Sugeno and M. Nishida. Fuzzy Control of Model Car. *Fuzzy Sets and Systems*, 16:103–113, 1985.
Sugeno93	M. Sugeno and T.Yasukawa. A Fuzzy-Logic-Based Approach to Qualitative Modeling. *IEEE Trans. on Fuzzy Systems*, 1:7–31, 1993.
Takagi85	H. Takagi and M. Sugeno. Fuzzy Identification of Systems and its Application to Modelling and Control. *IEEE Trans. on Systems, Man, and Cybernetics*, 15:116–132, 1985.
Takagi91a	H. Takagi and I. Hayashi. NN-Driven Fuzzy Reasoning. *Int. J. of Approximate Reasoning*, 5:191–212, 1991.
Takagi91b	H. Takagi, T. Konda, and Y. Kojima. Neural Networks Based on Approximate Reasoning Architecture. *Japanese J. Fuzzy Theory and Systems*, 3:63–74, 1991.
Tanaka87	H. Tanaka. Fuzzy Data Analysis by Possibilistic Linear Models. *Fuzzy Sets and Systems*, 24:363–375, 1987.
Tanaka88	H. Tanaka and Watada. Possibilistic Linear Systems and their Application to Linear Regression Models. *Fuzzy Sets and Systems*, 27:275–289, 1988.
Tanaka92	K. Tanaka and M. Sugeno. Stability Analysis and Design of Fuzzy Control Systems. *Fuzzy Sets and Systems*, 45:135–156, 1992.
Taunton89	C. Taunton. Expert Systems in Process Control: An Overview. *ISA'89 Advanced Control Conference*, 51–55. Birmingham, 1989.
Terano91	T. Terano, K. Asai, and M. Sugeno. *Fuzzy System Theory and its Applications.* Academic Press, Boston, MA, 1991.
Terano92	T. Terano, M. Sugeno, M. Mukaidono and K. Shigemasu, eds. *Fuzzy Engineering toward Human Friendly Systems.* IOS Press, Amsterdam, 1992.
Tong80	R.M. Tong. Some Properties of Fuzzy Feedback Systems. *IEEE Trans. on Systems, Man, and Cybernetics*, 10:327–331, 1980.
Tong84	R.M. Tong. A Retrospective View of Fuzzy Control Systems. *Fuzzy Sets and Systems*, 14:199–210, 1984.
Toth92	H. Toth. Reconstruction Possibilities for Fuzzy Sets: Towards a New Level of Understanding? *Fuzzy Sets and Systems*, 52:283–304, 1992.

Trillas84   E. Trillas and L. Valverde. An Inquiry into Indistinguishability Operators. In: H.J. Skala, S. Termini, and E. Trillas, eds., *Aspects of Vagueness*, 231–256. Reidel, Dosdrecht, 1984.

Trillas85   E. Trillas and L. Valverde. On the Implication and Indistinguishability in the Setting of Fuzzy Logic. In: J. Kacprzyk and R.R. Yager, eds., *Management Decision Support Systems Using Fuzzy Sets and Possibility Theory*, 198–212. Verlag TÜV Rheinland, Köln, 1985.

Turksen91   I.B. Turksen. Measurement of Membership Functions and their Acquisition. *Fuzzy Sets and Systems*, 40:5–38, 1991.

Umano82   M. Umano. FREEDOM-0 — A Fuzzy Database System. In: M.M. Gupta and E. Sanchez, eds., *Fuzzy Information and Decision Processes*, 339–347. North-Holland, Amsterdam, 1982.

Umano87   M. Umano. Fuzzy Set Prolog. *Proc. 2nd IFSA Congress*, 750–753, Tokyo, Japan, 1987.

Valverde85   L. Valverde. On the Structure of F-Indistinguishability Operators. *Fuzzy Sets and Systems*, 17:313–328, 85.

Wang80   P.P. Wang and S.K. Chang, eds. *Fuzzy Sets: Theory and Application to Policy Analysis and Information Systems*. Plenum Press, New York, 1980.

Wang83   P.Z. Wang. From the Fuzzy Statistics to the Falling Random Subsets. In: P.P. Wang, ed., *Advances in Fuzzy Sets, Possibility and Applications*, 81–96. Plenum Press, New York, 1983.

Wang92   Z. Wang and G.J. Klir. *Fuzzy Measure Theory*. Plenum Press, New York, 1992.

Wang93   P.Z. Wang and K.F. Loer, eds. *Mind and Computer — Fuzzy Science and Engineering*. World Scientific, Singapore, 1993.

Weber83   S. Weber. A General Concept of Fuzzy Connectives, Negation and Implication Based on t-Norms and t-Conorms. *Fuzzy Sets and Systems*, 11:115–134, 1983.

Whalen85   T. Whalen and B. Scott. Alternative Logics for Approximate Reasoning in Expert Systems: A Comparative Study. *Int. J. of Man Machine Studies*, 22:327–346, 1985.

White92   D.A. White and D.A. Sofge, eds. *Handbook of Intelligent Control*, Van Nostrand Reinhold, New York, 1992.

Yager80   R.R. Yager. On a General Class of Fuzzy Connectives. *Fuzzy Sets and Systems*, 4:235–242, 1980.

Yager80b   R.R. Yager. Aspects of Possibilistic Uncertainty. *Int. J. of Man–Machine Studies* 12:283–298, 1980.

Yager82   R.R. Yager. Some Procedures for Selecting Fuzzy Set Theoretic Operators. *Int. J. of General Systems*, 8:115–124, 1982.

Yager83   R.R. Yager. An Introduction to Applications of Possibility Theory. *Human Systems Management*, 3:246–269, 1983.

Yager85   R.R. Yager. Inference in a Multivalued Logic System. *Int. J. of Man Machine Studies*, 23:27–44, 1985.

Yager87   R.R. Yager, S. Ovchinnikov, R.M. Tong, and H.T. Nguyen. *Fuzzy Sets and Applications*. Selected Papers by L.A. Zadeh. John Wiley and Sons, New York, 1987.

Yager92a   R.R. Yager and L.A. Zadeh. *An Introduction to Fuzzy Logic Applications in Intelligent Systems*. Kluwer, Boston, MA, 1992.

Yager92b   R.R. Yager. On the Specificity of a Possibilistic Distribution. *Fuzzy Sets and Systems* 50:279–292, 1992.

Yager93a   R.R. Yager and D.P. Filev. SLIDE: A Simple Adaptive Defuzzification Method. *IEEE Transactions on Fuzzy Systems* 1, 1:69–78, 1993.

Yager93b   R.R. Yager and D.P. Filev. On the Issue of Defuzzification and Selection Based on a Fuzzy Set. *Fuzzy Sets and System*, 55:255–271, 1993.

Yager94   R.R. Yager, M. Fedrizzi, and J. Kacprzyk, eds. *Advances in the Dempster–Shafer Theory of Evidence*. Wiley & Sons, New York, 1994.

Yasunubo83   S. Yasunubo, S. Miyamoto and H. Ihara. Fuzzy Control for an Automatic Train Operation System. *Proc. 4th IFAC/IFIP/IFORS Int. Conf. on Control in Transportation Systems*, 33–39, Baden-Baden, Germany 1983.

Yasunubo85   S. Yasunubo and S. Miyamoto. Automatic Train Operation by Predictive Fuzzy Control. In: M. Sugeno, ed., *Industrial Applications of Fuzzy Control*, 1–18. North-Holland, Amsterdam, 1985.

Zadeh65   L.A. Zadeh. Fuzzy Sets. *Information and Control*, 8:338–353, 1965.

Zadeh68   L.A. Zadeh. Probability Measures of Fuzzy Events. *J. of Mathematical Analysis and Applications*, 10:421–427, 1968.

Zadeh71a   L.A. Zadeh. Similarity Relations and Fuzzy Orderings. *Information Sciences*, 3:177–200, 1971.

Zadeh71b   L.A. Zadeh. Towards a Theory of Fuzzy Systems. In: R.E. Kalman and N. de Claris, eds. *Aspects of Networks and System Theory*, 469–490, Rinehart and Winston, NY, 1971.

Zadeh72a   L.A. Zadeh. A Fuzzy-Set-Theoretic Interpretation of Linguistic Hedges. *J. of Cybernetics*, 2:4–34, 1972.

Zadeh72b   L.A. Zadeh. A Rationale for Fuzzy Control. *J. Dynamic Systems, Measurement and Control, Series 6*, 94:3–4, 1972.

Zadeh73   L.A. Zadeh. Outline of a New Approach to the Analysis of Complex Systems and Decision Processes. *IEEE Trans. on Systems, Man, and Cybernetics*, 3:28–44, 1973.

Zadeh75   L.A. Zadeh. The Concepts of a Linguistic Variable and its Application to Approximate Reasoning. *Information Sciences*, 8:199–249, 301–357, 9:43–80, 1975.

Zadeh76   L.A. Zadeh. A Fuzzy–Algorithmic Approach to the Definition of Complex or Imprecise Concepts. *Int. J. of Man Machine Studies*, 8:249–291, 1976.

Zadeh78a   L.A. Zadeh. Fuzzy Sets as a Basis for a Theory of Possibility. *Fuzzy Sets and Systems*, 1:3–28, 1978.

Zadeh78b   L.A. Zadeh. PRUF — A Meaning Representation Language for Natural Languages. *Int. J. of Man Machine Studies*, 10:395–460, 1978.

Zadeh79   L.A. Zadeh. A Theory of Approximate Reasoning. In: J.E. Hayes, D. Michie, and L.I. Mikulich, eds., *Machine Intelligence (9)*, 149–194. John Wiley and Sons, New York, 1979.

Zadeh81   L.A. Zadeh. Possibilistic Theory and Soft Data Analysis. In: L. Cobb and R. Thrall, eds., *Mathematical Frontiers of Social and Policy Sciences* 69–129, Westview Press, Boulder, 1981.

Zadeh83a   L.A. Zadeh. A Computational Approach to Fuzzy Quantifiers in Natural Language. *Comput. and Maths. with Applications*, 9:149–184, 1983.

Zadeh83b   L.A. Zadeh. The Role of Fuzzy Logic in the Management of Uncertainty in Expert Systems. *Fuzzy Sets and Systems*, 11:199–227, 1983.

Zadeh84   L.A. Zadeh. A Theory of Commonsense Knowledge. In: H.J. Skala, S. Termini, and E. Trillas, eds., *Aspects of Vagueness*, 257–296. Reidel, Dordrecht, 1984.

Zadeh85   L.A. Zadeh. Syllogistic Reasoning in Fuzzy Logic and its Application to Usuality and Reasoning with Dispositions. *IEEE Trans. on Systems, Man, and Cybernetics*, 15:754–763, 1985.

Zadeh86   L.A. Zadeh. Test-Score Semantics as a Basis for a Computational Approach to the Representation of Meaning. *Literary and Linguistic Computing*, 1:24–35, 1986.

Zadeh92   L.A. Zadeh and J. Kacprzyk, eds. *Fuzzy Logic for the Management of Uncertainty.* John Wiley and Sons, New York, 1992.

Zadeh94   L.A. Zadeh. Fuzzy Logic, Neural Networks, and Soft Computing. *Communications of the ACM*, 37:77-84, 1994.

Zemankova-Leech84 M. Zemankova-Leech and A. Kandel. *Fuzzy Relational Data Bases — A Key to Expert Systems.* Verlag TÜV Rheinland, Köln, 1984.

Zimmermann80   H.J. Zimmermann and P. Zysno. Latent Connectives in Human Decision Making. *Fuzzy Sets and Systems*, 4:37–51, 1980.

Zimmermann84   H.J. Zimmermann, L.A. Zadeh, and B.R. Gaines, eds. *Fuzzy Sets and Decision Analysis.* North-Holland, Amsterdam, 1984.

Zimmermann87   H.J. Zimmermann. *Fuzzy Sets, Decision Making and Expert Systems.* Kluwer, Boston, 1987.

Zimmermann91   H.J. Zimmermann. *Fuzzy Set Theory and its Applications.* Kluwer, Boston, 2nd edition, 1991.

Index

L-fuzzy set, 13
α-cut of a fuzzy set, 15
  strict, 35
i-th restrictions, 99, 103

acceptance degree, 28
and, 28
at least as specific as, 92

Bayesian networks, 144
belief function, 151
belief revision, 92, 143
biimplication, 58

centre of area method, 173
centre of gravity method, 173
characteristic function, 8
closed world assumption, 85, 95
COA, 173
COG, 173
concept-oriented view, 84
construction sequence, 120
context, 44
contour function, 45, 69
contradiction-free, 101
control function, 161
control variable, 160, 161
cycle, 118
cylindrical extension, 96, 101

data base, 164
decision logic, 164
decision making, 76
decomposability, 117
defuzzification interface, 164
degree of consistency, 136
degree of inconsistency, 136
degree of membership, 8, 28, 84
degree of possibility, 84
dependency hypergraph, 116
disturbance variable, 160

dot method, 184

elementary event, 93
elementary linguistic value, 74
equality relation with respect to, 47
evidence, 92
  total, 98, 102
evidence system, 98
  non-contradictory, 102
  possibilistic, 102
extension, 31
extension principle, 31
extensional hull, 52

F-expert system, 99
  equivalence, 120
  possibilistic, 92, 103
falling shadow, 69
focal sets, 44
focusing, 81, 92
fuzzification interface, 163
fuzzy controller, 159
fuzzy convex, 32
fuzzy interval
  of L-R-type, 41
fuzzy intervals, 32
fuzzy knowledge base, 139
fuzzy measure, 150
fuzzy numbers, 32
fuzzy probability value, 70
fuzzy quantifier, 70
fuzzy relation, 83, 210
fuzzy set
  Cartesian product, 214
  core, 11
  distance, 76
  epistemic interpretation, 12, 85
  extensional, 52
  horizontal representation, 15
  normal, 14

physical interpretation, 12
subset, 20
support, 11
vertical representation, 15
fuzzy truth value, 70
fuzzy-numbers, 14

Gödel relation, 110
gradual truth, 28
Graham reduction, 121

Hausdorff pseudometric, 76
hyperedge fusion, 127
hyperedges, 115
hypergraph, 115
hypertree, 120

implication
  Lukasiewicz, 57, 146
  Gödel, 57, 146
  Goguen, 57
  Kleene–Dienes, 57
  Kleene-Dienes, 146
  QL, 146
  R, 146
  Reichenbach, 57, 146
  S, 145
  Zadeh, 57, 146
implication expression, 139
implication rule, 83
index set, 96
indistinguishability operators, 218
inference, 98
inference rules, 82
influence diagram, 144
input variable, 161
interval arithmetics, 34

knowledge base, 92, 95, 97, 164
  possibilistic, 102
  uncertain, 133
knowledge map, 144
knowledge-based analysis, 163

linguistic approximation, 76
linguistic hedge, 70
linguistic quantifier, 72
linguistic rules, 163

max criterion method, 172
max dot method, 184

mean of maximum method, 172
measure of belief, 150
measure of uncertainty, 81
modularization, 97
MOM, 172
more specific than, 92

necessity measure, 132
non-redundant, 116, 157
non-truth-functional, 133
normal fuzzy sets, 32
normalization condition, 85

one-point coverage, 69
or, 28
output variable, 160

plausibility function, 151
plausibility measure, 150
Poincaré paradox, 49
possibilistic logic, 132
possibilistic resolution principle, 134
possibility distribution, 81, 84, 85
possibility measure, 133
probability, 86
product space, 96
projection, 96
  pointwise, 96
propagation algorithm, 114

random set, 44
  consonant, 151
reduction sequence, 121
representation theorem, 17
rule base, 97, 164
  non-contradictory, 102
  non-redundant, 116
  possibilistic, 101

semantics, 42
set representation, 34
similarity relations, 218
simple hypertree, 122
singleton, 52
specification phase, 74
state, 103
  consistent, 103

t-conorm, 23
  Archimedian, 26
t-norm, 23

Archimedian, 26
truth functionality, 22
twig, 120

uncertain clause, 133
uncertain datum, 81, 84

universe, 85, 96

vague concept, 81
vertex fusion, 127
vertices, 115